Lecture Notes in Mathematics

Edited by A. Dold and B. Eckmann

Subseries: USSR
Adviser: L. D. Faddeev, Leningrad

1100

Victor Ivrii

Precise Spectral Asymptotics for Elliptic Operators Acting in Fiberings over Manifolds with Boundary

Springer-Verlag
Berlin Heidelberg New York Tokyo 1984

Author

Victor Ivrii
Dept. of Mathematics, Institute of Mining and Metallurgy
Magnitogorsk 455000, USSR

Consulting Editor

Olga A. Ladyzhenskaya
Leningrad Branch of the V.A. Steklov Mathematical Institute
Fontanka 27, Leningrad 191011, USSR

AMS Subject Classification (1980): 35P20, 58G17, 58G20, 58G25

ISBN 3-540-13361-5 Springer-Verlag Berlin Heidelberg New York Tokyo
ISBN 0-387-13361-5 Springer-Verlag New York Heidelberg Berlin Tokyo

Printing and binding: Beltz Offsetdruck, Hemsbach/Bergstr.
2146/3140-543210

TABLE OF CONTENTS

INTRODUCTION

This book is devoted to the determination of the precise asymptotics for eigenvalues of certain elliptic selfadjoint operators acting in fiberings over compact manifolds with boundary and for more general elliptic selfadjoint spectral problems. The precise asymptotics for restriction to the diagonal of the Schwartz kernels of the corresponding spectral projectors is derived too. These asymptotics for closed manifolds were determined in author's paper [57] using the same method; therefore one can consider [57] as a simple and short introduction to the methods and ideas of this book. More general results were obtained, for example, in [32 - 35], but with much weaker remainder estimates.

In Part I we derive these asymptotics for the second-order elliptic selfadjoint differential operators acting in fiberings over manifolds with boundary on which an elliptic boundary condition is given; this condition must be either the Dirichlet condition or the generalized Neumann condition; the latter means that the boundary value of the derivative of the function with respect to the direction transversal to the boundary is expressed through the boundary value of the function by means of first-order pseudo-differential operator on the boundary; examples show that for more general elliptic boundary conditions our results may not be valid.

We shall derive the following asymptotics for the eigenvalue distribution function:

$$N(k) = \varkappa_0 k^d + O(k^{d-1}),$$

$$N(k) = \varkappa_1 k^{d-1} + O(k^{d-2}) \quad \text{or}$$

$$N(k) = O(1)$$

where d is the dimension; the second asymptotics may occur only if the boundary is not empty. It is impossible to derive stronger estimates for remainder without some condition of global nature. For the Laplace - Beltrami operator this condition is:

The set of all points periodic with respect to the geodesic flow with reflection at the boundary has measure zero.

In a general case if certain strong conditions involving some global condition are satisfied then the following asymptotics hold:

$$N(k) = \varkappa_0 k^d + \varkappa_1 k^{d-1} + o(k^{d-1}) \quad \text{or} \quad N(k) = \varkappa_1 k^{d-1} + \varkappa_2 k^{d-2} + o(k^{d-2}).$$

Under some conditions the asymptotics for the restriction to the diagonal of the Schwartz kernels of the spectral projectors are derived too. Inside these asymptotics have the same nature: $\varkappa_0(x)k^d + O(k^{d-1})$ or $O(1)$ but near the boundary the leading part of the asymptotics contains the term of boundary-layer type of the order k^d. If and only if this term exhausts the leading part of the asymptotics then we have the asymptotics $N(k) = \varkappa_1 k^{d-1} + O(k^{d-2})$ for the eigenvalue distribution functions.

In Part I the methods and ideas of [56, 57] are generalized and improved upon and in Part II these methods and ideas are applied to certain new situations and the results of Part I are extended to these situations. In §6 the asymptotics for the elliptic selfadjoint first-order operators are derived. In §7 we derive the asymptotics for the spectral problem

$$(-\mu \mathcal{J} + A)u = 0 \tag{*}$$

where A is the elliptic selfadjoint second-order positive definite operator acting in the fibering over a manifold with boundary and $\mathcal{J}$ is non-degenerate Hermitian matrix, acting in the fibers of this fibering. The opposite case when $\mathcal{J}$ is positive definite A is not necessarily positive or negative definite may be reduced to the case $\mathcal{J} = I$ investigated in Part I. In §8 we study the problem (*), when A and $\mathcal{J}$ differ from $(-\Delta)^p$, $(-\Delta)^q$ respectively only by lower order terms where $p > q \geqslant 0$ and Δ is the Laplace - Beltrami operator. We must note that the latter case is the most complicated and in this case we need some improvements and modifications of our methods.

We intend to write a few papers in future; in these papers the spectral asymptotics for global and partially global operators [62] and the quasiclassical spectral asymptotics for h-(pseudo)-differential operators [63] will be derived and applications of these results to the problem (*) where $\mathcal{J}$ may degenerate in a definite manner, will be given.

ACKNOWLEDGEMENTS. I express my sincere thanks to B.M.Levitan and M.A.Šubin who drew my attention to the spectral asymptotics (to H.Weyl conjecture), to M.S.Agranovič, V.M.Babič, L.Fridlender, G.Grubb, V.B.Lidskii, G.V.Rosenblum, M.Z.Solomjak and to M.A.Šubin again for useful, stimulating and clarifying discussions and to my wife Olga for her understanding, encouragement and support. I express my sincere thanks to Yu.N.Kovalik for assistance in translation.

Part I. THE ASYMPTOTICS FOR SECOND-ORDER OPERATORS

§ 0. Main theorems

0.1. Let X be a compact d-dimensional C^∞-manifold with the boundary $Y \in C^\infty$, $d \geq 2$, dx a C^∞-density on X, E a Hermitian D-dimensional C^∞-fibering over X. Let $\mathcal{A}: C^\infty(X,E) \to C^\infty(X,E)$ be a second-order elliptic differential operator, formally selfadjoint with respect to inner product in $L_2(X,E)$; if $Y=\varnothing$, then $\mathcal{A}$ may be a classical pseudo-differential operator. Let ν be a C^∞ vector field transversal to Y at every point, τ the operator of restriction to Y, $B=\tau$ or $B=\tau\nu + B_1\tau$ a boundary operator, where $B_1: C^\infty(Y,E) \to C^\infty(Y,E)$ is a first-order classical pseudo-differential operator on Y; certainly, only B but not its components is an invariant object. We suppose that (E). Operator $\{\mathcal{A}, B\}: C^\infty(X,E) \to C^\infty(X,E) \oplus C^\infty(Y,E)$ satisfies the Šapiro - Lopatinskii condition; (S) $\mathcal{A}_B$ - the restriction of $\mathcal{A}$ to the $\operatorname{Ker} B$ (i.e. operator with the domain $D(\mathcal{A}_B) = C^\infty(X,E) \cap \operatorname{Ker} B$) is a symmetric operator in $L_2(X,E)$.

Let $A_B: L_2(X,E) \to L_2(X,E)$ be a closure of $\mathcal{A}_B$ in $L_2(X,E)$; then A_B is selfadjoint, its spectrum is discrete, with finite multiplicity and tends either to $\pm\infty$, or to $+\infty$, or to $-\infty$. Without the loss of generality one can suppose that 0 is not an eigenvalue of A_B. Then, if X is a closed manifold, then $\Pi^\pm$ - selfadjoint projectors to positive(negative) invariante subspaces of A_B are zero-order classical pseudo-differential operators on X. If X is a manifold with boundary, then $\Pi^\pm$ belong to Boutet de Monvel algebra, i.e. $\Pi^\pm = \Pi^{0\pm} + \Pi'^{\pm}$ where $\Pi^{0\pm}$ are zero-order classical pseudo-differential operators with the transmission property on X and $\Pi'^{\pm}$ are zero-order classical singular Green operators (see [14] or § 1 of this book). $\pi^\pm$ - the principal symbols of $\Pi^\pm$ - are selfadjoint projectors to positive and negative invariant subspaces of a; a is the principal symbol of $\mathcal{A}$. If $\pi^\pm=0$ then $\Pi^\pm$ is a zero-order classical singular Green operator. It should be pointed out that for a more general boundary operator B even satisfying the Šapiro - Lopatinskii condition, these statements may not be valid: $\Pi^\pm$ may be operators of a more general nature;

are likely to belong to Rempel - Schulze algebra [83].

Let H be a closed subspace in $L_2(X,E)$ such that Π -selfadjoint projector to H - belongs to Boutet de Monvel algebra (if X is a closed manifold then Π is a classical zero-order pseudo-differential operator). We suppose that $\Pi A_B \subset A_B \Pi$; this means that H is an invariant subspace of A_B. Then $A_{B,H} : H \to H$, $\mathcal{D}(A_{B,H}) = \mathcal{D}(A_B) \cap H$ - the restriction of A_B to H - is a selfadjoint operator; its spectrum is discrete, with finite multiplicity and tends either to $\pm\infty$, or to $+\infty$, or to $-\infty$.

Let us introduce the eigenvalue distribution functions for $A_{B,H}$: $N^{\pm}(k) = N^{\pm}_H(k)$ is the number of eigenvalues of $A_{B,H}$ lying between 0 and $\pm k^2$.

Let $E(s)$ be spectral selfadjoint projectors of A_B, $E_H(s) = \Pi E(s)$ and $e^{\pm}_H(s) = \pm(E_H(\pm k^2) - E_H(0))$; let $e^{\pm}_H(x,y,k)$ be Schwartz kernels of $e^{\pm}_H(k)$; then

$$N^{\pm}_H(k) = \operatorname{tr} e^{\pm}_H(k) = \int_X e^{\pm}_H(x,x,k)\,dx .$$

We are interested in the asymptotics of $N^{\pm}_H(k)$ and $e^{\pm}_H(x,x,k)$ as k tends to ∞ assuming, of course, that $\pm A_{B,H}$ is not semibounded from above.

0.2. Let a and a^s be the principal and subprincipal symbols of $\mathcal{A}$, $\pi^{\pm}$ and π the principal symbols of $\Pi^{0\pm}$ and Π^0, $\varepsilon(s)$ spectral selfadjoint projectors of a, $\varepsilon^{\pm} = \pm(\varepsilon(\pm 1) - \varepsilon(0))$.

0.3. We start from the results of [57] concerning asymptotics on closed manifolds. One can assume that $\pi\pi^{\pm} \neq 0$ on some connective component of $T^*X\setminus 0$; otherwise $\Pi\Pi^{\pm}$ is of order (-1), therefore $\Pi\Pi^{\pm}$ is finite-dimensional projector and $\pm A_{B,H}$ is semibounded from above.

THEOREM 0.1. If X is closed manifold then the following asymptotics hold:

$$(0.1)\quad N^{\pm}_H(k) = \varkappa^{\pm}_0 k^d + O(k^{d-1}) \quad \text{as} \quad k \to \infty ,$$

$$(0.2)\quad e^{\pm}_H(x,x,k) = \alpha^{\pm}_0(x) k^d + O(k^{d-1})$$

as $k \to \infty$ uniformly with respect to $x \in X$,

where

$$(0.3)\quad \alpha^{\pm}_0(x) = \int_{T^*_xX} \varepsilon^{\pm}\, d\xi ,$$

$đ\xi=(2\pi)^{-d}d\xi$, $d\xi$ is the measure on $T^*_x X$, generated by the natural measure $dx\,d\xi$ on T^*X and by fixed measure dx on X; $\alpha_0^{\pm}(x)>0$ on Ω -connective component of X , if $T^*\Omega \cap$ cone $\operatorname{supp}\Pi\Pi^{\pm}\neq\emptyset$; otherwise

(0.2) $\quad e_H^{\pm}(x,x,k) = O(1)$ as $k\to\infty$

uniformly with respect to $x\in\Omega$;

(0.4) $$\varkappa_0^{\pm}=\int_X \operatorname{tr}\alpha_0^{\pm}(x)\,dx=\int_{T^*X}\operatorname{tr}\varepsilon^{\pm}\,dx\,đ\xi.$$

REMARK. If X is a manifold with the boundary Y then the asymptotics (0.2) holds for $x\in X\setminus Y$ uniformly on every compact subset of $X\setminus Y$.

REMARK. It is obvious that $\alpha_0^{\pm}\in C^\infty$; all coefficients $\alpha_{\cdot}^{\pm}(x)$ which will be introduced in future will also be smooth.

To obtain the second term of the asymptotics for the eigenvalue distribution function we need some condition of a global character. Let $0<\sigma_1^2\leqslant\sigma_2^2\leqslant\ldots\leqslant\sigma_{D_\pm}^2$ be all the positive eigenvalues of $\pm a$; $D_\pm$ may depend on connective component of $T^*X\setminus 0$. Let $\Sigma_j^{\pm}$ be a conical with respect to ξ closed nowhere dense subsets of $T^*X\setminus 0$ such that for every j σ_j^2 has the constant multiplicity outside of $\Sigma_j^{\pm}$; this multiplicity may depend on the connective component of $T^*X\setminus 0\setminus\Sigma_j^{\pm}$. These subsets exist without question.

Consider the <u>bicharacteristics</u> of σ_j - the curves along which

$$\frac{d\rho}{dt}=H_{\sigma_j}(\rho),\qquad \rho=\rho(t)\in T^*X\setminus 0\setminus\Sigma_j^{\pm},$$

where $H_f=\left\langle\frac{\partial f}{\partial\xi},\frac{\partial}{\partial x}\right\rangle-\left\langle\frac{\partial f}{\partial x},\frac{\partial}{\partial\xi}\right\rangle$ is the Hamiltonian field generated by f . a bicharacteristic is called periodic if there exists $T\neq 0$ such that $\rho(t+T)=\rho(t)$; T is called the <u>period</u> of bicharacteristic.

THEOREM 0.2. Let X be a closed manifold and the following condition hold:

(H1). There exists a set $\Lambda^{\pm}$ of measure zero, $\bigcup_j\Sigma_j\subset\Lambda^{\pm}\subset$ $\subset T^*X\setminus 0$, such that through each point of $T^*X\setminus 0\setminus\Lambda^{\pm}$ for every j there passes a nonperiodic bicharacteristic of σ_j lying in $T^*X\setminus 0\setminus\Sigma_j^{\pm}$, which is infinite in both directions.

Then the following asymptotics holds:

(0.5) $$N_H^{\pm}(k)=\varkappa_0^{\pm}k^d+\varkappa_1^{\pm}k^{d-1}+o(k^{d-1})\qquad k\to\infty ,$$

where

$$(0.6)\quad \varkappa_1^{\pm} = \mp \frac{d}{2} \int_{T_x^* X} \operatorname{tr} \varepsilon^{\pm} (\pm a)^{-1/2} \Pi\Pi^{\pm} a^{s} \, dx \, d\xi + \bar{\varkappa}^{\pm}.$$

REMARK. If A is a differential operator and $\Pi(x,-\xi) = \Pi(x,\xi)$, then $\varkappa_1^{\pm} = 0$ where $\bar{\varkappa}_1^{\pm}$ depends only on a, Π (see Appendices D, F_∞)

0.4. Now let X be a manifold with the boundary. Let $x_1 \in C^\infty$, $x_1 = 0$ and $dx_1 \neq 0$ at Y, $x_1 > 0$ in $X \setminus Y$. Consider the characteristic symbol $g_{\pm}(\rho, \tau) = \det(\tau^2 \mp a(\rho))$.

DEFINITION. The point $\rho \in T^* X \setminus Y$ has the <u>multiplicity</u> $\ell = \ell_{\pm}(\rho)$ if

$$\left(\frac{\partial}{\partial \tau}\right)^{i} g_{\pm}(\rho,\tau)\Big|_{\tau=1} = 0 \qquad \forall i < \ell ,$$

$$\theta_{\pm}(\rho) = \left(\frac{\partial}{\partial \tau}\right)^{\ell} g_{\pm}(\rho,\tau)\Big|_{\tau=1} \neq 0 .$$

DEFINITION. We shall say that the point $\rho \in T^* X|_Y \setminus 0$ with $\ell = \ell_{\pm}(\rho) \geq 1$

(i) is <u>positive</u> if

$$\theta_{\pm}^{-1}(\rho) \left(\frac{\partial}{\partial \tau}\right)^{i} \left(\frac{\partial}{\partial \xi_1}\right)^{\ell - i} g_{\pm}(\rho,\tau)\Big|_{\tau=1} > 0$$

$$\forall i = 0, \ldots, \ell - 1 ;$$

(ii is <u>negative</u> if

$$\theta_{\pm}^{-1}(\rho) \left(\frac{\partial}{\partial \tau}\right)^{i} \left(-\frac{\partial}{\partial \xi_1}\right)^{\ell - i} g_{\pm}(\rho,\tau)\Big|_{\tau=1} > 0$$

$$\forall i = 0, \ldots, \ell - 1 ;$$

(iii) is <u>tangential</u> if

$$\left(\frac{\partial}{\partial \tau}\right)^{i} \left(\frac{\partial}{\partial \xi_1}\right)^{\ell - i} g_{\pm}(\rho,\tau)\Big|_{\tau=1} = 0$$

$$\forall i = 0, \ldots, \ell - 1 ;$$

(iv) is <u>indefinite</u> in the remaining cases.

Here ξ_1 is the dual to x_1 variable, $\frac{\partial}{\partial \xi_1} = -H_{x_1}$.

Let $j : T^* X|_Y \longrightarrow T^* Y$ be a natural mapping.

THEOREM 0.3. Let $\Pi\Pi^{\pm}$ be not a singular Green operator (i.e.

$\Pi\Pi^{\pm} \neq 0$).

Then the asymptotics (0.1) holds with the same coefficient $\varkappa_0^{\pm}$.

THEOREM 0.4. Let $\Pi\Pi^{\pm}$ be a singular Green operator. Then the following asymptotics holds:

$$(0.7) \quad N_H^{\pm}(k) = \varkappa_1'^{\pm} k^{d-1} + O(k^{d-2})^{*)} \quad \text{as} \quad k \to \infty \qquad {}^{*)}$$

where $0 < \varkappa_1'^{\pm}$ depends only on $a|_Y$ and on $(\Pi\Pi^{\pm})_0'$ - the principal symbol of singular Green operator $(\Pi\Pi^{\pm})'$; b is the principal symbol of B.

0.5. To obtain the second term of the asymptotics for the eigenvalue distribution function we need some condition of a global nature.

Suppose first that

$$(H.2) \qquad g_{\pm}(\rho,\tau) = (\tau^2 - \mu(\rho))^{\ell} h(\rho,\tau),$$

where $\mu(\rho) = \mu(x,\xi)$ is a positive definite quadratic form in ξ (i.e. a Riemannian metric on X) and $h(\rho,\tau)$ does not vanish.

Then on S^*X -tangent spheres bundle of the Riemannian manifold X (more precisely on its subset of the complete measure) one can define a continious measure - preserving geodesic flow with reflection at the boundary. More particulary: consider S^*X with identified points ρ and ρ' such that $j\rho = j\rho'$.

By a geodesic we mean a curve lying in S^*X , along which

$$\frac{d\rho}{dt} = \frac{1}{2} H_{\mu}(\rho)$$

(i.e. the bicharacteristic of $\sigma = \sqrt{\mu}$); the geodesic, minus, perhaps, its endpoints, must lie in $S^*(X \setminus Y)$; t (the length) is the natural parameter along the geodesic.

By a geodesic billiard we mean a curve lying in S^*X consisting of segments of geodesics; the endpoint of the precedent segment and the starting point of the consequent must belong to $S^*X|_Y$ and be equivalent (i.e. identified); t (the length) is the natural parameter along geodesic billiards.

It is easy to show that there exists a set $\Sigma \subset S^*X$, of first Baire category and measure zero, such that through each point of $S^*X \setminus \Sigma$ a geodesic billiard of infinite length in both directions can be passed in such a way that each of its intervals of fi-

---

*) Here $O(k^0) \stackrel{def}{=} O(\ln k)$

nite length contains a finite number of segments, and all the geodesics included in it are transversal to the boundary. Thus, on $S^*X\setminus\Sigma$ a continuous measure - preserving flow $\Phi(t)$ is given. *)

We call a point $\rho\in S^*X\setminus\Sigma$ periodic if there exists a $T\neq 0$ such that $\Phi(T)\rho=\rho$; T is the period of point ρ.

THEOREM 0.5. Let $\Pi\Pi^{\pm}$ be not a singular Green operator and condition (H.2) be satisfied. Assume that the set of periodic points has measure zero. Then the following asymptotics holds:

$$(0.5)\qquad N_H^{\pm}(k)=\varkappa_0^{\pm}k^d+(\varkappa_1^{\pm}+\varkappa_1''^{\pm})k^{d-1}+o(k^{d-1})\qquad k\to\infty,$$

where $\varkappa_0^{\pm}$, $\varkappa_1^{\pm}$ are given by (0.4), (0.6), $\varkappa_1''^{\pm}$ depends only on $a|_Y$, b, $\Pi\Pi^{\pm}|_Y$, $(\Pi\Pi^{\pm})'_0$.

0.6. As we mentioned above, the asymptotics of $e_H^{\pm}(x,x,k)$ near boundary has a more complicated character. Let us identify some neighbourhood of Y with $[0,\delta)\times Y$; then point x will be identified with (x_1,x'), where x_1 is such as above, $x'\in Y$. If condition (H.2) is satisfied then X is a Riemannian manifold and this identification is canonical, $x_1=dist(x,Y)$.

THEOREM 0.6. Let $\Pi\Pi^{\pm}$ be not a singular Green operator and condition (H.2) be satisfied. Then in neighbourhood of Y the following asymptotics holds:

$$(0.8)\qquad e_H^{\pm}(x,x,k)=(\alpha_0^{\pm}(x)+Q_0^{\pm}(x',x_1k))k^d+O(k^{d-1})$$

as $k\to\infty$ uniformly with respect to x , where $\alpha_0^{\pm}(x)$ is given by (0.3),

$$Q_0^{\pm}\in C^{\infty}(Y\times\bar{\mathbb{R}}^+),$$

$$(0.9)\qquad \mathcal{D}_{x'}^{\alpha'}\mathcal{D}_s^{i}Q_0^{\pm}(x',s)=O(s^{-\frac{d+1}{2}})$$

as $s\to\infty$ uniformly with respect to $x'\in Y$ for every α', i.

THEOREM 0.7. Let $\Pi\Pi^{\pm}$ be not a singular Green operator and the following conditions be satisfied:

(H.3) For every point $\rho^*\in T^*Y\setminus 0$ we can find $\mu\in\mathbb{R}$ such that at all points of $j^{-1}\rho^*\cap\{g_{\pm}(\rho,1)=0\}$ the following inequalities hold:

---

*) The simple proof of these statements is published in [21] ; it coincides with the unpublished proof by the author [56].

(0.10) $$\theta_{\pm}^{-1}(\rho)\left(\frac{\partial}{\partial\tau}\right)^{i}\left(\frac{\partial}{\partial\tau}+(\xi_1-\mu)\frac{\partial}{\partial\xi_1}\right)^{l-i} g_{\pm}(\rho,\tau)\Big|_{\tau=1}>0$$

$$\forall i=0,\dots,l-1\ ;\ l=l_{\pm}(\rho)\ ;$$

(H.4) $\xi_1=\lambda_i(x,\xi')$ - all real roots of the $g_{\pm}(x,\xi',\xi_1,1)$ - have constant multiplicities in the neighbourhood of $N^*Y=\{x_1=\xi'=0\}$ and

(0.11) $$\frac{\partial\lambda_i}{\partial\xi_k}\Big|_{x_1=\xi'=0}$$ do not depend on i ,

(0.12) $$\operatorname{sign}\lambda_i\left(\frac{\partial^2\lambda_i}{\partial\xi_k\,\partial\xi_l}\Big|_{x_1=\xi'=0}\right)_{k,l=2,\dots,d}$$

are negative definite $(d-1)\times(d-1)$-matrices.

Then for $d\geqslant 3$ the asymptotics (0.9) holds; for $d=2$ the following asymptotics holds:

(0.8) $$e^{\pm}(x,x,k)=(\alpha_0^{\pm}(x)+Q_0^{\pm}(x',x_1k))k^2+O(k^{3/2})$$

as $k\to\infty$

uniformly with respect to x, where $Q_0^{\pm}$ satisfies (0.9).

REMARK. In theorems 0.6, 0.7 $Q_0^{\pm}(x',\cdot)$ depends only on $a(0,x',\cdot)$, $b(x',\cdot)$, $\Pi\Pi^{\pm}(0,x',\cdot)$, $(\Pi\Pi^{\pm})'_0(x',\cdot)$. Moreover

$$\varkappa_1''^{\pm}=\int_Y\int_0^{\infty}\operatorname{tr}Q^{\pm}(x',s)\,ds\,dx'$$

where $dx'=\frac{dx}{dx_1}\Big|_Y$ is a C^{∞}-density on Y .

COROLLARY. If the conditions of the theorem 0.6 hold or if the conditions of the theorem 0.7 hold and $d\geqslant 3$ then the asymptotics (0.2) holds outside the boundary layer $x_1\leqslant k^{-(d-1)/(d+1)}$.

THEOREM 0.8. Let $\Pi\Pi^{\pm}$ be a singular Green operator but not a smoothing operator. Then the following asymptotics holds:

(0.13) $$e_H^{\pm}(x,x,k)=k^d Q_0^{\pm}(x',x_1k)+O(\min(k^{d-1},x^{-d+1}))$$

as $k\to\infty$

uniformly with respect to x

where $Q_0^{\pm}\in C^{\infty}(Y\times\bar{\mathbb{R}}^+)$,

(0.14) $$\mathcal{D}_{x'}^{\alpha'}\mathcal{D}_s^{i}Q_0^{\pm}(x',s)=O(s^{-d-i})$$

as $s\to\infty$ uniformly with respect to $x'\in Y$ for every α', i ;

$Q_0^{\pm}(x',\cdot)$ depends only on $a(0,x',\cdot)$, $b(x',\cdot)$, $(\Pi\Pi^{\pm})_0(x',\cdot)$. Moreover,

$$\varkappa_1'^{\pm}=\int_Y\int_0^{\infty} \operatorname{tr} Q_0^{\pm}(x',s)\,ds\,dx' .$$

It is interesting that on Y $e^{\pm}(x,x,k)$ has degree-like asymptotics again.

THEOREM 0.9. Let $\Pi\Pi^{\pm}$ be not a smoothing operator and $B=i\tau\mathcal{D}_1+$ $+B_1\tau$. Then the following asymptotics holds:

$$e_H^{\pm}(x,x,k)=\varkappa_0'^{\pm}(x)k^{d}+O(k^{d-1}) \tag{0.15}$$

as $k\to\infty$ uniformly with respect to $x\in Y$ where $\varkappa_0'^{\pm}(x)$ depends only on $a(x,\cdot)$, $b(x,\cdot)$, $\Pi\Pi^{\pm}(x,\cdot)$, $(\Pi\Pi^{\pm})_0'(x,\cdot)$.

REMARK. In reality theorems 0.6-0.9 have the local character and therefore if we wish to derive the asymptotics of $e^{\pm}(x,x,k)$ on some subset $\Omega\subset X$ then we need the fulfillment of conditions of these theorems only on $\Omega\cap Y$.

0.7. As an illustration we shall consider the Laplace - Beltrami operator (or its certain generalization) on the Riemannian manifold with the boundary.

Thus, let X be a compact Riemannian manifold with the boundary Y, dx and dx' be natural measures on X and Y respectively, E a Hermitian fibering over X. Let $\mathcal{A}: C^{\infty}(X,E)\to C^{\infty}(X,E)$ be a formally selfadjoint differential operator with the principal part $-\Delta Id$,*) where Δ is the Laplace - Beltrami on X, $Id: E\to E$ is the identical mapping. We denote by A_D, A_N, A_G an operator A_B if the boundary operator is $B=\tau$, $B=\tau\frac{\partial}{\partial n}+i\varphi\tau$, $B=\tau\frac{\partial}{\partial n}+(i\varphi+B_1)\tau$ respectively where n is the interior unit normal to Y, φ is a smooth matrix such that A_N is the selfadjoint operator, $B_1: C^{\infty}(Y,E)\longrightarrow C^{\infty}(Y,E)$ is a first-order classical pseudo-differential operator on Y with the principal symbol β. Then A_G is selfadjoint if and only if B_1 is symmetric. Let $\beta_k (k=1,\dots,D)$ be the eigenvalues of β Then Šapiro-Lopatinskii condition means precisely that

$$\beta_k\neq 1 \quad \text{on } S^*Y \quad \forall k=1,\dots,D. \tag{0.16}$$

When (0.16) holds then A_G is semibounded from below if and only if

$$\beta_k<1 \quad \text{on } S^*Y \quad \forall k=1,\dots,D. \tag{0.17}$$

Let $\Pi=I$. One can concretize the statements of the theorems 0.3, 0.5, 0.6, 0.9:

<u>THEOREM 0.10 (i).</u> The following asymptotics hold as $k\to\infty$;:

*) Certainly, $-\Delta Id$ is not an invariant object ixcept in the case of $E=X\times\mathbb{C}^D$.

$$N^+(k) = \varkappa_0^+ + k^d + O(k^{d-1}),$$

$$e^+(x,x,k) = (\alpha_0^+ + Q_0^+(x', z x_1 k))k^d + O(k^{d-1}),$$

where $\varkappa_0^+ = (2\pi)^{-d}\omega_d D \operatorname{vol} X$, $\alpha_0^+ = (2\pi)^{-d}\omega_d Id$, ω_ℓ is the volume of the unit ball in $\mathbb{R}^\ell$,

$$Q_0^+(x', s) = (2\pi)^{-d+1} \int_{S^{d-2}} \mathcal{K}_d(\beta, s)\, d\theta,$$

$d\theta$ is the Lebesque measure on

$$S^{d-2} \ni \theta, \quad \beta = \beta(x', \theta),$$

$$\mathcal{K}_d(\beta, s) =$$

$$= -\frac{1}{2\pi}\int_\gamma (-i\omega + \beta\sqrt{1-\omega^2})^{-1} (i\omega + \beta\sqrt{1-\omega^2})(1-\omega^2)^{\frac{d-3}{2}} (-i\omega)^{-d} \left(\frac{\partial}{\partial s}\right)^{d-1} \frac{1}{s}(e^{-is\omega} - 1)\, d\omega$$

for $A = A_G$,

$$\mathcal{K}_d(s) = \pm\frac{1}{2\pi}\int_\gamma (1-\omega^2)^{\frac{d-3}{2}} (-i\omega)^{-d} \left(\frac{\partial}{\partial s}\right)^{d-1} \frac{1}{s}(e^{-is\omega} - 1)\, d\omega$$

for $A = A_N$ and $A = A_D$; here and below, the signs (+) and (-) correspond to $A = A_N$ and $A = A_D$ respectively, $\operatorname{Re}(1-\omega^2)^{1/2} > 0$, γ is a contour with the starting point -1 and the endpoint +1 lying in the lower complex half-plane $\operatorname{Re}\omega < 0$ below all points $\omega_k = -i\beta_k/\sqrt{1-\beta_k^2}$ with $0 \leqslant \beta_k < 1$.

(ii) Assume that the set of points periodic with respect to geodesic flow with reflection at the boundary has measure zero. Then the following asymptotics holds: $N^+(k) = \varkappa_0^+ k^d + \varkappa_1^+ k^{d-1} + o(k^{d-1})$ as $k \to \infty$, where

$$\varkappa_1^+ = -\frac{(2\pi)^{-d}}{2i(d-1)} \int_{S^*Y}\int_\gamma \operatorname{tr}(-i\omega + \beta\sqrt{1-\omega^2})^{-1}(i\omega + \beta\sqrt{1-\omega^2})(1-\omega^2)^{\frac{d-3}{2}} \omega^{-1}\, d\omega\, d\theta\, dy$$

for $A = A_G$, $\varkappa_1^\pm = \pm\frac{1}{4}(2\pi)^{-d+1}\omega_{d-1}\operatorname{vol} Y$ for $A = A_N$ and $A = A_D$.

COROBLAIRES AND REMARKS.

(i) As before

$$\mathcal{K}_d(\beta, s) = O(s^{-\frac{d+1}{2}}) \qquad \text{as } s \to +\infty.$$

(ii) The following asymptotics holds as $s \to +\infty$:

$$\mathcal{K}_d(\beta,s)=\frac{1}{4\pi}\Gamma\left(\frac{d-1}{2}\right)2^{\frac{d+1}{2}}\cos\left(s-\frac{\pi}{4}(d-3)\right)+O\left(s^{-\frac{d+2}{2}}\right)$$

for $A=A_G$,

$$\mathcal{K}_d(s)=\pm\frac{1}{4\pi}\Gamma\left(\frac{d-1}{2}\right)2^{\frac{d+1}{2}}\cos\left(s-\frac{\pi}{4}(d-3)\right)+O\left(s^{-\frac{d+3}{2}}\right)$$

for $A=A_N$ and $A=A_D$; therefore on the greater part of boundary layer we can simplify the expression for Q_0^+.

(iii) If d is odd then for $A=A_N$ and $A=A_D$

$$Q_0^+(s)=\pm 2(2\pi)^{-\frac{d+1}{2}}\left(-\frac{1}{s}\frac{\partial}{\partial s}\right)^{\frac{d-1}{2}}\frac{\sin s}{s}.$$

(iv). If d is even then for $A=A_N$ and $A=A_D$

$$Q_0^+(s)=\pm 2(2\pi)^{-\frac{d}{2}-1}\left(-\frac{1}{s}\frac{\partial}{\partial s}\right)^{\frac{d-1}{2}}\int_{-1}^{1}\left(\frac{\sin ws}{ws}+\frac{\cos ws-1}{w^2s^2}\right)\left(1-w^2\right)^{-1/2}dw.$$

(v) The asymptotics (0.15) holds with

$$\alpha_0'^{+}=(2\pi)^{-d}\left\{\omega_d-\frac{1}{d}\int_{S^{d-2}}\int_{\gamma}\left(-iw+\beta\sqrt{1-w^2}\right)^{-1}\left(iw+\beta\sqrt{1-w^2}\right)\left(1-w^2\right)^{\frac{d-3}{2}}dw\,d\theta\right\};$$

in particular, $\alpha_0'^{+}=2(2\pi)^{-d}\omega_d\,\mathrm{I}d$ for $A=A_N$.

(vi) Degree-like and boundary layer type terms in asymptotics for $e^+(x,x,k)$ generate the first and the second terms in the asymptotics for $N^+(k)$ (after integrating with respect to dx), but the asymptotics for $N^+(k)$ with the second term does not follow from the asymptotics for $e^+(x,x,k)$.

(vii) If β is odd with respect to θ , then

$$æ_1^+=-\frac{1}{4}(2\pi)^{-d+1}\omega_{d-1}\operatorname{vol}Y+\sum_{j:|\beta_j|<1}\frac{(2\pi)^{-d+1}}{2(d-1)}\int_{S^*Y}\left(1-\beta_j^2\right)^{-\frac{d+1}{2}}d\theta\,dx'.$$

One can concretize the statements of the theorems 0.4, 0.8 and 0.9 again.

THEOREM 0.11. The following asymptotics hold as $k\to\infty$:

$$N^-(k)=æ_1^-k^{d-1}+O(k^{d-2}),\qquad {}^{*)}$$

*) Here $O(k^0)=O(1)$.

$$e^-(x,x,k)=k^d Q_0^-(x',2x_1k)+O(\min(k^{d-1},x_1^{-d+1})),$$

where

$$\varkappa_1^-=\frac{(2\pi)^{-d+1}}{d-1}\int_{S^*Y}\operatorname{tr}(\beta^2-1)^{-\frac{d-1}{2}}\varepsilon\,d\theta\,dx',$$

$$Q_0^-(x',s)=2(2\pi)^{-d+1}\int_{S^{d-2}}\beta^{-d+1}\left(-\frac{\partial}{\partial s}\right)^{d-1}\frac{1}{s}\left(1-\exp(-s\beta/\sqrt{\beta^2-1})\right)\quad\varepsilon\,d\theta,$$

$\varepsilon=\varepsilon(x',\theta)$ is the selfadjoint projector to invariant subspace of $\beta(x',\theta)$ corresponding to eigenvalues $\beta_k>1$.

COROLLARIES AND REMARKS.

(i) Outside the boundary layer $x_1\leqslant Ck^{-d}\ln k$ $\quad e^-(x,x,k)=O(k^{d-1})$.

(ii) The asymptotics for $N^-(k)$ follows from the asymptotics for $e^-(x,x,k)$, but for $d=2$ one derives the remainder estimate $O(\ln k)$ instead of $O(1)$.

(iii) The asymptotics (0.15) holds with

$$\varkappa_0'^-=\frac{2}{d}(2\pi)^{-d+1}\int_{S^{d-2}}\beta(\beta^2-1)^{-d/2}\varepsilon\,d\theta.$$

It appears that one can obtain the second term of the asymptotics for $N^-(k)$. Really:

THEOREM 0.12. Let $0<\sigma_1^2\leqslant\sigma_2^2\leqslant\ldots\leqslant\sigma_n^2$, $\sigma_j=\sqrt{\beta_j^2-1}$ (we consider only $\beta_j>1$) satisfy the conditions of the theorem 0.2 on the manifold Then the following asymptotics holds as $k\to\infty$:

$N^-(k)=\varkappa_1^-k^{d-1}+\varkappa_2^-k^{d-2}+o(k^{d-2})$.

0.8. Describe the main ideas of this book, moments which seem to be essential and plan of Part I.

In §1 we shall give account of certain known results concerning L.Boutet de Monvel operator algebra and new results concerning selfadjoint projectors to positive and negative invariant subspaces of elliptic even-order differential operators acting in fiberings over manifolds with the boundary. We shall prove that if boundary conditions are elliptic and appropriate then these projectors belong to L. Boutet de Monvel operator algebra. For second-order operator $\mathcal{A}$ with positive definite principal symbol we shall derive necessary and sufficient condition (concerning elliptic boundary operator B) of the semiboundedness from below of operator A_B ; this condition generalizes the condition (0.17).

It is well-known that the most precise spectral asymptotics are derived by the method of hyperbolic operator; this method was used first by B.M.Levitan and V.G.Avakumovič. Namely, let $u(x,y,t)$ be the Schwartz kernel of the operator $A_B^{-1/2} \sin A_B^{1/2} t \cdot \Pi_+$ (we assume that $\Pi \subset \Pi\Pi^+$; otherwise we replace Π by $\Pi\Pi^+$; negative spectrum asymptotics can be obtained by replacing A_B by $-A_B$ and Π^+ by Π^-). Then $u(x,y,t)$ satisfies the boundary problem

(0.18) $\quad u_{tt} + \mathcal{A}_x u = 0$,

(0.19) $\quad B_x u = 0$,

(0.20) $\quad u|_{t=0} = 0, \quad u_t|_{t=0} = K_\Pi(x,y)$,

where $K_L(x,y)$ is the Schwartz kernel of the operator L ; $u(x,y,t)$ satisfies the analogous dual boundary problem (with respect to y,t). Let

$$\sigma(t) = \int_X \operatorname{tr} \Gamma u(x,t)\, dx ,$$

$$\Gamma u(x,t) = u(x,x,t) ;$$

then

$$\sigma(t) = \int_0^\infty k^{-1} \sin kt \, d N_H^+(k) ,$$

$$\Gamma u(x,t) = \int_0^\infty k^{-1} \sin kt \, d \Gamma e(x,k)^{*)} .$$

If we know $\sigma(t)$ and $\Gamma u(x,t)$ for small t (modulo smooth enough functions) then we can use the Tauberian theorems and derive the asymptotics for $N_H^+(k)$ and $\Gamma e_H^+(x,k)$. For obtained $\sigma(t)$ and $\Gamma u(x,t)$ both these asymptotics have remainders $O(k^{d-1})$ if Π is not a singular Green operator and remainders $O(k^{d-2})$ and $O(\min(k^{d-1}, x_1^{-d+1}))$ respectively if Π is a singular Green operator. To obtain the second term in asymptotics for $N_H^+(k)$ we need to show that all other singularities of $\sigma(t)$ are "weaker" then the "great" singularity (at $t=0$) which is isolated. Otherwise the spectrum of the operator may consist of separate clusters [18-20, 94, 95] or the cluster part of

---

*) Analogous formulas for first-order operators will be given in §6 .

the spectrum may have positive density with respect to whole spectrum [67]; there is no second term in asymptotics for $N_H^+(k)$ in all these ceses. Thus for obtaining the second term in the asymptotics for the $N_H^+(k)$ we need to prove that almost all singularities of solutions of the problem (0.18)-(0.19) propagate along certain trajectories (bicharacteristics, geodesics, geodesic billiards ets) almost all of which are nonperiodic. Therefore the conditions of the theorems 0.2, 0.5 are more restrictive then the conditions of other theorems nevertheless we are not interested in the exceptional propagation of singularities (along gliding rays for $A=-\Delta$ etc.). Note that we discuss here asymptotics with the remainder estimate $o(k^{d-1})$; to obtain more precise remainder estimates one need some more restrictive conditions (for example, see [12] where for the Laplace - Beltrami operator on closed manifold with the nonpositive sectional curvature remainder estimate $O(k^{d-1}/\ln k)$ was derived).

Note that the equation (0.18) is not necessarily hyperbolic; (0.18) is hyperbolic if and only if a is positive definite); all the more so problem (0.18)-(0.20) is not necessarily well-posed in common sense (it is well-posed if and only if A_B is semibounded from below); thus (0.18)-(0.19) is not necessarily solvable for arbitrary initial conditions. But we know that the problem (0.18)-(0.20) has the unique solution $u(x,y,t)$ and we can investigate it.

For example if we study asymptotics of $N^-(k)$ and $e^-(x,x,k)$ for operator $A=-\Delta$ (then the equation (0.18) is

$$(0.21)\qquad Pu \equiv u_{tt} + \Delta_x u = 0 \;;$$

this equation is elliptic; but the boundary operator $B=\tau\frac{\partial}{\partial n}+(i y+ B_1)\tau$ does not satisfy the Šapiro - Lopatinskii condition in the interesting case (B satisfies the Sapiro - Lopatinskii condition with respect to P if and only if A_B is semibounded from below) nevertheless B satisfies this condition with respect to $\mathcal{A}=-\Delta$. Thus solution of the equation (0.21) with the boundary condition

$$(0.22)\qquad \frac{\partial u}{\partial n} + (i y + B_1)u\Big|_Y = 0$$

is not in general smooth up to the boundary. Its singularities lies on Y and propagate along bicharacteristics of the symbols $\pm\sqrt{\beta^2(x,\xi')-|\xi'|^2}$ along which $\tau=\mp\sqrt{\beta^2(x,\xi')-|\xi'|^2}$ and $\beta(x,\xi')>|\xi'|$; we assume now that $E=X\times\mathbb{C}$.

At the other hand, if we study asymptotics of $N^+(k)$ and $e^+(x,x,k)$ for the same operator then the equation (0.18) is

(0.23) $\quad Pu \equiv u_{tt} - \Delta_x u = 0;$

it is usual wave equation; but in this case the problem (0.18)-(0.20) may be ill-posed too (if operator A_B is not semibounded from below, i.e. if the condition (0.17) is not fulfilled). Moreover, if A_B is semibounded from below, then the boundary operator B does not necessarily satisfy the Šapiro - Lopatinskii condition with respect to P in the elliptic zone $\{|\xi'|>|\tau|\}$; if $0<\beta(x,\xi')<|\xi'|$ then this condition fails and singularities propagate along geodesic billiards (and along generalized geodesic billiards with the branching in general) as in the case of Dirichlet and Neumann boundary conditions [53, 78, 79], but also along the bicharacteristics of the symbols $\pm\sqrt{|\xi'|^2-\beta^2(x,\xi')}$, along which $\tau=\mp\sqrt{|\xi'|^2-\beta(x,\xi')}, 0<\beta(x,\xi')<|\xi'|, x_1=0$ [54]. These bicharacteristics lie in the elliptic zone.

But in all these cases all singularities propagate with a finite speed with respect to x and ξ (or with respect to x and ξ' near the boundary; on the boundary ξ_1 varies by jump). It appears that in the general case singularities of solutions of the problem (0.18)-(0.19) propagate with a finite speed with respect to x and ξ (or with respect to x and ξ' near the boundary). This statement is proved in §2 by the method of energy estimates developed in the author's papers [49,50,53].

Note again that the presence of generalized geodesic billiards and non-standart bicharacteristics on Y is not the obstacle for obtaining the second term in the asymptotics for $N^+(k)$ because the generic singularities propagate only along usual geodesic billiards (i.e. along the trajectories of geodesic flow $\Phi(t)$).

One can note that in two considered model cases the singularities propagate with respect to x at a speed different from 0 ; therefore the singularity of $\sigma(t)$ at $t=0$ is isolated.

It appears that this statement remains true in the general case. Moreover one can improve this statement and prove that the great singularity of $\sigma(t)$ is normal, i.e. there exists $s=s(d)$ such that $t^n\sigma(t)\in H^{s+n}$ in the neighbourhood of $t=0$ for all $n\in\mathbb{Z}^+$.

Analogous statement holds for Γu on closed manifolds; moreover it appears that a differentiation with respect to x does not increase the order of the great singularity of Γu.

For manifolds with the boundary the situation is more complicated. If Π is a singular Green operator then Γu has a normal singularity at $t=x_1=0$.

But if Π is not a singular Green operator then the singularity of Γu is not normal. Thus, for $A=-\Delta$ and $\Pi=I$ for small t

$$\text{sing supp } \Gamma u=\{t=0\}\cup\{|t|=2x_1\}.$$

It is connected with the fact that in points near the boundary there exists a unique direction (that is to say, the direction tangent to the geodesic along which the minimal distance from x to Y is reached; this geodesic is normal to Y and this distance equals x_1) such that the singularity starting from x in this direction returns in x running this geodesic there and back and reflecting at the boundary; the returning time equals $2x_1$.

Thus the presence of normal waves prevents from the normality of the great singularity of Γu near the boundary and we need to cut these waves. Really, if Π is not a singular Green operator, condition (H.3) is fulfilled and $q(x, D_x', D_t)$ is a pseudo-differential operator with the symbol vanishing near $\xi'=0$ then the singularity of $\Gamma q u$ at $t=0$ is normal.

Note that the differentiation of Γu (or $\Gamma q u$) with respect to any vector field tangent to the boundary does not increase the order of singularity, but the differentiation with respect to x increases it. This nonsmoothness of Γu with respect to x_1 is closely connected with the term of boundary-layer type in asymptotics for $\Gamma e_H^+(x,k)$.

Note finally that all the statements concerning Γu on closed manifolds remain true for $\tau\Gamma u$ on manifolds with the boundary.

Thus in §3 we prove the normality of the great singularity of $\sigma(t)$, Γu (in the conditions of theorems 0.1, 0.8), $\Gamma q u$ (in condition (H.3)) and $\tau\Gamma u$; there we use the method of the energy integrals, microlocalization and the fact that for small t the singularities of u lie in the neighbourhood of the diagonal $x=y$.

In all the papers known to the author and using the method of the hyperbolic operator *) [8, 10-12, 18-20, 23, 24, 36-39, 43, 94, 95 etc] $u(x,y,t)$ was obtained by explicit construction in the form of a sum of oscillatory integrals. But in a general case such construction is complicated or even impossible because of either nonhyperbolicity of the equation (0.18), or ill-posedness of the problem (0.18)-(0.20) or variable multiplicity of characteristics or tangent to the boundary rays etc. In certain situations these difficulties was overcome; thus for the Laplace - Beltrami operator on manifolds with

*) With the exception of [82,87,88] where $u(x,y,t)$ was constructed for $|t|<max(x_1^\delta, y_1^\delta), 0<\delta<1$ by means of some variant of geometric optics.

strictly concave or convex boundary $u(x,y,t)$ was constructed and in the first case the asymptotics for $N(k)$ was derived (for Dirichlet and Neumann conditions) [10, 11, 76]; R.Melrose communicated to the author that he had derived the asymptotics for $N(k)$ in the second case too. On this way certain results were obtained for closed manifolds if the eigenvalues of the principal symbol are smooth and are not in involution [84]. But all these oscillatory integrals have an ill nature and in any case they are not Fourier oscillatory integrals [42].

At the same time $\sigma(t)$ and $\Gamma u(x,t)$ have a good nature for small t (this fact causes the normality of the great singularity of $\sigma(t)$, $\Gamma u(x,t)$, $\Gamma q u(x,t)$). The main idea of this book is to construct asymptotics (with respect to smoothness) for $\sigma(t)$, $\Gamma u(x,t)$ etc without construction of such asymptotics for $u(x,y,t)$. Moreover, the constructions of the asymptotics for $\sigma(t)$, $\tau\Gamma u(x,t)$, $\Gamma u(x,t)$ are parallel and independent and we have no need of asymptotics for $\Gamma u(x,t)$ to derive the asymptotics for $\sigma(t)$, $\tau\Gamma u(x,t)$.

In §4 we freeze coefficients of operators A, B in the point y (or in the point $(0,y')$) and construct the successive approximation series for $u(x,y,t)$; the terms of this series have the increasing orders of singularity but contain the increasing powers of t; the n-th term belong to the space $\sum_{j=0}^{M} t^j H_{loc}^{s(d)+n-j}$.

It appears that the terms of the corresponding series for $\sigma(t)$ and $\Gamma u(x,t)$ have the decreasing orders of singularity and thus we obtain the formal asymptotics for $\sigma(t)$ and $\Gamma u(x,t)$. We justify these asymptotics for small t by means of the theorems about normality of the great singularities of $\sigma(t)$ and $\Gamma u(x,t)$ respectively.

In the normal rays zone the asymptotics for $\Gamma u(x,t)$ is derived by separate construction: $u(x,y,t)$ is constructed modulo smooth function as a sum of Fourier oscillatory integrals by means of geometric optics (for this construction we need the first part of condition (H.4)) and then the asymptotics for $\Gamma u(x,t)$ is derived.

In §5 we prove the main theorems of part I by means of Tauberian theorem of Hörmander and its simple generalizations.

In appendices A, B we prove auxiliary statements, in appendix C we formulate one and in appendices D, E we make certain auxiliary calculations.

0.9. In conclusion we should like to give the sketch of another proof of theorems 0.11, 0.12; this proof seems to clarify the nature of condition " $B=\tau$ or

$$B=\tau\frac{\partial}{\partial n}+(i\varphi+B_1)\tau".$$

Let v be an eigenfunction of A_G with the negative eigenvalue $-k^2$; then from the elliptic with a parameter k equation

$$(\mathcal{A}+k^2)\,v=0$$

one can express v and $\tau\frac{\partial v}{\partial n}$ by means of $w=\tau v$. Exactly

$$v=h_k w,\quad \tau\frac{\partial v}{\partial n}=-q_k w$$

where h_k is Poisson (i.e. coboundary) operator belonging to the L.Boutet de Monvel algebra with a parameter k, q_k is first-order pseudo-differential with a parameter k operator on Y and modulo lower order terms $q_k\sim(|D'|^2+k^2)^{1/2}$,

$$h_k\sim exp(-x_1 q_k),$$

$$\|v\|_{L_2(X)}\sim\|(2q_k)^{-1/2}w\|_{L_2(Y)}.$$

Substituting $\tau\frac{\partial v}{\partial n}=q_k w$ into the boundary condition we obtain the "spectral" problem

(0.24) $$(-q_k+B_1+i\varphi)\,w=0.$$

Then modulo lower order terms

$v\sim exp(-x_1 B_1)w,\ \|v\|_{L_2(X)}\sim\|(2B_1)^{-1}w\|_{L_2(Y)}$.

It is easy to show that problem (0.24) is equivalent to problem

(0.25) $$(C-k^2)\,w=T_k w,$$

(0.26) $$(I-\Pi)\,w=T'_k w,$$

where $C \sim B_1^2 - |D'|^2$ is an elliptic second-order pseudo-differential operator, Π is commutating with C selfadjoint projector - pseudo-differential operator on Y with the principal symbol $\mathcal{E}$ (recall that $\mathcal{E}$ is a selfadjoint projector to invariant subspace of β, corresponding to eigenvalues greater then $|\xi'|$; β is the principal symbol of B_1), T_k and T_k' are pseudo-differential with a parameter k operators on Y of orders 1 and -1 respectively.

It is clear that problem (0.25) - (0.26) is closely related to problem

(0.27) $$(C-k^2)w=0, \quad w\in H=\Pi H^1(Y,E),$$

where in H the inner product $\langle w, w'\rangle = (2B_1 w, w)$ is introduced.

Therefore one can easily show that modulo lower order terms

$$K_{\Pi^-}(x,y) \sim 2B_1 \exp(-(x_1+y_1)B_1) K_\Pi(x',y'),$$

$$e^-(x,y,k) \sim 2B_1 \exp(-(x_1+y_1)B_1) e_H(x',y',k),$$

$$N^-(k) \sim N_H(k)$$

where $N_H(k)$, $e_H(x',y',k)$ are derived for C $H=\Pi H^1(X,E)$.

One can specify this sketch of proof and obtain the same remainder estimates as in theorems 0.11, 0.12.

Note that the hyperbolic operator for C equals $(-D_t^2+B_1^2+|D'|^2)$; the same (modulo lower order terms) operator can be obtained reducing problem (0.21) - (0.22) to the boundary by the method described above (i.e. one expresses $r\frac{\partial u}{\partial n}$ by means of ru from the elliptic equation (0.21), substitutes $r\frac{\partial u}{\partial n}$ into the boundary condition (0.22) and saves the derived equation from radical).

By the described method one can obtain the negative spectrum asymptotics in another case. Let B_1 be a pseudo-differential operator of order $m>1$, $m\in\mathbb{Z}^+$. Then the Šapiro - Lopatinskii condition holds is and only if $\beta_j \neq 0$ for all j, and A_G is semibounded from below if and only if $\beta_j<0$ for all j; recall that β_j

are eigenvalues of β_j, principal symbol of B_1. One can prove that modulo lower order terms

$$K_{\Pi^-} \sim 2 B_1 \exp(-(x_1+y_1)B_1)\, K_{\Pi}$$

where Π is the selfadjoint projector to positive invariant subspace of B_1 ; Π is a pseudo-differential operator with the principal symbol ε , the selfadjoint projector to positive invariant subspace of β ; now Π^- is not a classical singular Green operator.

The following asymptotics can be derived;

$$N^-(k) = \varkappa_1^- k^{\frac{d-1}{m}} + O(k^{\frac{d-2}{m}}),$$

$$e^-(x,x,k) = k^{\frac{d-1}{m}+1} \bar{Q}(x',x_1 k) + O(\min k^{\frac{d-1}{m}}, x^{-\frac{d-1}{m}}) \quad \text{as } k \to \infty,$$

where

$$\varkappa_1^- = \frac{(2\pi)^{-d+1}}{d-1} \int_{S^*Y} \operatorname{tr} \beta^{-\frac{d-1}{m}} \varepsilon \, d\theta \, dx',$$

$$Q^-(x',s) = \frac{2(2\pi)}{m} \int_{S^{d-2}} \beta^{-\frac{d-1}{m}} \varepsilon \, d\theta \, dx' \int_0^1 \alpha^{\frac{d-1}{m}} e^{-s\alpha} \, d\alpha.$$

Moreover, if the positive $\sigma_j = \beta_j$ satisfy conditions of theorem 0.2, then the following asymptotics holds:

$$N^-(k) = \varkappa_1^- k^{\frac{d-1}{m}} + \varkappa_2^- k^{\frac{d-2}{m}} + o(k^{\frac{d-2}{m}}) \qquad \text{as } k \to \infty .$$

Moreover, one can prove by this method that if $d=2$, a is positive definite and $\Pi = I$ then the asymptotics (0.7) holds for $N^-(k)$ with a remainder estimate $O(1)$ instead of $O(\ln k)$.

1. L.Boutet de Monvel operator algebra and selfadjoint projectors

1.0. We start from L.Boutet de Monvel results ([14], see [6, 86] too) which are of interest for us. Poisson (coboundary), boundary (trace) and singular Green operators considered here have type 0 with respect to L.Boutet de Monvel's terminology. We consider only the classical operators, i.e. the operators whose symbols can be expanded in series of positively homogeneous terms. Then we consider selfadjoint projectors to positive and negative invariant subspaces of even order elliptic operators.

1.1. At first let $X=\overline{\mathbb{R}}^+\times\mathbb{R}^{d-1}\ni x=(x_1,x')=(x_1,\ldots,x_d)$, $Y=0\times\mathbb{R}^{d-1}=\partial X$, $\overset{\circ}{X}=X\setminus Y=\mathbb{R}^+\times\mathbb{R}^{d-1}$. Let $H^s(\mathbb{R}^{d-1})$, $H^s(\mathbb{R}^d)$ be Sobolev spaces, $\mathcal{H}^{s,m}(X)=\{u,\ D_1^j u\in L_2(\mathbb{R}^+,H^{s-j}(\mathbb{R}^{d-1})),\ \forall j=0,\ldots,m\}$, where $s\in\mathbb{R}\cup\infty$, $m\in\mathbb{Z}^+\cup\infty$. The spaces $\mathcal{H}^{s,m}(X\times Y)$, $\mathcal{H}^{s,m}(X\times X)$ are introduced in a similar way; in the latter case m denotes the general number of derivatives with respect to x_1, y_1.

Let $OPS^m(\mathbb{R}^d)$, $OPS^m(\mathbb{R}^{d-1})$ be the spaces of properly supported classical pseudo-differential operators of the order m, $OPS^{m'}(X)=C^\infty(\overline{\mathbb{R}}^+,OPS^m(\mathbb{R}^{d-1}))$. The spaces of operators $OPS^{m'}(X\times Y)$, $OPS^{m'}(X\times X)$ are introduced in a similar way.

For $u\in\mathcal{H}^{t,n}(X)$ let us definite $WF_b^{s,n}(u)\subset T^*Y\setminus 0$: $\rho\in WF_b^{s,n}(u)$ if $a\in OPS^{0'}(X)$, $au\in\mathcal{H}^{s,n}(X)$ implies $a_0(\rho)=0$ where a_0 is the principal symbol of a; let us definite the full wave front sets

$$WF_f^{s,n}(u)=WF^s(u)\cup j^{-1}WF_b^{s,n}(u).$$

For $u(x,y)\in\mathcal{H}^{t,n}(X\times Y)$ we can introduce in a similar way $WF_b^{s,n}(u)\subset T^*(Y\times Y)\setminus 0$ and $WF_f^{s,n}(u)\subset T^*(X\times Y)\setminus 0$; for $u\in\mathcal{H}^{t,n}(X\times X)$, we can introduce

$$WF_{b,}^{s,n}(u)\subset T^*(Y\times\overset{\circ}{X})\setminus 0,$$

$$WF_{,b}^{s,n}(u)\subset T^*(\overset{\circ}{X}\times Y)\setminus 0,$$

$$WF_{b,b}^{s,n}(u)\subset T^*(Y\times Y)\setminus 0$$

and full wave front sets $WF_f^{s,n}(u) \subset T^*(X \times X)\setminus 0$.

In future we shall omit the sign ∞ either for s or n in the notations for the wave front sets.

1.2. A pseudo-differential operator $P \in OPS^m(\mathbb{R}^d)$ has the transmission property (with respect to Y) if $\forall j \in \mathbb{Z}^+$ $p_{m-j}(x,-\xi) - (-1)^{m-j} p_{m-j}(x,\xi)$ vanish with all their derivatives on NY; where p_{m-j} is positively homogeneous of degree $(m-j)$ term in asymptotic expansion of symbol P and NY is a normal bundle of Y.

Let $OPS^m(X)$ be a class of pseudo-differential operators of the same class on $\mathbb{R}^d$ with the transmission property with respect to Y. For $P \in OPS^m(X)$ we define an operator $P_X = r^+ P c^- : r^+ C_0^\infty(\mathbb{R}^d) \longrightarrow C^\infty(\mathring{X})$ where r^+ is the operator of restriction to X, c^- is the operator of continuation by zero outside of X.

Let $\mathcal{T}^m_{hom}(\mathbb{R}^{d-1} \times \overline{\mathbb{R}}^{+n} \times \mathbb{R}^{d-1})$ be the space of functions $k(x', \omega, \xi') \in C^\infty(\mathbb{R}^{d-1} \times \overline{\mathbb{R}}^{+n} \times (\mathbb{R}^{d-1} \setminus 0))$ such that estimates

$$|D^\alpha_{x'} D^\beta_{\xi'} D^\gamma_\omega k| \leq C_{\alpha\beta\gamma l} |\xi'|^{m-|\beta|+|\gamma|} (1+|\omega||\xi'|)^{-l} \qquad \forall \alpha, \beta, \gamma, l$$

hold and

$$k(x', \lambda^{-1}\omega, \lambda\xi') = \lambda^m k(x', \omega, \xi') \qquad \forall \lambda > 0 .$$

Let $\mathcal{T}^m(\mathbb{R}^{d-1} \times \overline{\mathbb{R}}^{+n} \times \mathbb{R}^{d-1})$ be the space of functions $k(x', \omega, \xi') \in C^\infty(\mathbb{R}^{d-1} \times \overline{\mathbb{R}}^{+n} \times \mathbb{R}^{d-1})$ such that estimates

$$|D^\alpha_{x'} D^\beta_{\xi'} D^\gamma_\omega k| \leq C_{\alpha\beta\gamma l} (1+|\xi'|)^{m-|\beta|+|\gamma|} (1+|\omega|+|\omega||\xi'|)^{-l} \qquad \forall \alpha, \beta, \gamma, l$$

hold, expanding into asymptotic series $k \sim \sum_{j=0}^{\infty} k_j$, $k_j \in \mathcal{T}^{m-j}_{hom}(\mathbb{R}^{d-1} \times \overline{\mathbb{R}}^{+n} \times \mathbb{R}^{d-1})$ in the sense that for $|\xi'| \geq 1$ estimates

$$| D^{\alpha}_{x'} D^{\beta}_{\xi'} D^{\gamma}_{\omega} (k - \sum_{j=0}^{N-1} k_j) | \leq$$

$$\leq C_{\alpha\beta\gamma\ell N} |\xi'|^{m-|\beta|+|\gamma|-N} (1+|\omega||\xi'|)^{-\ell} \quad \forall \alpha, \beta, \gamma, \ell, N$$

hold.

<u>Poisson (coboundary) operator</u> of order m is the operator $H: C_0^{\infty}(Y) \longrightarrow C^{\infty}(X)$ with the Schwartz kernel expressed modulo $C^{\infty}(X \times Y)$ by oscillatory integral

(1.1)
$$K_H(x, y') \sim (2\pi)^{-d+1} \int e^{i\langle x'-y', \xi'\rangle} k_H(x', x_1, \xi')\, d\xi' =$$

$$=(2\pi)^{-d} \int \exp i(\langle x'-y', \xi'\rangle + x_1 \xi_1)\, h(x', \xi)\, d\xi \quad ,$$

where $h(x', \xi) = \int_0^{\infty} e^{-ix_1\xi_1} k_H(x, \xi')\, dx_1$ is the symbol of H, $k_H \in \mathcal{T}^m(\mathbb{R}^{d-1} \times \overline{\mathbb{R}}^+ \times \mathbb{R}^{d-1})$,

$$h \sim \sum_{j=0}^{\infty} h_j \quad , \quad h_j = \int_0^{\infty} e^{-ix_1\xi_1} k_j(x, \xi')\, dx_1$$

are positively homogeneous of degrees $(m-j-1)$; h and h_j are analytic functions of $\xi_1 \in \mathbb{C}_- \setminus \mathbb{R}$.

h_0 is the <u>principal symbol</u> of H .

<u>Boundary (trace) operator</u> of order m is the operator $T: r^+ C_0^{\infty}(\mathbb{R}^d) \longrightarrow C^{\infty}(Y)$ with the Schwartz kernel expressed modulo $C^{\infty}(Y \times X)$ by oscillatory integral

(1.2)
$$K_T(x', y) \sim$$

$$\sim (2\pi)^{-d+1} \int e^{i\langle x'-y',\xi'\rangle} k_T(x',y_1,\xi')\,d\xi' =$$

$$= (2\pi)^{-d} \int \exp i(\langle x'-y',\xi\rangle - y_1\eta_1)\, t(x',\xi',\eta_1)\, d\xi'\, d\eta_1 ,$$

where $t(x',\eta_1,\xi') = \int_0^\infty e^{iy_1\eta_1} k_T(x',y_1,\xi')\,dy_1$
is the symbol of T, $k_T \in \mathcal{T}^{m+1}(\mathbb{R}^{d-1} \times \overline{\mathbb{R}}^+ \times \mathbb{R}^{d-1})$

$$t \sim \sum_{j=0}^{\infty} t_j , \quad t_j = \int_0^\infty e^{iy_1\eta_1} k_j(x',y_1,\xi')\,dy_1$$

are positively homogeneous of degrees $(m-j)$ with respect to (η_1,ξ') ; t and t_j are analytic functions of $\eta_1 \in \mathbb{C}_+ \setminus \mathbb{R}$.
t_0 is the principal symbol of T.

Singular Green operator of order m is the operator $G: r^+ C_0^\infty(\mathbb{R}^d) \longrightarrow C^\infty(X)$ with the Schwartz kernel expressed modulo $C^\infty(X \times X)$ by oscillatory integral

$$K_G(x,y) \sim (2\pi)^{-d+1} \int e^{i\langle x'-y',\xi'\rangle} k_G(x',x_1,y_1,\xi')\,d\xi' = \tag{1.3}$$

$$= (2\pi)^{-d-1} \int \exp i(\langle x'-y',\xi'\rangle + x_1\xi_1 - y_1\eta_1)\, g(x',\xi_1,\eta_1,\xi')\, d\xi\, d\eta_1$$

where

$$g(x',\xi_1,\eta_1,\xi') = \int_0^\infty\int_0^\infty k_G(x',x_1,y_1,\xi')\, e^{-ix_1\xi_1 + iy_1\eta_1}\, dx_1\, dy_1$$

is the symbol of G,

$k_G \in \mathcal{T}^{m+1}(\mathbb{R}^{d-1} \times \overline{\mathbb{R}}^{+2} \times \mathbb{R}^{d-1})$,

$$g \sim \sum_{j=0}^{\infty} g_j ,$$

$$g_j = \int_0^\infty\int_0^\infty k_j(x', x_1, y_1, \xi') e^{-ix_1\xi_1 + iy_1\eta_1}\, dx_1\, dy_1$$

are positively homogeneous of degrees $(m-j)$ with respect to (ξ, η_1); g and g_j are analytic functions of $\xi_1 \in \mathbb{C}_- \setminus \mathbb{R}$, $\eta_1 \in \mathbb{C}_+ \setminus \mathbb{R}$.

g_o is the principal symbol of G.

Let $OP^m(X)$, $OT^m(X)$, $OGS^m(X)$ be the spaces of properly supported Poisson, boundary and singular Green operators respectively.

PROPOSITION 1.1. Classes $OPS^m(X)$, $OP^m(X)$, $OT^m(X)$, $OGS^m(X)$ are invariant with respect to a change of coordinates preserving X and Y. The principal symbols of these operators are invariant functions on $T^*X \setminus 0$, $j^{-1}(T^*Y \setminus 0)$, $j^{-1}(T^*Y \setminus 0)$, $j_x^{-1} j_y^{-1}\, diag\, (T^*Y \setminus 0)^2$ respectibely.

PROPOSITION 1.2. If $P \in OPS^m(X)$, $H \in OP^m(X)$, $T \in OT^m(X)$, $G \in OGS^m(X)$ then

(i) There exist operators $Q_j \in OPS^{m-j}(Y)$ such that

$$r P_X - \sum_{j=0}^{m} Q_j r D_1^j \in OT^m(X).$$

(ii) Operators $C_o^\infty(Y) \ni v \longrightarrow r^+ P(v\delta(x_1))$, $v \longrightarrow r^+ G(v\delta(x_1))$ belong to $OP^{m+1}(X)$.

(iii) $rH \in OPS^m(Y)$; operator $C_o^\infty(Y) \ni v \longrightarrow T(v\delta(x_1))$ belong to $OPS^{m+1}(Y)$.

PROPOSITION 1.3. If $P \in OPS^m(X)$, $H \in OP^m(X)$, $T \in OT^m(X)$, $G \in OGS^m(X)$ then the following mappings are continuous:

(i) $P_X : \mathcal{H}_{loc}^{s,n}(X) \longrightarrow \mathcal{H}_{loc}^{s-m,\, n-m}(X)$ for $n \geq m$,

(ii) $H : H_{loc}^{s}(Y) \longrightarrow \mathcal{H}_{loc}^{s-m+1/2,\, \infty}(X) \cap C^\infty(\mathring{X})$,

(iii) $T : \mathcal{H}^{s,0}_{loc}(X) + \mathcal{E}'_{loc}(\mathring{X}) \longrightarrow H^{s-m-1/2}_{loc}(Y)$,

(iv) $G : \mathcal{H}^{s,0}_{loc}(X) + \mathcal{E}'_{loc}(\mathring{X}) \longrightarrow$

$$\mathcal{H}^{s-m,\infty}_{loc}(X) \cap C^{\infty}(\mathring{X}),$$

where $\mathcal{E}'_{loc}(\mathring{X})$ is the space of distributions $f \in \mathcal{D}'(\mathbb{R}^d)$ such that $\operatorname{supp} f \subset \mathring{X}$.

PROPOSITION 1.4. If $P \in OPS^m(X)$, $H \in OP^m(X)$, $T \in OT^m(X)$, $G \in OGS^m(X)$, then $H^* \in OT^{m-1}(X)$, $T^* \in OP^{m+1}(X)$, $G^* \in OGS^m(X)$ and for $m \leq 0$

$$(P_X)^* = (P^*)_X .$$

PROPOSITION 1.5. Let $P_j \in OPS^{m_j}(X)$, $Q_j \in OPS^{m_j'}(X)$, $R_j \in OPS^{m_j}(Y)$, $H_j \in OP^{m_j}(X)$, $T_j \in OT^{m_j}(X)$, $G_j \in OGS^{m_j}(X)$; then

(i) $P_{1X} P_{2X} - (P_1 P_2)_X \in OGS^{m_1+m_2}(X)$ for $m_2 \leq 0$,

(ii) $Q_1 H_2$, $P_{1X} H_2$, $H_1 R_2 \in OP^{m_1+m_2}(X)$.

(iii) $R_1 T_2$, $T_1 Q_2 \in OT^{m_1+m_2}(X)$;

$$T_1 P_{2X} \in OT^{m_1+m_2}(X) \quad \text{for} \quad m_2 \leq 0 .$$

(iv) $G_1 G_2$, $H_1 T_2$, $P_{1X} G_2$, $Q_1 G_2$,

$$G_1 Q_2 \in OGS^{m_1+m_2}(X); \quad G_1 P_{2X} \in OGS^{m_1+m_2}(X) \quad \text{for } m_2 \leq 0 .$$

(v) $T_1 H_2 \in OPS^{m_1+m_2}(Y)$.

(vi) $x_1^j H_1 \in OP^{m_1-j}(X)$, $T_1 x_1^k \in OT^{m_1-k}(X)$,

$$x_1^j G_1 x_1^k \in OGS^{m_1-j-k}(X);$$

moreover for $m_2 \leq k$

$$T_1 x_1^k P_{2X} \in OT^{m_1+m_2-k}(X),$$

$$G_1 x_1^K P_{2X} \in OGS^{m_1+m_2-k}(X).$$

REMARK. We consider only operators of the type 0 ; therefore we need the condition $m_2 \leq 0$ (or $m_2 \leq k$) in formulas containing P_2.

PROPOSITION 1.6. If $P \in OPS^m(X)$, $G \in OGS^m(X)$, $H \in OP^m(X)$, $T \in OT^m(X)$ then the following inclusions hold:

(i) ${}^T WF'^{,k}_f(K_P) \subset diag(T^*X\setminus 0)^2 \cup j_x^{-1} j_y^{-1} diag(T^*Y\setminus 0)^2$ for $m \leq -k - d/2$;

(ii) ${}^T WF_f(K_H) \subset j_x^{-1}\, diag(T^*Y\setminus 0)^2$;

(iii) ${}^T WF_f(K_T) \subset j_y^{-1}\, diag(T^*Y\setminus 0)^2$;

(iv) ${}^T WF_f(K_G) \subset j_x^{-1} j_y^{-1} diag(T^*Y\setminus 0)^2$,

where ${}^T M = \{(x,\xi,y,\eta) : (x,\xi,y,-\eta) \in M\}$.

PROPOSITION 1.7. If $P \in OPS^m(\mathbb{R}^d)$, $H \in OP^m(X)$, $T \in OT^m(X)$, $G \in OGS^m(X)$ then the next inclusions hold:

(i) $(x-y)^\alpha K_P \in C(\mathbb{R}^d_y, H^{-m+|\alpha|-\frac{d}{2}-\varepsilon}(\mathbb{R}^d_x))$,

(ii) $(x'-y')^{\alpha'} x_1^{\alpha_1} K_H \in C(\mathbb{R}^{d-1}_{y'}, \mathcal{H}^{1-m+|\alpha|-d/2-\varepsilon,\infty}(X_x))$,

(iii) $(x'-y')^{\alpha'} y_1^{\alpha_1} K_T \in C(\mathbb{R}^{d-1}_{y'}, \mathcal{H}^{-m+|\alpha|-d/2-\varepsilon,\infty}(\mathbb{R}^{d-1}_{x'} \times \overline{\mathbb{R}}^+_{y_1}))$,

(iv) $(x'-y')^{\alpha'} x_1^{\alpha_1} y_1^{\alpha_0} K_G \in C(\mathbb{R}^{d-1}_{y'}, \mathcal{H}^{-m+|\alpha|+1/2-d/2-\varepsilon,\infty}(\mathbb{R}^{d-1}_{x'} \times \overline{\mathbb{R}}^+_{x_1} \times \overline{\mathbb{R}}^+_{y_1}))$

for any $\varepsilon > 0$.

It goes without saying that we can permutate x and y, H and T and obtain new inclusions.

1.3. Propositions 1.1, 1.6 imply that one can define the operator classes $OPS^m(X,E,F)$, $OGS^m(X,E,F)$, $OP^m(X,E,F')$, $OT^m(X,E',F)$ if X is compact C^∞-manifold with the boundary $Y \in C^\infty$, E, F are C^∞-fiberings over X, E', F' are C^∞-fibering over Y. In what follows $E=F$, $E'=F'=E|_Y$ and we shall write simply $OPS^m(X,E)$,

$OGS^m(X,E)$, $OP^m(X,E)$, $OT^m(X,E)$.

Let us give an account of L.Boutet de Monvel's results concerning a parametrix in a very special case. Let $\mathcal{A}: C^\infty(X,E) \to C^\infty(X,F)$ be a differential operator of order m, $B = \{B_1, \dots, B_s\}$, $B_j = \sum_{k=0}^{m-1} B_{jk} \tau \nu^k$ be a boundary operator where $B_{jk} \in OPS^{\mu_j - k}(Y, F_j, E)$, ν is a C^∞ vector field transversal to Y in every point; certainly only B_j but not their components are invariant objects.

Consider the operator $\{\mathcal{A}, B\}: C^\infty(X,E) \to C^\infty(X,F) \oplus C^\infty(Y,F_1) \oplus \dots \oplus C^\infty(Y,F_s)$.

PROPOSITION 1.8. The following statements are equivalent:
(i) Operator $\{\mathcal{A}, B\}: H^{n+m}(X,E) \to H^n(X,F) \oplus H^{n+m-\mu_1-1/2}(Y,F_1) \oplus \dots \oplus H^{n+m-\mu_s-1/2}(Y,F_s)$

has the closed image and finite-dimensional kernel and cokernel.
(ii) There exist $Q \in OPS^{-m}(X, E, F)$, $G \in OGS^{-m}(X, E, F)$, $H_j \in OP^{-\mu_j}(X, E, F_j)$ such that operator

$$\mathcal{R}: H^n(X,F) \oplus H^{n+m-\mu_1-1/2}(Y,F_1) \oplus \dots \oplus H^{n+m-\mu_s-1/2}(Y,F_s) \to H^{n+m}(X,E),$$

$$\mathcal{R}(f, v_1, \dots, v_s) = (Q_X + G) f + H_1 v_1 + \dots + H_s v_s$$

is two-sided parametrix of $\{\mathcal{A}, B\}$.
(iii) $\{\mathcal{A}, B\}$ is elliptic, i.e. $\mathcal{A}$ is elliptic and $\{\mathcal{A}, B\}$ satisfies the Šapiro - Lopatinskii condition.

Moreover, Q is the parametrix of $\mathcal{A}$ on $\widetilde{X}$ -manifold without boundary containing X; $\widetilde{X}$ may be noncompact. The principal symbol of Q equals a^{-1} where a is the principal symbol of $\mathcal{A}$; the values of the principal symbols of G and H_k on $j_x^{-1} j_y^{-1}(x', \xi')$, $j^{-1}(x', \xi')$ respectively are defined by the values of principal symbols of $\mathcal{A}, B$ on $j^{-1}(x', \xi')$.

1.4. Let $E = F$ be a Hermitian fibering, dx density on X and $\{\mathcal{A}, B\}$ an elliptic operator. Let $\mathcal{A}_B$ be the restriction of $\mathcal{A}$ to $C_0^\infty(X, E) \cap \operatorname{Ker} B$. We suppose that $\mathcal{A}_B$ is a symmetric operator in $L_2(X, E)$. Let A_B be a closure of $\mathcal{A}_B$ in $L_2(X, E)$. Then A_B is selfadjoint operator in $L_2(X, E)$, its spectrum is discrete, with finite multiplicity, and tends either to $\pm\infty$, or to $+\infty$, or to $-\infty$. Assume that 0 is not an eigenvalue of A_B . Then

$$(1.4) \qquad \Pi^{\pm} = \frac{1}{2} \pm \frac{1}{\pi}\int_0^{\infty} (A_B^2 + \lambda^2)^{-1} A_B \, d\lambda$$

are selfadjoint projectors to positive and negative invariant subspaces of A_B.

THEOREM 1.9. Let X be a closed compact manifold, $A \in OPS^{m}(X,E)$ be elliptic and selfadjoint and 0 be not an eigenvalue of A . Then $\Pi^{\pm} \in OPS^{\circ}(X,E)$ and $\pi^{\pm}$ — the principal symbols of $\Pi^{\pm}$ are the selfadjoint projectors to positive and negative invariant subspaces of a (the principal symbol of A).

THEOREM 1.10. Let X be a compact manifold with the boundary, $\mathcal{A}$ differential operator of order $m = 2s$, $F_1 = \cdots = F_s = E|_Y$,

$$B_j = \gamma \nu^{\mu_j} + \sum_{k=0}^{\mu_j - 1} B_{jk} \gamma \nu^{k}$$

where $0 \leqslant \mu_1 < \mu_2 < \ldots < \mu_s \leqslant 2s-1$, $B_{jk} \in OPS^{\mu_j - k}(Y,E)$. Suppose that $\{\mathcal{A}, B\}$ is elliptic, $\mathcal{A}_B$ is symmetric and 0 is not an eigenvalue of A_B . Then $\Pi^{\pm} = \Pi^{\circ\pm} + \Pi'^{\pm}$ where $\Pi^{\circ\pm} \in OPS^{\circ}(X,E)$, $\Pi'^{\pm} \in OGS^{\circ}(X,E)$. Moreover, $\pi^{\pm}$ the principal symbols of $\Pi^{\circ\pm}$ — are the selfadjoint projectors to positive and negative invariant subspaces of a ; for every $(x,\xi') \in T^*Y \setminus 0$ the values of $\Pi_0'^{\pm}$ — the principal symbols of $\Pi'^{\pm}$ — on $j_x^{-1} j_y^{-1}(x',\xi')$ are defined by the values of the principal symbols of $\mathcal{A}, B$ on $j^{-1}(x',\xi')$.

PROOF. We change variable and rewrite (1.4)

$$(1.4') \qquad \Pi^{\pm} = \frac{1}{2} + \frac{m}{\pi} A_B \int_0^{\infty} (A_B^2 + z^{2m})^{-1} z^{m-1} \, dz .$$

If X is a closed manifold then $A^2 + z^{2m}$ is elliptic with a parameter z operator.

If X is a manifold with the boundary then $\{\mathcal{A} \pm i z^{m}, B\}$

are elliptic with a parameter z operators, i.e. the following estimates hold:

(1.5)
$$\sum_{l=0}^{m} (1+z)^{m-l} \|u\|_l \leq$$
$$C\|(\mathcal{A} \pm iz^m)u\|_0 + C\sum_{j=0}^{s} \|(\Lambda^2+z^2)^{(m-\mu_j-1/2)/2} B_j u\|_{Y,0}$$

for all u, $z>0$, where $\Lambda \in OPS^1(Y, E)$ is elliptic, selfadjoint and invertible operator, C does not depend on u, z. Really, (1.5) hold obviously for $u \in \operatorname{Ker} B$. To prove (1.5) for an arbitrary u it is sufficient (because of the special type of B) to construct operators $E_k(z): C^\infty(Y, E) \longrightarrow C^\infty(X, E)$ such that

$$\|E_k(z)v\|_j \leq C\|(\Lambda^2+z^2)^{-\frac{1}{4}-\frac{k}{2}+\frac{j}{2}} v\|_{Y,0} \quad \forall j \in \mathbb{Z}^+,$$
$$\gamma\nu^j E_k(z)=0 \quad \forall j<k,$$
$$\gamma\nu^k E_k(z) = I.$$

But if we identify the small neighbourhood of Y with $[0,\delta)\times Y \ni (x_1,x')$ then the following operators are appropriate:

$$E_k: v \longrightarrow \frac{x_1^k}{k!} \exp(-(\Lambda^2+z^2)^{1/2} x_1)\, \varphi(x_1)v$$

where $\varphi \in C^\infty(\mathbb{R})$, $\varphi = 1$ for $x_1 < \delta/3$, $\varphi=0$ for $x_1 > \delta/2$. Certainly, E_k are the Poisson operators of order k with a parameter z.

Therefore operators $\{\mathcal{A} \pm iz^m, B\}$ have parametrices $\mathcal{R}_\pm$, $\mathcal{R}_\pm(f, v_1,\dots,v_s)=(Q_{\pm X}+G_\pm)f+H_{\pm 1}v_1+\dots+H_{\pm s}v_s$, where $Q_\pm$ are pseudo-differential operators of order $-m$ with a parameter z, $G_\pm$ are singular Green operators of order $-m$ with a parameter z, $H_{\pm j}$ are Poisson operators of order $-\mu_j$ with a parameter z.

This means that

(i) Q is pseudo-differential operator with the symbol $q(x,\xi,z)$ satisfying the estimates

$$|D_x^\alpha D_\xi^\beta q(x,\xi,z)| \leq C_{\alpha\beta}(1+|\xi|+z)^{-m}(1+|\xi|)^{-|\beta|} \quad \forall \alpha,\beta^{*)}$$

(sign $\pm$ is omited) expanding in asymptotic series

$$q(x,\xi,z) \sim \sum_{n=0}^{\infty} q_n(x,\xi,z)$$

where q_n are positively homogeneous of degree $(-m-n)$ with respect to (ξ, z) , satisfy the estimates

$$|D_x^\alpha D_\xi^\beta q_n(x,\xi,z)| \leq C_{\alpha\beta n}(|\xi|+z)^{-m}|\xi|^{-n-|\beta|} \quad \forall \alpha,\beta,n ,$$

in the sense that for $|\xi| \geq 1$

$$|D_x^\alpha D_\xi^\beta (q - \sum_{n=0}^{N-1} q_n)| \leq$$

$$\leq C_{\alpha\beta N}(|\xi|+z)^{-m}|\xi|^{-N-|\beta|} \quad \forall \alpha,\beta,N ;$$

neglible operators (i.e. operators of order $-\infty$) are operators satisfying the following estimates:

$$(1+z)^m \|Qu\|_s \leq C_{st}\|u\| \quad \forall s,t,u,\ z>0 .$$

(ii) H has Schwartz kernel expressed modulo Schwartz kernel of neglible operator by oscillatory integral (1.1) with $k(x,\xi',z)$

*) One can improve these and following estimates for $n+|\beta|>0$; one can essentially improve these estimates for $n+|\beta|>0$ if $\mathcal{A}$ is a differential operator.

satisfying the estimates

$$|D_x^{\alpha} D_{\xi'}^{\beta} k(x,\xi',z)| \leq C_{\alpha\beta\ell}(1+|\xi'|+z)^{-\mu+\alpha_1}(1+|\xi'|)^{-|\beta|}(1+x_1(1+|\xi'|+z))^{-\ell} \quad \forall \alpha,\beta,\ell$$

(indices j, $\pm$ are omited), expanding in asymptotic series

$$k \sim \sum_{n=0}^{\infty} k_n$$

where $k_n(x', \lambda^{-1}x_1, \lambda\xi', \lambda z) = \lambda^{-\mu-n} k_n(x', x_1, \xi', z) \quad \forall \lambda > 0$,

$$|D_x^{\alpha} D_{\xi'}^{\beta} k_n(x,\xi',z)| \leq C_{\alpha\beta\ell n}(|\xi'|+z)^{-\mu+\alpha_1}|\xi'|^{-n-|\beta|}(1+x_1(|\xi'|+z))^{-\ell} \quad \forall \alpha,\beta,\ell,n,$$

in the sense that for $|\xi'| \geq 1$

$$|D_x^{\alpha} D_{\xi'}^{\beta}(k - \sum_{n=0}^{N-1} k_n)| \leq C_{\alpha\beta\ell N}(|\xi'|+z)^{-\mu+\alpha_1}|\xi'|^{-|\beta|-N}(1+x_1(|\xi'|+z))^{-\ell} \quad \forall \alpha,\beta,\ell,N;$$

neglible operators (i.e. operators of order $-\infty$) are operators satisfying the following estimates:

$$z^{\ell}\|x_1^j D_1^i H v\|_{s,0} \leq C_{ij\ell st}\|v\|_{Y,t} \quad \forall s,t \in \mathbb{R},\ \ell \in \overline{\mathbb{R}}^+,\ i,j \in \mathbb{Z}^+$$

such that $j+\mu+1/2 \geq \ell+i$.

(iii) G has Schwartz kernel expressed modulo Schwartz kernel of neglible operator by oscillatory integral (1.3) with $k(x', x_1, y_1, \xi', z)$ satisfying the estimates

$$|D_x^{\alpha} D_{y_1}^{\alpha_0} D_{\xi'}^{\beta} k(x,y_1,\xi',z)| \leq$$

$$C_{\alpha\alpha_0\beta\ell}(1+|\xi'|+z)^{1-m+\alpha_0+\alpha_1}(1+|\xi'|)^{-|\beta|}(1+(x_1+y_1)(1+|\xi'|+z))^{-\ell} \quad \forall \alpha,\alpha_0,\beta,\ell,$$

$$k \sim \sum_{n=0}^{\infty} k_n \quad ,$$

where $k_n(x', \lambda^{-1}x_1, \lambda^{-1}y_1, \lambda\xi'; \lambda z) = \lambda^{1-m-n} k_n(x', x_1, y_1, \xi', z) \quad \forall \lambda > 0$,

$$| D_x^{\alpha} D_{y_1}^{\alpha_0} D_{\xi'}^{\beta} k_n(x, y_1, \xi', z) | \leq$$

$$C_{\alpha\alpha_0\beta\ell n} (|\xi'|+z)^{1-m+\alpha_0+\alpha_1} |\xi'|^{-n-|\beta|} (1+(x_1+y_1)(|\xi'|+z))^{-\ell} \qquad \forall \alpha, \alpha_0, \beta, \ell, n,$$

in the sense that for $|\xi'| \geq 1$

$$| D_x^{\alpha} D_{y_1}^{\alpha_0} D_{\xi'}^{\beta} (k - \sum_{n=0}^{N-1} k_n) | \leq$$

$$C_{\alpha\alpha_0\beta\ell n} (|\xi'|+z)^{1-m+\alpha_0+\alpha_1} |\xi'|^{-N-|\beta|} (1+(x_1+y_1)(|\xi'|+z))^{-\ell} \qquad \forall \alpha, \alpha_0, \beta, \ell, N;$$

neglible operators (i.e. operators of order $-\infty$) are operators satisfying the following estimates:

$$z^{\ell} \| x_1^{j} D_1^{i} G x_1^{\rho} D_1^{q} v \|_{s,0} \leq C_{ij\ell pqst} \|v\|_t$$

$$\forall s,t \in \mathbb{R}, \ell \in \overline{\mathbb{R}}^+, \quad i,j,p,q \in \mathbb{Z}^+$$

$$j+p+m \geq i+\ell+q \qquad \forall v \in C_0^{\infty}(\mathring{X}).$$

Let

$$q = \int_0^{\infty} Q_+(z) Q_-(z) z^{m-1} dz ,$$

$$g = \int_0^{\infty} (Q_+(z) G_-(z) + Q_-(z) G_+(z) + G_+(z) G_-(z) z^{m-1} dz . \tag{1.6}$$

It is easy to prove that q is a classical pseudo-differential operator of order $-m$; if A is differential operator then $q_j(x, -\xi) = (-1)^j q_j(x, \xi)$ where $q_j(x, \xi)$ is positively homogeneous of degree $(-m-j)$ term in the expansion of q and therefore q is the operator with the transmission property for even m. Also it is proved that if X is either a closed manifold or a manifold with the boundary then $\Pi^{0\pm} = \frac{1}{2} \pm \frac{m}{2} \mathcal{A} q \in OPS^{0}(X, E)$. Replacing in formula (1.4) operators $\Pi^{\pm}, A$ and $(A_B^2 + z^{2m})^{-1}$ by their principal symbols $\Pi^{\pm}$, a and $(a^2 + z^{2m})^{-1}$ respectively we obtain that $\Pi^{\pm}$ are the selfadjoint projectors to positive and negative subspaces of a.

Theorem 1.9 has been proved. But it does not follow from the direct consideration of the right side of (1.6) that $g \in OGS^{-m}(X,E)$; it follows only that $g \in GOGS^{-m,m-2}(X, E)$ *), where $GOGS^{-\mu,h}(X,E)$ is a class of operators with Schwartz kernels expressed modulo Schwartz kernels of neglible operators by oscillatory integrals (1.3) with $k(x, y_1, \xi')$ satisfying the estimates $x_1^i y_1^j |D_x^{\alpha} D_{y_1}^{\alpha_0} D_{\xi'}^{\beta} k(x, y_1, \xi')| \leq C_{\alpha\alpha_0\beta ij\ell} (1+|\xi'|)^{\mu-|\beta|+1+\alpha_0+\alpha_1-i-j} (1+(x_1+y_1)(1+|\xi'|))^{-\ell}$ $\forall \alpha, \alpha_0, \beta, i, j, \ell$ such that $\alpha_0 + \alpha_1 \leq i + j + h$, expanding in asymptotic series

$$k \sim \sum_{n=0}^{\infty} k_n$$

where $k_n(x', \lambda^{-1} x_1, \lambda^{-1} y_1, \lambda\xi') = \lambda^{\mu+1-n} k_n(x', x_1, y_1, \xi') \quad \forall \lambda > 0$,

$$x_1^i y_1^j |D_x^{\alpha} D_{y_1}^{\alpha_0} D_{\xi'}^{\beta} k_n(x, y_1, \xi')| \leq$$

$$\leq C_{\alpha\alpha_0\beta ij\ell n} |\xi'|^{\mu-|\beta|+1+\alpha_0+\alpha_1-i-j-n} (1+(x_1+y_1)|\xi'|)^{-\ell}$$

$\forall \alpha, \alpha_0, \beta, i, j, \ell, n$ such that $\alpha_0 + \alpha_1 \leq i + j + h$, in the sense that for $|\xi'| \geq 1$

$$x_1^i y_1^j |D_x^{\alpha} D_{y_1}^{\alpha_0} D_{\xi'}^{\beta} (k - \sum_{n=0}^{N-1} k_n)| \leq$$

$$\leq C_{\alpha\alpha_0\beta ij\ell N} |\xi'|^{\mu-|\beta|+1+\alpha_0+\alpha_1-i-j-N} (1+(x_1+y_1)|\xi'|)^{-\ell}$$

---

*) Or some stronger conclusion.

$\forall\alpha,\alpha_0,\beta,i,j,\ell,N$ such that $\alpha_0+\alpha_1\leqslant i+j+h$; negligible operators (i.e. operators of order $-\infty$) are operators satisfying the following estimates:

$$\| x_1^i D_1^p g x_1^j D_1^q u \|_{s,0} \leqslant C_{ijpqst} \|u\|_{t,0}$$

$$\forall u \in C_0^\infty(\mathring{X}),\quad s,t\in \mathbb{R},\quad i,j,p,q\in \mathbb{Z}^+$$

such that $p+q\leqslant i+j+h$.

One can correctly define the described operator class in the case when X is a compact manifold with the boundary, E is a fibering over X . It is easy to prove that if $g\in GOGS^{\mu,h}(X,E)$ then $\operatorname{sing\,supp} K_g \subset Y\times Y$ but this inclusion is not valid for the product of two operators of the class $GOGS^{\mu,h}(X,E)$ and therefore this operator class is nonclosed with respect to multiplication.

Let us prove that operator g defined by (1.6) belongs to $GOGS^{-m,N}(X,E)$ for every N ; then $g\in OGS^{-m}(X,E)$ since in local coordinates all terms in the asymptotic expansion of K_g are defined by g uniquely.

Let $X_1=[0,\delta]\times Y$ be identified with some closed neighbourhood of Y in X, $\partial X_1 = Y\cup Y_1$, $E_1=[0,\delta]\times E'$, $E'=E|_Y$, E_1 identified with the restriction of E to X_1, $\nu = iD_1$.

Decompose E' in the sum of fiberings E'_+ and E'_- whose fibers are positive and negative invariant subspaces of $a_0(0,x')$ respectively, where $a_0(x)$ is the coefficient at D_1^m in $\mathcal{A}$. Then

$$(1.7)\quad \mathcal{A}=\begin{pmatrix} a_0^+(x') & 0\\ 0 & -a_0^-(x')\end{pmatrix} D_1^m + a_0'(x)x_1 D_1^m + a_1(x,D')D_1^{m-1}+\cdots+a_m(x,D'),$$

$$B_j=(iD_1)^{\mu_j}\tau + B_{j1}(iD_1)^{\mu_j-1}+\cdots+B_{j\mu_j}\tau ,$$

where a_k are differential operators of order k, $B_{jk}\in OPS^k(Y,E')$, $a_0^\pm(x')$ are positive definite.

Let $\bar{\mathcal{A}}^\pm = a_0^\pm(x')D_1^m + \Lambda^\pm$, $\Lambda^\pm(x,D')$ are differential operators of order m , $\bar{B}^\pm=(\bar{B}_1^\pm,\ldots,\bar{B}_s^\pm, C_1^\pm,\ldots,C_s^\pm)$, $\bar{B}_j^\pm=\tau(iD_1)^{\mu_j}$, $C_j^\pm=\tau_1(iD_1)^{\mu_j}$, τ and τ_1 are operators of the restriction to Y and Y_1 respectively; then for appropriate $\Lambda^\pm$ operators $\{\bar{A}^\pm,\bar{B}^\pm\}$ are elliptic and operators $\bar{\mathcal{A}}^\pm_{\bar{B}^\pm}$ are symmetric and positive definite. Let

$$\bar{\mathcal{A}} = \begin{pmatrix} \bar{\mathcal{A}}^+ & 0 \\ 0 & -\bar{\mathcal{A}}^- \end{pmatrix}, \quad \bar{B} = \begin{pmatrix} \bar{B}^+ & 0 \\ 0 & \bar{B}^- \end{pmatrix}.$$

Then $\mathcal{A} = \bar{\mathcal{A}} + \mathcal{A}'$, $\quad B_j = \bar{B}_j + B_j'$,

$$\mathcal{A}' = a_0'(x)x_1 D_1^m + a_1' D_1^{m-1} + \dots + a_m',$$

$$B_j' = B_{j1}\tau (iD_1)^{\mu_j - 1} + \dots + B_{j\mu_j}\tau$$

where $\quad a_k' = a_k'(x, D')$ $\quad$ are differential operators of order k.

Since for every f

$$(\bar{\mathcal{A}} \pm iz^m)(A_B \pm iz^m)^{-1} f = f - \mathcal{A}'(A_B \pm iz^m)^{-1} f,$$

$$\bar{B}_j (A_B \pm iz^m)^{-1} f = -B_j'(A_B \pm iz^m)^{-1} f$$

(and some conditions on Y_1 are given) then

$$(A_B \pm iz^m)^{-1} = (\bar{A}_{\bar{B}} \pm iz^m)^{-1} - (\bar{A}_{\bar{B}} \pm iz^m)^{-1} \mathcal{A}'(A_B \pm iz^m)^{-1} -$$

$$\sum_{j=1}^{s} \bar{H}_j^+(z) B_j' (A_B \pm iz^m)^{-1} + \dots$$

where $\bar{H}_j^{\pm}(z)$ $\quad$ are Poisson operators constructed for $\{\bar{\mathcal{A}}, \bar{B}\}$ and acting from Y, points denote terms containing Poisson operators acting from Y_1.

Repeating this procedure N_2 times and multiplying the expressions obtained it follows that $(A_B^2 + z^{2m})^{-1}$ is the sum of the terms where each term is either product in some order of n factors $(\bar{A}_{\bar{B}} \pm iz^m)^{-1}\mathcal{A}'$, $\quad \bar{H}_j^{\pm}(z) B'$ $\quad$ and two factors $(\bar{A}_{\bar{B}} \pm iz^m)^{-1}$ or product in some order of $n > N$ factors $(\bar{A}_{\bar{B}} \pm iz^m)^{-1}\mathcal{A}'$, $\bar{H}_j^{\pm}(z) B'$ $\quad$ and two factors $(\bar{A}_{\bar{B}} \pm iz^m)^{-1}$, $(A_B \pm iz^m)^{-1}$ *) or the term which one can estimate with all its derivatives for $x_1 + y_1 < \delta$ $\quad$ by $const(1 + z^2)^{-l}$ $\forall l$.

Multiplying by z^{m-1} and integrating we obtain that every

---

*) Certainly the coefficients at these products are absolute constants.

product of $(n+1)$ factors gives an operator belonging to $OPS^{-m}(X,E) + GOGS^{-m,\, m+n-2}(X,E)$. Since we want to prove that $g \in GOGS^{-m,N}(X,E)$ we need only to consider only the sum of terms which appear after integrating of products of $n \leq N$ factors and to prove that this sum belongs to $OPS^{-m}(X,E)+GOGS^{-m,N}(X,E)$.

Let $\bar{\bar{\mathcal{A}}} = \begin{pmatrix} \bar{\mathcal{A}}^{+} & \underline{0} \\ 0 & \mathcal{A}^{-} \end{pmatrix}$. Then for "small" $\mathcal{A}', \underline{B}'$ operator $\{\bar{\bar{\mathcal{A}}} + \mathcal{A}', \bar{B} + B'\}$ is elliptic and if operator $(\bar{\bar{\mathcal{A}}} + \mathcal{A}')_{\bar{B}+B'}$ is symmetric then it is positive definite and operator Π^{-} constructed for it equals 0 ; therefore

$$\int_0^\infty [(\bar{\bar{\mathcal{A}}} + \mathcal{A}')^2_{\bar{B}+B'} + z^{2m}]^{-1} z^{m-1} dz =$$

$$\frac{\pi}{2m} (\bar{\bar{\mathcal{A}}} + \mathcal{A}')^{-1}_{\bar{B}+B'} \in OPS^{-m}(X,E) + OGS^{-m}(X,E) .$$

One can apply the procedure described above for the operator $(\bar{\mathcal{A}} + \mathcal{A}')_{\bar{B}+B'}$ to the operator $(\bar{\bar{\mathcal{A}}} + \mathcal{A}')_{\bar{B}+B'}$. Therefore: If $\mathcal{A}', B'$ are "small" and $(\bar{\bar{\mathcal{A}}} + \mathcal{A}')_{\bar{B}+B'}$ is symmetric then the sum of terms oftained by means of this procedure belongs to $OPS^{-m}(X,E)+GOGS^{-m,N}(X,E)$.

We want to discard the conditions of smallness and symmetricity in this statement. The operator $(\bar{\bar{\mathcal{A}}} + \mathcal{A}')_{\bar{B}+B'}$ is called ℓ-symmetric if one can symmetrize this operator by the change of a'_r, $B_{j,r-1}$ for $r > m-\ell$. The operator is 0-symmetric if and only if it is a symmetric; the arbitrary operator of the type described is $(m+1)$-symmetric. We state that the following statement holds:

Let $(\bar{\bar{\mathcal{A}}} + \mathcal{A}')_{\bar{B}+B'}$ be ℓ-symmetric. Let us choose only the products containing precisely p_r factors with a'_r $(m \geq r > m-\ell)$ and q_{jr} factors with $B_{j,r-1}$ $(m \geq r > m-\ell)$ and replace a'_r in its s-th entry by α_{r,i_s} and replace $B_{j,r-1}$ in its t-th entry by $\beta_{j,r-1,k_t}$ $(s=1,\dots,p_r,\ t=1,\dots,q_{jr})$ where α_{ri} are differential with respect to x' operators of order r, $\beta_{j,r-1,k} \in OPS^{r-1}(Y,E')$, $\{i_1,\dots,i_{p_r}\}$ and $\{k_1,\dots,k_{q_{jr}}\}$ are the permutions of the numbers $\{1,\dots,p_r\}$ and $\{1,\dots,q_{jr}\}$. Let us symmetrize the derived expressions with respect to all permutations and sum with respect to all selected terms. Then after integrating with respect to $z^{m-1}dz$ we obtain an operator belonging to $OPS^{-m}(X,E)+GOGS^{-m,N}(X,E)$.

We use the unduction on l. For $l=0$ this statement is proved. Let this statement be proved for $l=\bar{l}$. Note that $\bar{l}$-symmetricity of $(\bar{\bar{\mathcal{A}}} + \mathcal{A}')_{\bar{B}+B'}$ remains if we replace $a'_{m-\bar{l}}$ by

$$a'_{m-\bar{l}} + \lambda_1 \alpha_{m-\bar{l},1} + \dots + \lambda_{p_{m-\bar{l}}} \alpha_{m-\bar{l},p-\bar{l}}$$

and replace $B_{j,m-\bar{l}-1}$ by

$$B_{j,m-\bar{l}-1} + \lambda_{j1} \beta_{j,m-\bar{l}-1,1} + \dots + \lambda_{j,q_{j,m-\bar{l}}} \beta_{j,m-\bar{l}-1,q_{j,m-\bar{l}}}$$

where $\alpha_{m-\bar{l},s}$ are differential with respect to x' operators of order $m-\bar{l}$, $\beta_{j,m-\bar{l}-1,t} \in OPS^{m-\bar{l}-1}(Y,E)$, $\alpha_{m-\bar{l},s}$ and $\beta_{j,m-\bar{l}-1,t}$ are symmetric and $\lambda_s, \lambda_{jt} \in \mathbb{R}$. Therefore if we choose in every sum of previous step only the products containing precisely $p_{m-\bar{l}}$ factors with $a'_{m-\bar{l}}$ and $q_{j,m-\bar{l}}$ factors with $B_{j,m-\bar{l}}$ and replace $a'_{m-\bar{l}}$, $B_{j,m-\bar{l}}$ by $\alpha_{m-\bar{l},i_s}$, $\beta_{j,m-\bar{l},k_t}$, symmetrize with respect to all permutations $\{i_1,\dots,i_{p_{m-\bar{l}}}\}$, $\{k_1,\dots,k_{j,m-\bar{l}}\}$ and sum with respect to all selected terms, then after integrating with respect to $z^{m-1}dz$ every new (sub) sum gives an operator belonging to $OPS^{-m}(X,E) + GOGS^{-m,N}(X,E)$ provided operators $\alpha_{m-\bar{l},s}$, $\beta_{j,m-\bar{l}-1,t}$ are symmetric.

Sinse every new sum is linear with respect to each operator $\alpha_{m-\bar{l},s}$, $\beta_{j,m-\bar{l}-1,t}$ (and polylinear with respect to their totality) then we can discard the assumption that these operators are "symmetric". The step of induction is made.

Set now $\alpha_{\tau s} = a'_{\tau}$, $\beta_{j,\tau-1,t} = B_{j,\tau-1}$ and sum all derived subsums with appropriate coefficients which are absolute constants; we obtain the original sum again.

Thus, the sum derived by the canonical procedure for $(\bar{\bar{\mathcal{A}}}+\mathcal{A}')_{\bar{B}+B'}$ belongs to $OPS^{-m}(X,E) + GOGS^{-m,N}(X,E)$ even if $\mathcal{A}', B$ are not small and $(\bar{\bar{\mathcal{A}}}+\mathcal{A}')_{\bar{B}+B'}$ is not symmetric.

But this sum for $(\bar{\bar{\mathcal{A}}}+\mathcal{A}')_{\bar{B}+B'}$ coincides with the similar sum for $(\bar{\bar{\mathcal{A}}}+J\mathcal{A}')_{\bar{B}+JB'}$ right-multiplied by $J=\begin{pmatrix} I & 0 \\ 0 & -I \end{pmatrix}$ and therefore this sum belongs to $OPS^{-m}(X,E)+GOGS^{-m,N}(X,E)$ too. Therefore $g \in OGS^{-m}(X,E)$ and $\Pi^{\pm} \in OPS^{0}(X,E) + OGS^{0}(X,E)$.

It is clear that the principal symbols of g and $\Pi'^{\pm}$ depend only on the principal symbols of $\mathcal{A}, B$. Theorem 1.10 has been proved.

1.5. Assume now that a is positive definite; this means precisely that $\Pi^{-}=0$ (and then $\Pi^{0-}\sim 0$); we want to know when $\Pi'^{-}\sim 0$, i.e. when A_B is semibounded from below. We consider only secon-order operators.

We suppose that $B=\tau\nu+B_1\tau$ because of for $B=\tau$ operator $A_B = A_D$ is undoubtedly semibounded from below. One can assume without a loss of generality that operator A_D is positive definite and invertible. Then there exists an operator $M\in OPS^1(Y,E): \tau\nu u = M\tau u$ for every u such that $\mathcal{A}u=0$. One can define this operator modulo infinitely smoothing operator by factorization of $\mathcal{A}$:

$$
(1.8)\qquad \mathcal{A}=-a_0\nu^2+2i\mathcal{A}_1\nu+\mathcal{A}_2\sim(\nu+\overset{*}{M})a_0(\nu-M)
$$

where $\mathcal{A}_j$ are differential operators of order j with respect to x', $a_0=\mathcal{A}_0$, $M\in OPS^{1'}(X, E)$ and all eigenvalues of m – the principal symbol of M – have negative real parts. Since $\mathcal{A}$ is formally selfadjoint then a_j – the principal symbols of $\mathcal{A}_j$ – are Hermitian matrices. The following formula for the principal symbol of M holds:

$$
m=\{-\pi a_0^{-1}+i\imath\imath ai\int_{-\infty}^{\infty} z\,a^{-1}(z)\,dz\}\{\int_{-\infty}^{\infty} a^{-1}(z)\,dz\}^{-1}
$$

where $a(z)=a_0z^2+2a_1z+a_2$.

It is easy to show that $\mathcal{A}_B$ is symmetric if and only if $a_0(M+B_1)$ is symmetric and $\{\mathcal{A}, B\}$ is elliptic if and only if $a_0(M+B_1)$ is elliptic.

Similarly for elliptic with a parameter k operator $\mathcal{A}+k^2$ one can definite the first-order pseudo-differential with a parameter k operator $M(k)$ with the principal symbol $m(k)$. Then $u\in Ker\,(A_B+k^2)$ implies $\tau=v\in Ker\,a_0(M(k)+B_1)$ and $v.v.: v\in Ker\,a_0(M(k)+B_1)$ implies that $u=H(k)v\in Ker\,(A_B+k^2)$ where $H(k)$ is a Poisson operator with a parameter k : $u=H(k)v$ if $(\mathcal{A}+k^2)u=0$, $\tau u=v$.

PROPOSITION 1.11. The following statements are equivalent:

(i) A_B is semibounded from below.

(ii) $\{\mathcal{A} + k^2, B\}$ is elliptic with a parameter k.
(iii) $a_0(M(k) + B_1)$ is elliptic with a parameter k.
(iv) $a_0(m + b)$ is negative definite where b is the principal symbol of B_1.

PROOF. It is easy to prove the equivalence of (i), (ii), (iii) and we leave it for the reader. Let $m(k)$ be the principal symbol of $M(k)$ (with respect to ξ', k). Since $m(k) = -|k| a_0^{-1} + O(1)$ as $k \to \infty$ then (iii) implies (iv); moreover, we have proved that $a_0(M(k) + B_1)$ is elliptic with a parameter k if $a_0(m(k) + b)$ is invertible for every k.

Note that (1.8) implies equalities $a_0 m(k) - m^* a_0 = -2a_1 i$, $m^*(k) a_0 m(k) = a_2 + k^2$; therefore $n(k) = a_0 \frac{d}{dk} m(k)$ is Hermitian and $m^*(k)n(k) + n(k)m^*(k) = 2k$.

Since the eigenvalues of $m(k)$ lie in the left complex halfplane then this equality implies that

$$k^{-1} n(k) = -2 \int_0^\infty e^{m^*(k)z} e^{m(k)z} dz ;$$

certainly the left side is a negative definite Hermitian matrix; therefore for every v $\langle a_0(m(k) + b) v, v \rangle$ is a decreasing function of k^2 and (iv) implies (iii).

Thus if a is positive definite but A_B is not semibounded from below then the investigation of negative spectrum of A_B is equivalent to the investigation of "spectrum" of operator $a_0(M(k) + B_1)$ depending on "spectral" parameter k by complicated way.

We wish to show that in certain special cases it is possible to obtain reasonable formulas for Π^-. Let H, $H(k)$ be Poisson operators: $\tau u \to u$ if $\mathcal{A} u = 0$, $(\mathcal{A} + k^2) u = 0$ respectively. Let $\nu x_1 = 1$.

PROPOSITION 1.12. Let $a|_Y$ be a scalar symbol; then $H \sim \sum_{n=0}^{\infty} H_n$ where $H_n \in OP^{-n}(X, E)$ have Schwartz kernels expressed in local coordinates in the neighbourhood of Y by oscillatory integrals

$$K_{H_n}(x, y') =$$

$$= (2\pi)^{-d+1} \int \exp\{i\langle x' - y', \xi' \rangle + m(x', \xi') x_1\} \sum_{j=0}^{2n} \alpha_{nj}(x', \xi') x_1^j \, d\xi',$$

d_{nj} are symbols of order $j-n$, $d_{n0}=\delta_{n0}$. The similar formula holds for $H(k)$.

PROOF can be made a recurrent explicit construction of d_{nj} such that $\mathcal{A}H \sim 0$, $\tau H \sim I$.

PROPOSITION 1.13. Let $a|_Y$, b be scalar symbols; then for every $L \in \mathbb{Z}^+$ one can decompose $H(k)$:

$$H(k) = H' + H''(k)[M(k) + B_1] + H'''(k) \tag{1.9}$$

where $H''(k)$ is a Poisson operator of order (-1) uniformly with respect to k, $k^L H'''(k)$ is infinitely smoothing operator uniformly with respect to k, $H' \in OP^0(X,E)$, $H' \sim \sum_{n=0}^{\infty} H'_n$, $H'_n \in OP^{-n}(X,E)$ and in local coordinates in the neighbourhood of Y Schwartz kernel of H'_n are expressed by oscillatory integrals

$$K_{H'_n}(x,y') = \tag{1.10}$$

$$= (2\pi)^{-d+1} \int \exp\{i\langle x'-y', \xi'\rangle - b(x',\xi')x_1\} \sum_{j=0}^{2n} d'_{nj}(x',\xi')\, x_1^j \, d\xi' ,$$

d'_{nj} are symbols of order $j-n$, $d_{n0}=\delta_{n0}$, $\Omega=\{m+b>0\} \subset \{\operatorname{Re} b<0\}$.

PROOF. On $T^*Y \setminus 0 \setminus \Omega$ operator $M(k)+B_1$ is elliptic with a parameter k; on Ω we can apply formula

$$e^{m(k)x_1} - e^{-bx_1} = \int_0^1 \exp\{(-sb + (1-s)m(k))x_1\}\, ds\, (m(k)+b)\, x_1$$

and obtain decomposition (1.9) - (1.10) but with $L=0$ and with d'_{nj} depending on k. To exclude k from d'_{nj} and to obtain an arbitrary L is possible by means of formulas

$$|m(k) - b|^2 a_0 = a_2 - a_0 |b|^2 + k^2 ,$$

$$(M(k) - B_1^*) a_0 (M(k) + B_1) \sim A_2 - B_1^* A_0 B_1 + k^2$$

modulo first-order pseudo-differential with a parameter k operator.

REMARK. One can prove that if $a|_Y$ is scalar symbol and on every connected component of $T^*Y\setminus 0$ $m+b$ is either positive definite or negative definite then the decomposition (1.9) takes place but (1.10) holds only for $n=0$.

PROPOSITION 1.14. Let $a|_Y, b$ be scalar symbols; then $\Pi^- \sim \sum_{n=0}^{\infty} G_n$, $G_n \in OGS^{-n}(X, E)$ and in local coordinates in the neighbourhood of Y Schwartz kernels of G_n are expressed by oscillatory integrals

$$K_{G_n}(x,y) =$$

$$=(2\pi)^{-d+1}\int_{\Omega} exp\{i\langle x'-y',\xi'\rangle-b(x',\xi')x_1-\bar{b}(x',\xi')y_1\}\sum_{j,l=0}^{2n}\omega_{njl}(x',\xi')x_1^j y_1^l\, d\xi'$$

where ω_{njl} are symbols of order $j+l-n+1$, $\omega_{00}=2\,\mathrm{Re}\, b$.

PROOF. Since

$$K_{\Pi^-}(x,y) = \sum_s \varphi_s(x)\otimes\varphi_s^{\dagger}(y)$$

where $(\dagger)$ means Hermitian conjugation, $(\psi_1\otimes\psi_2^{\dagger})\psi = \psi_1\langle\psi,\psi_2\rangle$ and we sum with respect to all orthonormal eigenfunctions of A_B corresponding to negative eigenvalues $-k_s^2 \searrow -\infty$ and since $\varphi_s = H(k_s)\tau\varphi_s$, $(M(k_s)+B_1)\tau\varphi_s = 0$ then

$$K_{\Pi^-}(x,y) = \sum_s (H'+H'''(k_s))\tau\varphi_s(x)\otimes[(H'+H'''(k_s))\tau\varphi_s(y)]^{\dagger}$$

because of proposition 1.12.

Since $\|\varphi_s\|_n \leqslant Ck_s^n$, $\|H'''(k_s)\tau\varphi\|_n \leqslant Ck_s^{n-L}$ and since $k_s \geqslant \varepsilon|s|^{1/d}$, $\varepsilon>0$ (see [2] for example) then the series

$$\sum_s H'''(k_s)\tau\varphi_s(x)\otimes(H'\tau\varphi_s(y))^{\dagger},$$

$$\sum_s H' \tau \varphi_s(x) \otimes (H'''(k_s) \tau \varphi_s(y))^\dagger ,$$

$$\sum_s H'''(k_s) \tau \varphi_s(x) \otimes (H'''(k_s) \tau \varphi_s(y))^\dagger$$

converge to infinitely smooth functions (recall that L is arbitrary and H does not depend on L); therefore

$$K_{\Pi^-}(x,y) \equiv \sum_s H' \tau \varphi_s(x) \otimes (H' \tau \varphi_s(y))^\dagger (mod\, C^\infty)$$

where the series in the right side converges.

Setting $x_1 = y_1 = 0$ (one can prove that it is a correct operation) it follows that $\sum_s \tau \varphi_s(x') \otimes (\tau \varphi_s(y'))^\dagger \equiv$ $\equiv K_{\Pi^-}(0, x', 0, y') = K_\omega(x', y') \, (mod\, C^\infty)$ where the series in the left side converges and operator $\omega : v \to \tau \Pi^-(v \delta(x_1))$ belongs to $OPS^1(Y, E)$.

Then modulo infinitely smoothing operator $\Pi^- \sim H' \omega H'^*$ and in neighbourhood of $Y \times Y$ K_{Π^-} is expressed by described oscillatory integrals. We need only to prove that $\omega_{00} = 2 Re\, b$ on Ω. Let Σ be some connected component of Ω ; since ω_{00} on Σ depends only on a on $j^{-1}\Sigma$ and b on Σ then one can assume without a loss of generality that $\Omega = \Sigma$. Equality $\Pi^{-2} = \Pi^-$ implies that $\omega_{00}^2 (2 Re\, b)^{-1} = \omega_{00}$ i.e. that either $\omega_{00} = 2 Re\, b$ or $\omega_{00} = 0$; but in the latter case Π^- would be smoothing, i.e. finite-dimensional and A_B would be semibounded from below; but it is impossible because of proposition 1.11.

One can prove the following simple statement:

PROPOSITION 1.15. Let $a|_Y$ be a scalar symbol; then $\Pi^- \equiv g (mod\, OGS^{-1}(X, E))$ where

$$K_g(x,y) =$$

$$= (2\pi)^{-d+1} \int exp(i<x'-y', \xi'> - b(x',\xi')x_1 - b^*(x',\xi')y_1)(b+b^*)^{-1}(x',\xi')\varepsilon(x',\xi')d\xi' ,$$

ε is selfadjoint projector to invariant subspace of b corresponding to the eigenvalues greater then $- Re\, m = (a_0^{-1} a_2 - a_0^{-2} a_1^2)^{1/2}$.

The following statement does not appear difficult to prove:

CONJECTURE. In a general case $\Pi^- \equiv g \,(mod\ OGS^{-1}(X,E))$ where in local coordinates in the neighbourhood of Y Schwartz kernel of g is expressed by oscillatory integral

$$K_g(x,y) =$$

$$=(2\pi)^{-d+1}\int exp(i\langle x'-y',\xi'\rangle - b(x',\xi')x_1)\,\omega(x',\xi')\,exp(-b^*(x',\xi')y_1)\,d\xi',$$

ω is a Hermitian matrix, $Ran\,\omega$ coincides with the linear hull of the set $\{v : \exists k;\ a_0(m(k)+b)\,v = 0\}$ and

$$\omega\int_0^\infty e^{-b^*(x',\xi')x_1}\, e^{-b(x',\xi')x_1}\, dx_1\,\omega = \omega$$

(these conditions define ω uniquely).

1.6. Let us consider two pathological examples. Let $X = \bar{\mathbb{R}}^+ \times \mathbb{R}^{d-1}$, A_B be the operator $-\Delta$ with the boundary condition $(iD_1 + b(D'))\,u|_Y = 0$, where $b(D') = F^{-1}_{\xi'\to x'}\ b(\xi')\,F_{x'\to\xi'}$, $F_{x'\to\xi'}$ and $F^{-1}_{\xi'\to x'}$ denote the Fourier transform and the inverse Fourier transform. The Šapiro - Lopatinskii condition means precisely that $b(\xi') = |\xi'|$ for $|\xi'| > C$ and the symmetricity condition for A_B is $b(\xi') \in \mathbb{R}$.

It is easy to show that

$$K_{\Pi^-}(x,y) =$$

(1.11)

$$=(2\pi)^{-d+1}\int_\Omega 2b(\xi')\,exp(i\langle x'-y',\xi'\rangle - b(x',\xi')(x_1+y_1))\,d\xi',$$

$$\Omega = \{b(\xi') > |\xi'|\}.$$

(i) Let $B = \alpha(D')\tau iD_1 + \beta(D')\tau$, $b(D') = \alpha^{-1}(D')\,\beta(D')$. If $\alpha(\xi')$, $\beta(\xi')$ are positively homogeneous of degrees $0, 1$ respectively then the Šapiro - Lopatinskii condition means precisely that $\alpha(\xi')|\xi'| \neq \beta(\xi')$. If $\alpha(\xi')$ vanishes then Π^- is not a singular Green operator; if $\alpha \in \mathbb{R}^+$ then Π^- is infinite-dimensional projector and if $\alpha \in \bar{\mathbb{R}}^-$ then $\Pi^- = 0$. Thus in class of general boundary operators satisfying the Šapiro - Lopatinskii condition nonnegative definite operators are not separated from nonsemibounded from below.

(ii) If $b(\xi')$ is classical symbol positively homogeneous of degree $m \in \mathbb{Z}^+$, $m \geq 2$ then the Šapiro - Lopatinskii condition means precisely that $b(\xi') \neq 0$. But operator Π^- given by formula (1.11) does not belong to $OGS^0(X)$. Certainly one can construct another calculation of singular Green operators containing Π^-. Note that with these boundary conditions the different asymptotics of eigenvalues are obtained (see §0).

1.7. Selfadjoint projectors to kernels of elliptic operators (without boundary conditions) are undoubtedly singular Green operators. Selfadjoint projectors to kernels of underdetermined elliptic operators on closed manifolds are pseudo-differential operators. In particular, selfadjoint projectors to kernels of d, δ are pseudodifferential operators.

On manifolds with the boundary selfadjoint projectors to kernels of d, δ belong to $OPS^0 + OGS^0$.

1.8. In future we need the following statement:

PROPOSITION 1.6. The following estimate holds:

$$Re(\mathcal{A}\Pi^+ v, \Pi^+ v) \geq c_0 \| \Pi^+ v \|_1^2 - C \| Bv \|^2_{Y, 1/2-\mu}, \quad c_0 > 0,$$

for every $v \in H^2(X, E)$ where μ is the order of the boundary operator B.

PROOF. Consider Hilbert spaces: $\Pi^+ L_2(X, E)$, $\Pi^+ \mathcal{D}(A_B^{1/2})$, $\Pi^+ \mathcal{D}(A_B)$ with norms $\|w\|$, $(A_B w, w)^{1/2}$, $\|A_B w\|$ respectively; the second space can be obtained by interpolation between first and third [73].

Consider Hilbert spaces $L_2(X,E)$, $\overset{1}{H}(X,E)$, $\overset{2}{H}(X,E)$ with the usual norms; the second space can also be obtained by interpolation between first and third.

But $\Pi^+ \mathcal{D}(A_B) \subset \overset{2}{H}(X, E)$ since $\{\mathcal{A}, B\}$ is elliptic; therefore $\Pi^+ \mathcal{D}(A_B^{1/2}) \subset \overset{1}{H}(X, E)$ because of interpolation theorem and hence (1.12) holds for all $v \in \mathcal{D}(A_B)$.

Let $F: H^{1/2-\mu}(Y, E) \longrightarrow H^1(X, E)$ be an operator such that $BF = I$. Aplying (1.12) to $w = v - FBv \in \mathcal{D}(A_B)$ we obtain (1.12) for arbitrary $v \in H^2(X, E)$.

§2. The "Wave" equation: finite speed of propagation of singularities

2.1. We assume that the main hypotheses are fulfilled: $\mathcal{A}$ is a second-order operator, $\{\mathcal{A}, B\}$ is elliptic, A_B is selfadjoint,

$B=\tau$ or $B=\tau\nu+B_1\tau$. Let $\Pi\in OPS^\circ(X,E)+OGS^\circ(X,E)$ be a selfadjoint projector, $\Pi A_B\subset A_B\Pi$; then $\Pi\Pi^\pm=\Pi^\pm\Pi$; we assume that $\Pi\Pi^+=\Pi$; otherwise we replace Π by $\Pi\Pi^+$ (we study only the positive spectrum asymptotics).

Since $A_{B,H}$ – the restriction of A_B to $\Pi L(X,E)$ is a positive selfadjoint operator – then we can introduce the operator $U(t)=A_{B,H}^{-1/2}\sin A_{B,H}^{1/2}t\cdot\Pi$; for every $v_0\in L_2(X,E)$ $v=U(t)v_0$ is the solution of the problem

$$P_B v\equiv(-D_t^2+A_B)v=0\,, \tag{2.1}$$

$$v|_{t=0}=0\,,\quad v_t|_{t=0}=\Pi v_0\,. \tag{2.2}$$

If $\Pi v_0\in\mathcal{D}(A_B^{n/2})$ then $U(t)v_0\in\bigcap_j^{n+1} C^j(\mathbb{R},\mathcal{D}(A_B^{(n+1-j)/2}))$.

Let $u(x,y,t)\in\mathcal{D}'(X\times X\times\mathbb{R},\mathrm{Hom}(E))$ *) be Schwartz kernel of $U(t)$; then $u(x,y,t)$ is the solution of the problem

$$P_x u\equiv(-D_t^2+\mathcal{A}_x)u(x,y,t)=0, \tag{2.3}$$

$$B_x u(x,y,t)=0\,, \tag{2.4}$$

$$u|_{t=0}=0\,,\quad u_t|_{t=0}=K_\Pi(x,y) \tag{2.5}$$

where $\mathcal{A}_x$ is a natural lifting of operator $\mathcal{A}$ acting on sections of E to an operator acting on sections of $\mathrm{Hom}(E)$ etc.

Since $U(t)$ is selfadjoint then

$$u^\dagger(x,y,t)=u(y,x,t) \tag{2.6}$$

where $(\dagger)$ denotes the Hermitian conjugation. Moreover $u(x,y,t)$

---

*) If E,F are fiberings over X,X_1 respectively then $\mathrm{Hom}(E,F)$ is a fibering over $X\times X_1$; the fiber of $\mathrm{Hom}(E,F)$ lying over point (x,y) is the space of linear mappings from F_y to E_x; $\mathrm{Hom}(E)=\mathrm{Hom}(E,E)$.

is the solution of the dual problem

$$(2.3)^{\dagger} \qquad u P_y^{\dagger} \equiv u(x,y,t)(-D_t^2 + \mathcal{A}_y^{\dagger}) = 0,$$

$$(2.4)^{\dagger} \qquad u(x,y,t)\, B_y^{\dagger} = 0$$

and (2.5) where $\mathcal{A}_y^{\dagger}$ is a natural lifting of operator $\mathcal{A}^{\dagger}$ acting on sections of $E^{\dagger}$ to an operator acting on sections of $\mathrm{Hom}(E)$ etc; here $E^{\dagger}$ is dual to E fibering, $v\mathcal{A}^{\dagger} \overset{def}{\equiv} (\mathcal{A}^* v^{\dagger})^{\dagger}$, $vB^{\dagger} \overset{def}{\equiv} (B^* v^{\dagger})^{\dagger}$ where $\mathcal{A}^*$ is formally adjoint to $\mathcal{A}$ operator, $\{\mathcal{A}^*, B^*\}$ is formally adjoint to $\{\mathcal{A}, B\}$ problem (recall that $\mathcal{A}^* = \mathcal{A}$, $B^* = B$); if v is section of E then $v^{\dagger}$ is section of $E^{\dagger}$ and v.v. It is convenient for us to write dual operators to the right of the function; this notations contradicts the usual notation of the operator theory but corresponds to the notations of the matrix theory; recall that $u(x,y,t)$ is $D \times D$ -matrix as well as $\mathcal{A}, \mathcal{A}^{\dagger}, B, B^{\dagger}$.

Obviously, $\{\mathcal{A}^{\dagger}, B^{\dagger}\}$ is elliptic since $\{\mathcal{A}, B\}$ is elliptic and $\mathcal{A}^{\dagger}_{B^{\dagger}}$ is symmetric since $\mathcal{A}_B$ is symmetric; therefore (2.3), (2.4), $(2.3)^{\dagger}$, $(2.4)^{\dagger}$ imply that

$$u \in C^{\infty}(X \times \mathbb{R}, \mathcal{D}'(X, \mathrm{Hom}(E))) \cap C^{\infty}(X \times X, \mathcal{D}'(\mathbb{R}, \mathrm{Hom}(E))).$$

Therefore we can correctly define

$$\Gamma u \equiv u(x,x,t) \in C^{\infty}(X, \mathcal{D}'(\mathbb{R}, \mathrm{hom}(E))) \quad \text{where} \quad \mathrm{hom}(E) \overset{def}{\equiv} \Gamma\, \mathrm{Hom}(E).$$

REMARK. Certainly, problem (2.3) - (2.5) as well as dual problem $(2.3)^{\dagger}$, $(2.4)^{\dagger}$, (2.5) may be ill-posed. But we know that $u(x,y,t)$ exists.

One can assume without a loss of generality that there is a Riemannian metric on X . Let $\mathrm{dist}(x,y)$ be the corresponding distance between points x, y and $x_1 = \mathrm{dist}(x, Y)$. We assume that $0 < \delta$ is small enough, $X \cap \{x_1 \leqslant 6\delta\}$, $E \cap \{x_1 \leqslant 6\delta\}$ are identified with $[0,6\delta] \times Y$, $[0,6\delta] \times E'$, $E' = E|_Y$ respectively. Let $dx' = \frac{dx}{dx_1}\Big|_Y$ be a density on Y.

Near $Y \times Y \times \mathbb{R}$ $u(x,y,t)$ depends on three "inconvenient" variables: x_1, y_1 and t (we will cut with respect to t); therefore we need to observe the smoothness with respect to these variables.

PROPOSITION 2.2. There exists $s = s(d) \in \mathbb{R}$ such that $u \in \mathcal{H}^{s;0}_{x',y';x_1,y_1,t}(X \times X \times \mathbb{R} \cap \{x_1 < 6\delta, y_1 < 6\delta\}, \mathrm{Hom}(E))_{loc}$.

PROOF. Since for every $\tau \in L_2([0,6\delta])$, $w \in L_2(Y,E')$

$$U(t)(\tau w) \in C(\mathbb{R}_t, D(A_B^{1/2})) \subset C(\mathbb{R}_t \times \overline{\mathbb{R}}^+_{x_1}, L_2(Y,E'))$$

then there exists $s = s(d)$ such that for every $\tau \in L_2([0,6\delta])$

$$\int u(x,y,t)\tau(y_1)dy_1 \in C(\mathbb{R}_t \times \overline{\mathbb{R}}^+_{x_1}, H^s(Y \times Y, \mathrm{Hom}(E))$$

and

$$u(x,y,t) \in C(\mathbb{R}_t \times \overline{\mathbb{R}}^+_{x_1}, \mathcal{H}^{s;0}_{x',y';y_1}(\overline{\mathbb{R}}^+_{y_1} \times Y \times Y, \mathrm{Hom}(E)).$$

2.2. We need certain rough estimates of wave front set of u. (2.3), $(2.3)^\dagger$ and ellipticity of $\mathcal{A}$, $\mathcal{A}^\dagger$ imply that

$$\mathrm{WF}(u) \cup j_y^{-1}\,\mathrm{WF}_{,b}(u) \subset \{C_1^{-1}|\tau| \leqslant |\xi| \leqslant C_1|\tau|\}, \tag{2.7}$$

$$\mathrm{WF}(u) \cup j_x^{-1}\,\mathrm{WF}_{b,}(u) \subset \{C_1^{-1}|\tau| \leqslant |\eta| \leqslant C_1|\tau|\} \tag{$2.7^\dagger$}$$

provided C_1 is large enough.

Since for $x \in Y$, $|\xi'| > C_2|\tau|$ point $(x,\xi',\tau) \in T^*Y \times \mathbb{R}$ is elliptic with respect to $P = -D_t^2 + \mathcal{A}$ (i.e. $j^{-1}(x,\xi',\tau) \cap \mathit{Char}\, P = \emptyset$) and since the Šapiro - Lopatinskii condition holds for $\{P, B\}$ in (x,ξ',τ) provided C_2 is large enough because of this condition holds for $\{\mathcal{A}, B\}$ (i.e. for $\{P, B\}$ at $\tau = 0$) then (2.3) - (2.4), $(2.3)^\dagger$ - $(2.4)^\dagger$ imply that

$$\mathrm{WF}_{b,}(u) \cup j_y^{-1}\,\mathrm{WF}_{b,b}(u) \subset \{|\xi'| \leqslant C_2|\tau|\}, \tag{2.8}$$

$$\mathrm{WF}_{,b}(u) \cup j_x^{-1}\,\mathrm{WF}_{b,b}(u) \subset \{|\eta'| \leqslant C_2|\tau|\}. \tag{$2.8^\dagger$}$$

REMARK. Certainly the ellipticity of point (x,ξ',τ) with respect to P does not imply the fulfilment of the Šapiro - Lopatinskii condition in this point for $\{P, B\}$; therefore C_2 depends on B as well as $\mathcal{A}$.

Finally the equality

$$\mathcal{A}_x u = u \mathcal{A}_y^{\dagger} \tag{2.9}$$

implies that

$$WF(u) \cup WF(u^{\pm}) \subset \{C_3^{-1}|\xi| \leq |\eta| \leq C_3|\xi|\} \tag{2.10}$$

where $u^{\pm} = u$ for $\pm t \geq 0$, $u^{\pm} = 0$ for $\pm t < 0$, $u^{\pm}$ satisfy equations

$$P_x u^{\pm} = \pm\delta(t) K_{\Pi}(x,y) , \tag{2.11}$$

$$u^{\pm} P_y^{\dagger} = \pm\delta(t) K_{\Pi}(x,y) \tag*{(2.11)†}$$

with the boundary conditions (2.4), (2.4)† .

THEOREM 2.2. Let X be a closed manifold; then

$$WF(u) \subset \{ dist(x,y) + |\xi+\eta|(|\xi|^{-1} + |\eta|^{-1}) \leq C|t|\}, \tag{2.12}$$

$$WF(u^{\pm}) \subset \{ dist(x,y) + |\xi+\eta|(|\xi|^{-1} + |\eta|^{-1}) \leq \pm ct\} \cup \{t = \xi = \eta = 0\}. \tag{2.13}$$

THEOREM 2.3. Let X be a manifold with the boundary, $t_0 > 0$ small enough; then

(i) Inclusions (2.12), (2.13) hold for

$$WF(u) \cap \{x_1 > \delta, y_1 > \delta, |t| \leq t_0\} ,$$

$$WF(u^{\pm}) \cap \{x_1 > \delta, y_1 > \delta, |t| \leq t_0\} .$$

(ii) $u \in C^{\infty}(\{x_1 > 3\delta, y_1 < 2\delta, |t| \leq t_0\} \cup \{x_1 < 2\delta, y_1 > 3\delta, |t| \leq t_0\}$.

(iii) $WF'^{1}_{f}(u^{\pm}) \cap \{x_1 < 4\delta, y_1 < 4\delta, |t| \leq t_0\} \subset \{dist(x,y) + |\xi' + \eta'|(|\xi'| + |\tau|)^{-1} + |\xi' + \eta'|(|\eta'| + |\tau|)^{-1} \leq$

$\pm ct\} \cup \{|\xi'| + |\eta'| \leq \pm c|\tau| t\} \subset \{c_4^{-1}|\xi'| \leq |\eta'| \leq c_4|\xi'|\} \cup \{|\xi'| + |\eta'| \leq c|t||\tau|\}$,

$WF_f(u) \cap \{x_1 < 4\delta, y_1 < 4\delta, |t| \leq t_0\} \subset \{dist(x,y) + |\xi' + \eta'||\tau|^{-1} \leq$

$c|t|\} \cup \{|\xi'| + |\eta'| \leq c|t||\tau|\}$.

(iv) If $\Pi \in OGS^0(X, E)$ then sing supp $u \subset Y \times Y \times \mathbb{R}$,

(2.14) $$WF_{b,b}(u) \subset \{c_5^{-1}|\tau| \leq |\xi'| \leq c_5|\tau|,\ c_5^{-1}|\tau| \leq |\eta'| \leq c_5|\tau|\},$$

(2.12)' $$WF_f(u) \subset \{(x,y) \in Y \times Y,\ dist(x,y) + |\xi' + \eta'|(|\xi'|^{-1} + |\eta'|^{-1}) \leq c|t|\},$$

(2.13)' $$WF_f(u^{\pm}) \subset \{(x,y) \in Y \times Y,\ dist(x,y) + |\xi' + \eta'|(|\xi'|^{-1} + |\eta'|^{-1}) \leq c|t|\} \cup$$

$$\{t = \xi = \eta = 0\} \cup \{y \in Y, t = \xi = \eta' = 0\} \cup$$

$$\{x \in Y, t = \xi' = \eta = 0\} \cup \{x \in Y, y \in Y, t = \xi' = \eta' = 0\}.$$

PROOF. We use the method of [49, 50, 53].

Since the proofs of theorems 2.2, 2.3(i) - (iii) are similar we dwell at length on the proof of theorem 2.3(iii) which is most complicated; the modifications which we need to prove statements (i),(ii) of theorem 2.3 will be explained as we go along. The proof of theorem 2.2 coincides with the proof of theorem 2.3(i) *).

---

*) In reality theorems 2.2, 2.3(i), (ii) follow from main theorem [49].

Let $\varphi(x,y,t,\xi',\eta',\tau)$ be real positively homogeneous of degree 0 with respect to (ξ',η',τ) function such that

$$\varphi>0 \quad \text{at the set } \{\xi'=0\}\cup\{t\geq 2t_0\}\cup\{x=y,\xi'=-\eta',t=0\}\cup \{x_1>5\delta,\ y_1>5\delta\}, \tag{2.15}$$

$$\varphi_t > C_0(|\varphi_x|+(|\xi'|+|\tau|)|\varphi_{\xi'}|) \tag{2.16}$$

where C_0 will be chosen later.

In the proof of (i) we take $\varphi=\varphi(x,y,t,\xi,\eta,\tau)$ where

$$\varphi>0 \quad \text{at the set } \{x_1<\delta/2\}\cup \{y_1<\delta/2\}\cup\{\xi=0\}\cup\{x=y,\xi=-\eta,t=0\}\cup\{t\geq 2t_0\}, \tag{2.15'}$$

$$\varphi_t > C_0(|\varphi_x|+|\xi|\,|\varphi_\xi|) \tag{2.16'}$$

and in the proof of (ii) we take $\varphi=\varphi(x,y,t,\xi,\tau)$ where

$$\varphi>0 \quad \text{at the set } \{x_1<\delta\}\cup \{y_1>5\delta\}\cup\{\xi=0\}\cup\{t\geq 2t_0\} \tag{2.15''}$$

and (2.16)$'$ holds.

In the proof of (i) we shall prove that $WF^{\ell+1}(u^+)\cap\{\varphi<0\}=\emptyset$ for every $\ell\in\mathbb{R}$; in the proof of (ii) we shall prove that $WF_f^{\ell+1,0}(u^+)\cap\{\varphi<0\}=\emptyset$ for every $\ell\in\mathbb{R}$; at last in the proof of (iii) we shall prove that $WF_f^{\ell+1,3}(V^+)\cap\{\varphi<0\}=\emptyset$ for every $\ell\in\mathbb{R}$ where $V=A^{-1}_{B,x}u=u\,(A_B^{-1})^+_y$. We shalll use the induction on $\ell\in\mathbb{R}$ since for some $\ell\in\mathbb{R}$ these statements hold.

Thus we shall prove that $WF(u^+)\cap\{\varphi<0\}=\emptyset$ for every φ satisfying (2.15)$'$, (2.16)$'$; this implies that

$$WF(u^+)\cap\{x_1>\delta,y_1>\delta,t\leq t_0\}\subset\{dist(x,y)+|\xi+\eta|\,|\xi|^{-1}\leq ct\}\cup\{t=\xi=0\}.$$

But x and y are equal in rights (see (2.6)) and u is odd with respect to t ; therefore we can permutate x and y as well as replace t by $-t$, u^+ by u^- and obtain (2.13); (2.13) and (2.8), (2.8) imply (2.12).

Moreover we shall prove that $WF_f^{',0}(u^+) \cap \{\varphi < 0\} = \emptyset$ for every φ satisfying (2.15)$''$, (2.16)$'$; therefore $WF_f^{',0}(u^+) \cap \{x_1 > 3\delta, y < 2\delta, t \leq t_0\} \subset \{t = \xi = 0\}$; then (2.7), (2.7)†, (2.8), (2.8)† and the fact that u is odd with respect to t imply that

$$WF_f^{',0}(u) \cap \{x_1 > 3\delta, y_1 < 2\delta, |t| \leq t_0\} = \emptyset ;$$

here $WF_f^{',0}(u)$ can be replaced by $WF_f(u)$ because of (2.3)†, since x and y are equal in rights we obtain (ii).

At last we shall prove that $WF_f^{',3}(V^+) \cap \{\varphi < 0\} = \emptyset$ for every φ satisfying (2.16) - (2.16). Therefore

$$WF_f^{',3}(V^+) \cap \{t \leq t_0\} \subset \{ dist(x,y) + |\xi' + \eta'|(|\xi'| + |\tau|)^{-1} \leq$$

$$ct\} \cup \{|\xi'| \leq ct|\tau|\} \subset \{ dist(x,y) + |\xi' + \eta'|(|\xi'| + |\tau|)^{-1} + |\xi' + \eta'|(|\eta'| + |\tau|)^{-1} \leq$$

$$c't|\tau|\} \cup \{ |\xi'| \leq c't|\tau|\};$$

since x and y are equal in rights, u, V are odd with respect to t and $u^\pm = \mathcal{A}_x V^\pm$, $u = \mathcal{A}_x V$ then the first inclusion (iii) holds; the second inclusion follows from the first inclusion and equations (2.3), (2.3)†.

Since V^+ satisfies the equations

$$\text{(2.17)} \qquad P_x V^+ = \delta(t) K_{A_B^{-1}\Pi}(x,y) ,$$

$$\text{(2.17)}^\dagger \qquad V^+ P_y^\dagger = \delta(t) K_{A_B^{-1}\Pi}(x,y)$$

then $V^+ \in \mathcal{H}^{s;3}(X \times X \times \mathbb{R}, Hom(E))$ for some $s = s(d)$.

Assume that

$$\text{(2.18)} \qquad WF_f^{l+1/2,3}(V^+) \cap \{\varphi < 0\} = \emptyset ,$$

Let $\chi \in C^\infty(\mathbb{R})$, $\chi(t) = 0$ for $t \geq 0$, $\chi'(t) \leq 0$ for $t < 0$, $OPS^{0'}(X \times X \times \mathbb{R}) \ni \Psi$ operator with the symbol $\chi(\varphi(x, y, t, \xi', \eta', \tau) + \alpha)$ *), $0 < \alpha$ small enough, $\Psi v \subset \{|t| \leq 3t_0,\ x_1 < 5\delta,\ y_1 < 5\delta\}$ for every v such that $\operatorname{supp} v \subset \{t \geq 0\}$. We shall prove that

$$(2.19)\qquad \Psi V^+ \in \mathcal{H}^{l+1,3}(X \times X \times \mathbb{R},\ Hom(E)).$$

Since $0 < \alpha$ is arbitrary then (2.19) implies that (2.18) with l replaced by $l + 1/2$ holds, i.e. the step of the induction will be made.

Obviously that

$$(2.20)\qquad \frac{\partial}{\partial t}\{\|D_t v\|^2 + Re(\mathcal{A}v, v)\} =$$

$$2Re(Pv; iD_t v) + Re\{(\mathcal{A}D_t v, v) - (D_t v, \mathcal{A}v)\}$$

where $\|\cdot\|$ and $(\cdot,\cdot)$ are the norm and the inner product in $L_2(X,E)$.

Since $\mathcal{A}_B$ is symmetric then

$$(2.21)\qquad (\mathcal{A}v_1, v_2) - (v_1, \mathcal{A}v_2) = (\beta B v_1, v_2)_Y - (v_1, \beta B v_2)_Y$$

where $\|\cdot\|_Y$ and $(\cdot,\cdot)_Y$ are the norm and the inner product in $L_2(Y,E)$, $\beta \in OPS^0(Y,E)$; we assume that $B = i\tau D_1 + B_1$; the case $B = \tau$ can be considered by the same way with inessential simplifications.

(2.20), (2.21) imply that

$$(2.22)\qquad 2k(\|e^{-kt}D_t v\|^2 + Re(\mathcal{A}e^{-kt}v, e^{-kt}v)) = -2Re\,i(e^{-kt}Pv, e^{-kt}D_t v) +$$

$$2Re\,i(e^{-kt}\beta Bv, e^{-kt}D_t v)_Y + 2k\,Re(e^{-kt}\beta Bv, e^{-kt}v)_Y$$

---

*) More precisely, we consider an operator $\Psi \in OPS^{0'}(X \times X \times \mathbb{R}, Hom(E))$ which is scalar in chosen local cards in X and in chosen trivialization of E over these cards. We need such operators only to obtain the appropriate formulas for principal symbols of commutators. We shall follows this agreement concerning notation $\Psi \in OPS(X \times X \times \mathbb{R})$ etc in future.

where now norms and inner products contain the integration with respect to $t \in \mathbb{R}$ too; $0 < k_+$ will be chosen later.

We set $v = \mathcal{J}_\varepsilon \Lambda^\ell \psi V^+$ where

$$\mathcal{J}_\varepsilon = (I + \varepsilon^2 (|D'_x|^2 + |D'_y|^2 + |D_t|^2)^{-1}, \quad \varepsilon > 0,$$

are mollifying operators,

$$\Lambda^\ell = (I + |D'_x|^2 + |D'_y|^2 + |D_t|^2)^{\ell/2},$$

and include in norms and inner products the integration with respect to $y \in Y$ too. Exactly for v_1, v_2 sections of $Hom(E)$ we set

$$(v_1, v_2) = \int_{X \times X \times \mathbb{R}} tr\, v_2^+ v_1 \, dx\, dy\, dt$$

and for v_1, v_2 - sections of $Hom(E|_Y, E)$ - we set

$$(v_1, v_2) = \int_{Y \times X \times \mathbb{R}} tr\, v_2^+ v_1 \, dx\, dy\, dt .$$

(2.18) implies that $v \in \mathcal{H}^{2,3}(X \times X \times \mathbb{R}, Hom(E))$ and we can substitute v into (2.22); moreover (2.18) implies that $\| v ; \mathcal{H}^{1/2;3}_{x',y',t;\, x_1,y_1} \|$ is bounded as ε tends to 0.

Note that $(I - \Pi^+_x) v = -[\Pi^+_x, \mathcal{J}_\varepsilon, \Lambda^\ell \psi] V^+$ since $(I - \Pi^+_x) V^+ = 0$. It is clear that $[\Pi^+_x, \mathcal{J}_\varepsilon \Lambda^\ell \psi]$ is an operator of order $(\ell - 1 ; 0)$ with respect to $(x', y', t; x_1)$ uniformly with respect to ε and is the sum if a pseudo-differential operator with the transmission property and a singular Green operator. Therefore $\| (I - \Pi^+_x) v, \mathcal{H}^{3/2}_{x',y',t;\, x_1,y_1} \|$ is bounded as ε tends to 0 and modulo terms bounded as ε tends to 0

$$(\mathcal{A}_x e^{-kt} v, e^{-kt} v) \sim (\mathcal{A}_x e^{-kt} \Pi^+_x v, e^{-kt} \Pi^+_x v).$$

Then proposition 1.16 implies that

$$\text{(2.23)} \qquad Re\,(\mathcal{A}_x e^{-kt} v, e^{-kt} v) \geqslant$$

$$c_0 |||e^{-kt}\Pi_x v|||_1^2 - C|||e^{-kt}B_x v|||^2_{Y,-1/2} - C \geq$$

$$c_0|||e^{-kt} v|||_1^2 - C|||e^{-kt}B_x v|||^2_{Y,-1/2} - C_1$$

where $|||\cdot|||_n$ and $|||\cdot|||_{Y,\sigma}$ are norms in $\mathcal{H}^{n;0}_{x;y,t}(X\times X\times\mathbb{R}, Hom(E))$ and $\mathcal{H}^{\sigma;0}_{x';y,t}(Y\times X\times\mathbb{R}, Hom(E))$ respectively (these notations are temporary), C_j do not depend on ε and $0 < c_0$ does not depend on ε, k, φ.

Since $B_x V^+_l = 0$ and $[B_x, \mathcal{J}_\varepsilon \Lambda^l \psi]$ is an operator of order l with respect to (x', y', t) uniformly with respect to ε and since cone $supp\,\psi \cap \{\xi' = 0\} = \emptyset$ then (2.18) implies that $|||e^{-kt}B_x v|||_{Y,-1/2}$ is bounded as ε tends to 0 and (2.23) implies the inequality

$$\text{(2.24)}\qquad Re(\mathcal{A}_x e^{-kt}v, e^{-kt}v) \geq \frac{1}{2}c_0|||e^{-kt}v|||_1^2 - C.$$

Consider now the right side of (2.22). Since $P_x V^+ = \delta(t)K_{A^{-1}_B\Pi}(x,y)$, $cone\ supp\,\psi\cap\{\xi'=0\}=\emptyset$, $cone\ supp\,\psi\cap\{x=y, \xi'=-\eta', t=0\}=\emptyset$ and $WF'^{1}_f(\delta(t)K_{A^{-1}_B\Pi}) \subset \{x=y, \xi'=-\eta', t=0\}$ then $\psi P_x V^+ \in \mathcal{H}^{\infty,1}(X\times X\times\mathbb{R}, Hom(E))$. Recall that $B_x V^+ = 0$.

Therefore the right side of (2.22) equals modulo terms bounded as ε tends to 0 to

$$\text{(2.25)}\quad -2Re\,i(e^{-2kt}[P_x,\mathcal{J}_\varepsilon\Lambda^l]\psi V^+, D_t v) - 2Re\,i(e^{-2kt}\mathcal{J}_\varepsilon\Lambda^l[P_x,\psi]V^+, D_t v) +$$

$$2Re\,i(e^{-2kt}\beta_x[B_x,\mathcal{J}_\varepsilon\Lambda^l]\psi V^+, D_t v)_Y + 2Re\,i(e^{-2kt}\beta_x\mathcal{J}_\varepsilon\Lambda^l[B_x,\psi]V^+, D_t v)_Y.$$

Note that

$$[P_x, \mathcal{J}_\varepsilon\Lambda^l] = \sum_{j=0}^{2}[A_{jx}, \mathcal{J}_\varepsilon\Lambda^l]D^{2-j}_{x_1},$$

$$[A_{j\varepsilon}, \mathcal{J}_\varepsilon] = -\varepsilon^2\mathcal{J}_\varepsilon[A_{jx}, |D'_x|^2]\mathcal{J}_\varepsilon,$$

$\varepsilon^2\mathcal{J}_\varepsilon[A_{jx}, |D'_x|^2]$ are operators of order $j-1$ uniformly with respect to ε ; therefore

$$[P_x, \mathcal{J}_\varepsilon \Lambda^l] = K_\varepsilon \mathcal{J}_\varepsilon \Lambda^l + D_{x_1} K'_\varepsilon \mathcal{J}_\varepsilon \Lambda^l + K''_\varepsilon P$$

where $K_\varepsilon, K'_\varepsilon, K''_\varepsilon$ are operators of orders 1, $0, l-1$ respectively uniformly with respect to ε (with respect to variables (x', y', t)); therefore the first term in (2.25) can be estimated by

$$c_1 |||e^{-kt} v|||_1^2 + C|||\Lambda^{l-1} P_x \psi V^+|||^2 + C$$

because of cone $supp\ \psi \cap \{\xi' = 0\} = \emptyset$, $c_1, c_2, \dots$ depend perhaps on $\mathcal{G}$ but do not depend on ε, k.

Since $\psi P_x V^+ \in \mathcal{H}^{\infty, 1}$ then (2.18) implies that the first term in (2.25) can be estimated by

$$c_1 |||e^{-kt} v|||_1^2 + C_1 .$$

Similarly, $[B, \mathcal{J}_\varepsilon \Lambda^l] = K'''_\varepsilon \mathcal{J}_\varepsilon \Lambda^l$ where K'''_ε is an operator of order 0 uniformly with respect to ε ; therefore the third term in (2.25) can be estimated by

$$c_1 |||e^{-kt} v|||_{Y, 1/2}^2 + C \leq c_2 |||e^{-kt} v|||_1^2 + C .$$

(2.18) implies that the second and the fourth terms in (2.25) are equal modulo terms bounded as ε tends to 0 to

$$-2 \operatorname{Re} i([P_x, \psi] w, \psi D_t w),$$

$$2 \operatorname{Re} i(\beta_x [B_x, \psi] w, \psi D_t w)_Y$$

respectively where $w = e^{-kt} \mathcal{J}_\varepsilon \Lambda^l V^+$.

Therefore (2.22), (2.24) and estimates made above imply that

$$\begin{aligned} (2.26) \qquad & (c_0 k - c_3) |||e^{-kt} v|||_1^2 - C \leq \\ & -\operatorname{Re} i([P_x, \psi] w, \psi D_t w) + \\ & \operatorname{Re} i(\beta_x [B_x, \psi] w, \psi D_t w)_Y = \end{aligned}$$

$$\mathrm{Re}\, i([D_t^2,\psi]w,\ \psi D_t w) -$$

$$\mathrm{Re}\, i([\mathcal{A}_x,\psi]w,\ \psi D_t w) +$$

$$\mathrm{Re}\, i(\ \beta_x[B_x,\psi]w,\ \psi D_t w)_Y .$$

We set $K = 2c_3 c_0^{-1}$. We want to prove that the right part of (2.26) is semibounded from above as ε tends to 0 . Let $\omega \in OPS^{0'}(X \times X \times \mathbb{R})$ be an operator with the symbol

$$\varphi_t(x,y,t,\xi',\eta',\tau)\,\chi_1(\varphi(x,y,t,\xi',\eta',\tau)+\alpha)$$

where $\chi_1 = \sqrt{-\chi\chi'} \in C^\infty$. Let

$$Q = Q_2 + Q_1 D_{x_1} + Q_0 D_{x_1}^2$$

be an operator with the principal symbol $-\varphi_t^{-1}\{a,\varphi\}$, $Q_j \in OPS^{j-1'}(X \times X \times \mathbb{R}, E)$. Let $q \in OPS^{0'}_{-1}(Y \times X \times \mathbb{R}, E)$ be an operator with the principal symbol $\beta_0 \varphi_t^{-1}\{\xi_1 + b,\ \varphi\}$ where β_0 , b are the principal symbols of β, B_1 respectively; here $a = a(x,\xi)$, $b = b(x',\xi')$, $\beta_0 = \beta_0(x',\xi')$.

Then the right part of (2.26) is equal modulo terms bounded as ε tends to 0 to

$$(2.27)\quad -2(D_t^2\omega w,\omega w) + \mathrm{Re}(Q\omega w, D_t\omega w) + \mathrm{Re}(q\omega w, D_t \omega w)_Y .$$

One can rewrite the operator Q in the form

$$Q = Q_2' + Q_1' D_{x_1} + Q_0' P_x .$$

(2.16) implies that the principal symbols of Q_2 , Q_1 , Q_0 can be estimated by $c_4 C_0^{-1}|\xi'|$, $c_4 C_0^{-1}$, $c_4 C_0^{-1}(|\xi'| + |\tau|)^{-1}$ respectively; therefore the principal symbols of Q_2' , Q_1' , Q_0' can be estimated by $c_5 C_0^{-1}(|\xi'|+|\tau|)$, $c_5 C_0^{-1}$, $c_5 C_0^{-1}(|\xi'|+|\tau|^{-1})^{-1}$ respectively; here c_4, c_5 do not depend on φ . Therefore the absolute value of the second term in (2.27) does not exceed

$$4c_5 C_0^{-1}(|||\omega w|||_1^2 + \|D_t \omega w\|^2) + C\|\Lambda^{-1} P_x \omega w\|^2 + C \leq$$

$$4c_5 C_0^{-1}(|||\omega w|||_1^2 + \|D_t \omega w\|^2) + C_1 .$$

Moreover (2.16) implies that the principal symbol of q can be estimated by $c_4 C_0^{-1}$; therefore the third term in (2.27) can be rewritten modulo terms bounded as ε tends to 0 in the form

$$2\,\mathrm{Re}(-i D_1 q \omega w,\ D_t \omega w)$$

and can be estimated by

$$c_5 C_0^{-1} |||\omega w|||_1 \, \|D_t \omega w\| + C .$$

On the other hand, the first term in (2.27) equals

$$-\|D_t \omega w\|^2 - \mathrm{Re}(\mathcal{A}_x \omega w, \omega w) + \mathrm{Re}(P_x \omega w, \omega w).$$

The third term here is bounded as ε tends to 0 .

Repeating the arguments used in the proof of (2.24) it follows that

$$\mathrm{Re}(\mathcal{A}_x \omega w, \omega w) \geq c_0 |||\omega w|||_1^2 - C .$$

Therefore the right part of (2.27) can be estimated from above by

$$-c_0 |||\omega w|||_1^2 - \|D_t \omega w\|^2 + 5c_5 C_0^{-1}(|||\omega w|||_1^2 + \|D_t \omega w\|^2) + C \leq C_1$$

provided C_0 is large enough, C_0 depends only on a, b .

Then (2.27) implies that $|||v|||_1 = |||\Lambda^{l} \mathcal{J}_\varepsilon \psi V^+|||_1$ is bounded as ε tends to 0 and hence

(2.28) $$\psi V^+ \in \mathcal{H}^{l+1;\,0}_{x',y',t;\ x_1,y_1}$$ and

$$D_{x_1}\psi \overset{+}{V} \in \mathcal{H}^{l;0}_{x',y',t;x_1,y_1}.$$

Note that

$$P_x \psi \overset{+}{V} = \psi(\delta(t)K_{A_B^{-1}\Pi}(x,y)) + [P_x,\psi]\overset{+}{V},$$

$$(\psi\overset{+}{V})P_y^\dagger = \psi(\delta(t)K_{A_B^{-1}\Pi}(x,y)) + \overset{+}{V}[P_y,\psi]^\dagger$$

and that $[P_x,\psi]\overset{+}{V},\ \overset{+}{V}[P_y,\psi]^{\dagger} \in \mathcal{H}^{l-1/2;1}_{x',y',t;x_1,y_1}$ because of (2.18) and $\psi(\delta(t)K_{A_B^{-1}\Pi}) \in \mathcal{H}^{\infty;1}_{x',y',t;x_1,y_1}$

Combining (2.28) with these equalities and inclusions we obtain that (2.19) holds.

Thus theorem 2.3(iii) has been proved.

Now we want to prove statement (iv). If $\Pi \in OGS^0(X,E)$ then $WF_f(u) \subset j_x^{-1} j_y^{-1}(T^*(Y\times Y\times \mathbb{R})\setminus 0)$ because of proposition 1.6 and we need to prove only that $WF_f(u) \subset \{|\tau| \leqslant C|\xi'|\}$; these two inclusions combined with the equality in rights of x and y, (2.8), $(2.8)^\dagger$ imply (2.14). Combining (2.14) with (iii) we obtain (2.12)', (2.13)'.

If a is negative definite then P is elliptic and one can check that operator $\{P, B\}$ satisfies the Šapiro - Lopatinskii condition in the zone $\{|\tau| \geqslant C|\xi'|\}$ provided C is large enough. Therefore $WF_f(u) \cap \{|\tau| \geqslant C|\xi'|\} = \emptyset$; the fact that B_1 is a pseudo-differential operator with respect to x' but not with respect to (x',t) is inessential: we can reduce the problem (2.3) - (2.4) to an elliptic equation on the boundary; the principal symbol of the corresponding operator is continuous and invertible at $\xi'=0$ and this operator commutes with D_t.

To prove (iv) in a general case we need to refer to the method of p.3.5. We can reduce (2.3) - (2.4) to boundary value problem for first-order system for $U=(D_t u,\ D_{x_1}u) \in \mathcal{D}'(X\times X\times \mathbb{R},\ \mathrm{Hom}(E\oplus E, E))$; this reduction is microlocal in the neighbourhood of the point $(x,0,\tau) \in T^*(Y\times\mathbb{R})\setminus 0$; this system is differential with respect to x_1 and pseudo-differential with respect to (x',t). By means of pseudo-differential operators one can reduce this system to four "systems"

$$(2.29)_j \qquad (D_{x_1} + M_{jx})U_j \in C^\infty \quad (j=1,\dots,4)$$

where $U_j \in \mathcal{D}'(X \times X \times \mathbb{R},\ \mathrm{Hom}(E_j, E))$, $E \oplus E = E_1 \oplus E_2 \oplus E_3 \oplus E_4$ *) is the decomposition of $E \oplus E$, $M_j \in OPS^{1'}(X \times \mathbb{R}, E_j)$ have the principal symbols $m_j(x, \xi', \tau)$ such that m_1 and m_2 are Hermitian matrices and eigenvalues of m_1, m_2, m_3, m_4 at $(x, 0, \tau)$ are equal to $\sqrt{\sigma_j}\,|\tau|$, $-\sqrt{\sigma_j}\,|\tau|$, $i\sqrt{-\theta_k}\,|\tau|$, $-i\sqrt{-\theta_k}\,|\tau|$ respectively, where σ_j and θ_k are respectively positive and negative eigenvalues of $a_0(x)$; therefore $\dim E_1 = \dim E_2 = D_+$, $\dim E_3 = \dim E_4 = D_-$, $D_\pm$ are numbers of positive and negative eigenvalues of $a_0(x)$.

Equations $(2.29)_{3,4}$ are both elliptic but $(2.29)_3$ needs no boundary condition and $(2.29)_4$ needs the Cauchy condition at Y. Therefore $(2.29)_3$ implies that $U_3 \in C^\infty$ in the neighbourhood of $(x, 0, \tau)$; we consider y only as a parameter.

Equations $(2.29)_{1,2}$ are both symmetric and hyperbolic with respect to x_1. The theory of propagation of the singularities for symmetric hyperbolic systems [49] combined with the fact that $U_1, U_2 \in C^\infty$ outside of Y implies that $U_1, U_2 \in C^\infty$ up to the boundary in the neighbourhood of $(x, 0, \tau)$.

Consider now boundary conditions. This conditions in general "mix" τU_1, τU_2, τU_3 and τU_4 but since $\tau U_1, \tau U_2, \tau U_3 \in C^\infty$ we obtain D equations with respect to D_--dimensional vector τU_4; certainly these equations are microlocal in the neighbourhood of $(x, 0, \tau)$. The obtained system is overdetermined in general; one can check that this system is elliptic in the neighbourhood of $(x, 0, \tau)$ and hence $\tau U_4 \in C^\infty$ in the neighbourhood of $(x, 0, \tau)$; then $(2.29)_4$ implies that $U_4 \in C^\infty$ in the neighbourhood of $(x, 0, \tau)$.

Thus we proved that $u \in C^\infty$ in the neighbourhood of $(x, 0, \tau)$ with respect to (x, t) ; y was considered as a parameter. But $(2.3)^\dagger - (2.4)^\dagger$ and ellipticity of $\{\mathcal{A}^\dagger, B^\dagger\}$ imply that the smoothness with respect to t is equivalent to the smoothness with respect to (microlocally) and hence $(x, 0, \tau) \notin WF_f(u)$. Statement (iv) has been proved.

2.3. It is interesting that there exist commutating semibounded operator A_B and selfadjoint projector Π which is a singular Green operator but not a smoothing operator. Exactly, let in notations of p.1.6. $\Omega = \{ |\xi'| > b(\xi') > 0 \}$ be a connected. component of $\mathbb{R}^{d-1} \setminus 0$. Consider an operator $\Pi \in OGS^0(\overline{\mathbb{R}}^+ \times \mathbb{R}^{d-1})$ with Schwartz kernel expressed by oscillatory integral (1.9). It is clear that Π is selfadjoint projector commuting with $(-\Delta)_B$, $\Pi\Pi^+ = \Pi$.

*) This decomposition is not necessarily orthogonal.

Certainly, singularities of $u(x,y,t)$ corresponding to this projector are absolutely "nonclassical".

2.4. Finally for small t we decompose $u(x,y,t)$ in two terms one of which corresponds to "free" propagation and another corresponds to reflecting at the boundary.

It is clear that there exist a compact C^∞-manifold $\tilde{X}$ with the boundary $\tilde{Y}$, a density $d\tilde{x}$ on $\tilde{X}$, a Hermitian C^∞-fibering $\tilde{E}$ over $\tilde{X}$, formally selfadjoint second-order elliptic differential operator $\tilde{\mathcal{A}}$ acting on sections of $\tilde{E}$ and a boundary operator $\tilde{B}$ such that $\tilde{A}_{\tilde{B}}$ is selfadjoint in $L_2(\tilde{X},\tilde{E})$, $\{\tilde{\mathcal{A}},\tilde{B}\}$ is elliptic and $X\subset\tilde{X}\setminus\tilde{Y}$, $dx = d\tilde{x}|_{\tilde{X}}$, $\tilde{E}=\tilde{\tilde{E}}|_X$, $\mathcal{A}=\tilde{\mathcal{A}}|_X$.

Let $\tilde{\Pi}^{\pm}$ be selfadjoint projectors to positive and negative invariant subspaces of $\tilde{A}_{\tilde{B}}$; then

$$K_{\tilde{\Pi}^{\pm}} \equiv K_{\Pi^{0\pm}} \ (\operatorname{mod} C^\infty(X\times X,\ \operatorname{Hom}(E)), \tag{2.30}$$

recall that $\Pi^{0\pm}$ are pseudo-differential components of $\Pi^{\pm}$.

Let $\tilde{u}(x,y,t)$ be Schwartz kernel of operator $\tilde{A}_{\tilde{B}}^{-1/2}\cdot\sin\tilde{A}_{\tilde{B}}^{1/2}t\cdot\tilde{\Pi}^{+}$; let $u_0(x,y,t) = \theta_x\,\tilde{u}(x,y,t)\theta_y^{\dagger}$ where $\theta\in OPS^0(\tilde{X},E)$ is an operator such that $K_\theta \equiv K_{\Pi^0}$ $(\operatorname{mod} C^\infty(X\times X,\operatorname{Hom}(E)))$; such operator exists without question.

Let $u_1(x,y,t) = u(x,y,t) - u_0(x,y,t)$. Then

$$P_x u_0 \equiv 0\ ,\quad P_x u_1 \equiv 0\ , \tag{2.31}$$

$$u_0 P_y^{\dagger} \equiv 0\ ,\quad u_1 P_y^{\dagger} \equiv 0 \tag{2.31†}$$

$$(\operatorname{mod} C^\infty(X\times X\times\mathbb{R},\ \operatorname{Hom}(E))),$$

$$u_0|_{t=0} = u_1|_{t=0} = 0\ , \tag{2.32}$$

$$i D_t u_0|_{t=0} \equiv K_{\Pi^0}(x,y)\,,$$

$$i D_t u_0|_{t=0} \equiv K_{\Pi'}(x,y)\ \ (\operatorname{mod} C^\infty(X\times X,\operatorname{Hom}(E)))\,,$$

$$B_x u_1 = -B_x u_0\ , \tag{2.33}$$

$$u_1 B_y^{\dagger} = -u_0 B_y^{\dagger}\ . \tag{2.33†}$$

Theorem 2.1(ii) implies that a change of $\tilde{\mathcal{A}}$ outside of the neighbourhood of X and a change of $\tilde{B}$ do not change $\tilde{u}(x,y,t)$, $u_0(x,y,t)$, $u_1(x,y,t)$ modulo $C^\infty(X\times X\times[-t_0,t_0], \mathrm{Hom}(E))$ where $0<t_0$ depends on the neighbourhood of X mentioned above.

THEOREM 2.4. $u_1 \in C^\infty(\{x_1+y_1 > c|t|, |t|<t_0\}, \mathrm{Hom}(E))$.

PROOF. Let $v = C^+ u_1 \in \mathcal{D}'(\tilde{X}\times X\times \mathbb{R}, \mathrm{Hom}(\tilde{E},E))$ where C^+ is the operator of the continuation from X to $\tilde{X}$, $C^+ w = 0$ outside of X. Then

$$\tilde{P}_x v = f \in C^\infty((\tilde{X}\setminus Y)_x \times X_y \times \mathbb{R}, \mathrm{Hom}(\tilde{E},E)),$$

$$v|_{t=0} = 0,$$

$$iD_t v|_{t=0} = g \in C^\infty((\tilde{X}\setminus Y)_x \times X_y, \mathrm{Hom}(\tilde{E},E)),$$

$$\tilde{B}_x v = 0 .$$

Using the Duhamel formula we can express $\tilde{\Pi}^+_x v$ by means of $\tilde{\Pi}^+_x f$, $\tilde{\Pi}^+_x g$ and $\tilde{u}(x,\cdot,\cdot)$. Then theorem 2.3(i) implies that $\tilde{\Pi}^+_x v \in C^\infty(\{x_1 > c|t|, |t|<t_0\})$. Similarly the same inclusion holds for $\tilde{\Pi}^-_x v$. Therefore $u_1 = v \in C^\infty$ at $\{x_1 > c|t|, |t|<t_0\}$.

The theorem is proved since x and y are equal in rights.

§3. The normality of the "great" singularity

3.1. Let us prove the theorems about normality of the great singularity. First let us examine the singularity of Γu outside of the boundary and this simplest situation we shall clarify our main ideas. The considerations near the boundary will be essentially more complicated: so we shall need to use certain additional ideas. As before $u(x,y,t)$ is the Schwartz kernel of operator $U(t) = A_B^{-1/2}\cdot \sin A_B^{1/2} t\cdot \Pi$ and $\Pi\Pi^+ = \Pi$.

THEOREM 3.1(i) Let X be a closed manifold and $A \in OPS^2(X,E)$ an elliptic selfadjoint operator. Then there exists $t_1>0$ such that $\Gamma u \equiv u(x,x,t) \in C^\infty$ for $0<|t|<t_1$ and

$$t^{|\alpha|+m+n} D_x^\alpha D_t^n \Gamma u \in L(X\times[-t_1,t_1], \mathrm{hom}(E)) \tag{3.1}$$

$$\forall n \in \mathbb{Z}^+, \ \alpha \in \mathbb{Z}^{+d}$$

where $m = m(\alpha)$.

(ii) Let X be a manifold with the boundary, $0<\delta$; then there exists $t_1 = t_1(\delta) > 0$ such that $\Gamma u \in C^\infty$ for $0<|t|<t_1$, $x_1 > \delta$ and (3.1) holds in the domain $\{\delta \leqslant x_1, \ |t| \leqslant t_1\}$.

We consider closed manifolds; manifolds with the boundary are considered by the same method with the help of theorem 2.3 (i), (ii).

Let $(y^*, \eta^*, \tau^*) \in (T^*X \setminus 0) \times (\mathbb{R} \setminus 0)$, $q \in OPS^0(X \times \mathbb{R})$ and cone supp q is contained in the small enough conic neighbourhood of $(y^*, \eta^*, \tau^*) \times \mathbb{R}_t$, $q_y = q(y, D_y, D_t)$.
Let $\chi \in C_0^\infty(\mathbb{R})$, $\operatorname{supp} \chi \subset (-t, t_0)$. Then theorem 2.2 and (2.10) imply that $WF(q_y \chi(t) u^+)$ is contained in the small enough conic neighbourhood of $(y^*, -\eta^*, y^*, \eta^*, 0, \tau^*)$ and our future investigation is microlocal.

Let $l(x)$ be a function such that

$$\tau^{*-1} \langle \{a, l\}(y^*, -\eta^*) v, v \rangle \geqslant c_1 |v| \tag{3.2}$$
$$\forall v \in \operatorname{Ker}(\tau^{*2} - a(y^*, -\eta^*)), \ c_1 > 0 .$$

Note that every function $l(x)$ such that $dl(y^*) = -\tau^{*-1}\eta^*$ satisfies (3.2).

Let us introduce functions $\varphi(x, y, t) = \sigma t + l(x) - l(y)$ and $\psi_\lambda(x, y, t) = \varphi_+^\lambda(x, y, t)$ where $z_+ = \max(z, 0)$, $\lambda \in \frac{1}{2}\mathbb{Z}^+$.

Then theorem 3.1 follows from (2.7), $(2.7)^+$ and the statement:

THEOREM 3.2. Let X be a closed manifold, $l(x)$ satisfy (3.2), cone supp q be contained in the small enough conic neighbourhood of $(y^*, \eta^*, \tau^*) \times \mathbb{R}_t$, $\chi \in C_0^\infty(\mathbb{R}_t)$, $\operatorname{supp} \chi \subset [-t_1, t_1]$, $0<t_1$, σ be small enough; then

$$D_x^\alpha D_y^\beta D_t^n \psi_{n+|\alpha|+|\beta|+m} \, q_y \chi u^+ \in L_2(X \times X \times \mathbb{R}, \operatorname{Hom}(E)) \tag{3.3}$$
$$\forall \alpha, \beta \in \mathbb{Z}^{+d}, \ n \in \mathbb{Z}^+$$

where $m = m(d)$.

REMARKS. Theorem 3.2 is a more refined statement that the singularities of $u(x, y, t)$ leave the diagonal $x = y$. By the same method one can prove that (3.3) holds even if l does not satisfy (3.2) but σ is a large enough negative number. The latter statement is a more refined one about the finite speed of the propagation of singularities (compare with theorem 2.2). A similar statement can be proved near the boundary.

(ii) Lemma 3.3 (see below) implies that (3.3) and hence (3.1) hold true not only for u but also for $Q'_x u Q''^+_y$ where $Q', Q'' \in OPS^0(X, E)$; in particular (3.1) holds for $u_0(x, y, t)$ up to the boundary.

(iii) Using the refined statement about the finite speed of the propagation of singularities (see above) one can prove that

$$(x_1+y_1)^{|\alpha|+|\beta|+n+m} D_x^\alpha D_y^\beta D_t^n \chi u_1 \in L_2(X \times X \times \mathbb{R} \cap \{x_1+y_1 \geq c|t|\}, Hom(E))$$

$$\forall \alpha, \beta \in \mathbb{Z}^{+d}, \quad n \in \mathbb{Z}^+$$

provided C is large enough; it is a more refined statement of theorem 2.4.

(iv) In §4 we shall prove that Γu has an essentially better nature near $t=0$ than is pointed out in theorem 3.1. That is to say Γu is a classical Fourier integral distribution with the Lagrangean manifold $N(X \times 0) = \{(x, 0, 0, \tau), \ x \in X, \ \tau \in \mathbb{R} \setminus 0\}$ near $t=0$ and therefore

$$t^{m+n} D_x^\alpha D_t^n \Gamma u \in L_2(X \times [-t_1, t_1], hom(E)) \tag{3.4}$$

$$\forall \alpha \in \mathbb{Z}^{+d}, \quad n \in \mathbb{Z}^+, \quad \text{here} \quad m = m(\alpha).$$

It is easy to show that (3.4) is a characteristic property of Fourier integral distributions with the Lagrangean manifold $N(X \times 0)$ and with the symbols belonging to Hörmander class $S_{1,0}$; at the same time one can construct many distribution of an ill nature satisfying (3.1);

PROOF OF THEOREM 3.2. Since (2.9) holds and cone supp $q \subset \{\eta \neq 0, \ \tau \neq 0\}$ then we need to prove (3.3) only for $\alpha = \beta = 0$; Let $\chi_0, \chi \in C_0^\infty(\mathbb{R}_t)$, $\chi_0 = 1$ on $[-2t_1, 2t_1]$, $\operatorname{supp} \chi \subset (-2t_1, 2t_1)$, $\chi = 1$ in the neighbourhood $[-t_1, t_1]$, $\chi'(t) \leq 0$ for $t \geq t_1$.

We shall prove that

$$\chi \Psi_{\lambda+m_1} \Lambda^{\lambda-m_2} v \in L_2(X \times X \times \mathbb{R}, Hom(E)) \tag{3.5}$$

$$\forall\lambda \in \tfrac{1}{2}\mathbb{Z}^+ ,$$

where $m_j = m_j(d)$, $v = q_y \chi_0 u^+$, $\Lambda^s = (I + |D_t|^2)^{s/2}$.
We shall use the induction on $\lambda \in \frac{1}{2}\mathbb{Z}^+$ decreasing $\operatorname{supp}\chi$ on every step. It is clear that for $\lambda = 0$ (3.5) is valid for appropriate $m_2 = m_2(d)$. Let (3.5) be valid for every $\lambda \leqslant \mu - \frac{1}{2}$, we shall prove (3.5) for $\lambda = \mu$.

Theorem 2.2, (2.11) and definitions of q, χ_0 imply that

$$\text{(3.6)} \qquad WF(v) \cap \{t < 2t_1\} \subset \Omega$$

where $\Omega = \Omega_{t_1}$ is the conic neighbourhood of $(y^*, -\eta^*, y^*, \eta^*, 0, \tau^*) \in (T^*X \setminus 0)^2 \times (T^*\mathbb{R} \setminus 0)$, Ω_{t_1} tends to $(y^*, -\eta^*, y^*, \eta^*, 0, \tau^*)$ cone as t_1 tends to 0,

$$\text{(3.7)} \qquad P_x v \equiv q_y \delta(t) K_\Pi(x, y) \pmod{C^\infty}$$

for $t < 2t_1$,

$$\text{(3.8)} \qquad v \equiv 0 \pmod{C^\infty} \qquad \text{for } t < 0 .$$

We use the formula

$$\text{(3.9)} \qquad -2\operatorname{Re} i(GPw, w) = i([P, G]w, w)$$

which is valid for arbitrary symmetric operators P, G and set $P = P_x$, $G = \Lambda^{\mu-1/2} \chi^2 \psi_{2m_1+2\mu+1} \Lambda^{\mu-1/2}$, $w = \Lambda^{-m_2} \mathcal{J}_\varepsilon v$ where $\mathcal{J}_\varepsilon = (I + \varepsilon^2 |D_t|^2)^{-N}$, N is large enough,

$$(w_1, w_2) = \int_{X\times X\times\mathbb{R}} \operatorname{tr} w_2^\dagger w_1 \, dx\, dy\, dt$$

is the inner product in $L_2(X\times X\times\mathbb{R}, \operatorname{Hom}(E))$; in what follows $\|\cdot\|$ denotes the corresponding norm.

(3.6) implies that $w \in H^N(X\times X\times\mathbb{R}, \operatorname{Hom}(E))$ and $\|w\|$ is bounded as ε tends to 0 for appropriate $m_2 = m_2(d)$; therefore the left of (3.9) equals modulo terms bounded as ε tends to 0 to

$$-2\operatorname{Re} i(G\mathcal{J}_\varepsilon \Lambda^{-m} q_y \delta(t) K_\Pi , w).$$

Obviously,

$$G = \Lambda^{-\nu} \psi_0 \sum_{j=0}^{\mu-\frac{1}{2}+\nu} \chi_j D_t^j \varphi^{2m_1+\mu+\frac{1}{2}-\nu+j} \Lambda^{\mu-\frac{1}{2}}$$

where $0 \leq \nu \leq 3/2$, $\mu - 1/2 + \nu \in 2\mathbb{Z}^+$, $\chi_j \in C^\infty(X\times X\times \mathbb{R})$, $\operatorname{supp}\chi_j \subset \operatorname{supp}\chi$; hence

$$G\mathcal{J}_\varepsilon \Lambda^{-m_2} q_y = \Lambda^{-\nu}\psi_0 \sum_{j=0}^{2m_1+2\mu+\frac{1}{2}} R_{j\varepsilon}\, q' \varphi^j$$

where $q' = q'(y,t,D_y,D_t) \in OPS^0(X\times\mathbb{R})$, cone supp $q' \subset$ cone supp $q \cap$ supp χ , $R_{j\varepsilon}$ are operators of order $(-2m_1+\nu-2)$ with respect to (t,y) uniformly with respect to ε.

Note that

$$\varphi^j \delta(t) K_\Pi(x,y) = (\ell(x)-\ell(y))^j \delta(t) K_\Pi(x,y)$$

and that propositions 1.6(i), 1.7(i) imply that

$$(\ell(x)-\ell(y))^j K_\Pi \in H_{x,y}^{j-\frac{d+1}{2}}(X\times X, \operatorname{Hom}(E)).$$

Therefore

$$\Lambda^{-1}(\ell(x)-\ell(y))^j \delta(t) K_\Pi \in H_{x,y;t}^{j-\frac{d+1}{2}}(X\times X\times\mathbb{R}, \operatorname{Hom}(E))$$

and consequently

$$q'_y \delta(t) \varphi^j K_\Pi \in H_{x,y,t}^{j-\frac{d+3}{2}}(X\times X\times\mathbb{R}, \operatorname{Hom}(E));$$

therefore for $2m_1 + m_2 \geq \frac{d+3}{2}$
$\| G\mathcal{J}_\varepsilon \Lambda^{-m_2} q_y \delta(t) K_\Pi \|$ is bounded as ε

tends to 0 ; therefore the left side of (3.9) is bounded as ε tends to 0.

Consider now the right side of (3.9). Obviously

$$[P_x, G] = \Lambda^{\mu-1/2}\left([P_x, \psi_{2\mu+2m_1+1}]\chi^2 + \psi_{2\mu+2m_1+1}[-D_t^2, \chi^2]\right)\Lambda^{\mu-1/2}.$$

If A is a differential operator then

$$(3.10)\qquad [P_x, \psi_{2\mu+2m_1+1}]\chi^2 =$$

$$(2\mu+2m_1+1)[P_x, \varphi]\psi_{2\mu+2m_1}\chi^2 + \chi_1\psi_{2\mu+2m_1-1} =$$

$$(2\mu+2m_1+1)\chi\psi_{\mu+m_1}[P_x, \varphi]\chi\psi_{\mu+m_1} + \chi_2\psi_{2\mu+2m_1-1}$$

where $\chi_1, \chi_2 \in C^\infty$, $\operatorname{supp}\chi_j \subset \operatorname{supp}\chi$.

If A is a pseudo-differential operator then we need more complicated considerations. Namely we need the following statement:

LEMMA 3.3. If $h \in C^{m+n}(X)$, $b \in OPS^n(X)$ then

$$bh = \sum_{|\alpha|\le n} \frac{1}{\alpha!} h_{(\alpha)} b^{(\alpha)} + \tilde{b},$$

$$hb = \sum_{|\alpha|\le n} \frac{1}{\alpha!} (-1)^{|\alpha|} b^{(\alpha)} h_{(\alpha)} + \tilde{b}'$$

where $OPS^{n-|\alpha|}(X) \ni b^{(\alpha)}$ are operators with symbols $(\partial/\partial\xi)^\alpha b(x,\xi)$, $b(x,\xi)$ is the symbol of b, $h_{(\alpha)} = D_x^\alpha h$, $\tilde{b}$ and $\tilde{b}'$ are bounded operators in $L_2(X)$, $m = m(\alpha)$.

Moreover if b depends on the parameter and is bounded uniformly with respect to this parameter as an element of $L^n_{1,0}(X)$ (see [46]) then $b^{(\alpha)}$ are bounded uniformly with respect to this parameter as elements of $L^{n-|\alpha|}_{1,0}(X)$ and $\tilde{b}, \tilde{b}'$ are bounded uniformly with respect to this parameter as bounded operators in $L_2(X)$.

PROOF of this lemma coincides in essential with the proof of the L_2 - boundedness and formula for the superposition for pseudo-differential operators (see [45, 46], th 3.3, 3.6) and we omit it.

Note that

$$\Lambda^{\mu-1/2} = \Lambda^{-2\nu} \sum_{j=0}^{\frac{\mu}{2}-\frac{1}{4}+\nu} D_t^{2j}$$

where $0 \leqslant \nu \leqslant \frac{3}{4}$, $\mu - \frac{1}{2} + 2\nu \in 2\mathbb{Z}^+$; then lemma 3.3, the boundedness of $\|w\|$ as ε tends to 0 and (3.6) imply that the right side of (3.9) will not change modulo terms bounded as ε tends to 0 if we replace

$$i\Lambda^{\mu-1/2} [A_x, \psi_{2\mu+2m_1+1}] \chi^2 \Lambda^{\mu-1/2}$$

by

$$\Lambda^{-2\nu} \sum_{j=0}^{\mu-\frac{1}{2}+2\nu} \sum_{k=0}^{2\mu-j+1} \psi_{2\mu+2m_1-j-k} R_{jkx} D_t^{\mu+2\nu-j-1/2} \Lambda^{\mu-1/2} \tag{3.11}$$

where $R_{jk} \in C^\infty(\mathbb{R}, OPS^{-k}(X, E))$, $\operatorname{supp} R_{jk} \subset \operatorname{supp} \chi$ and

$$R_{00} = i(2\mu + 2m_1 + 1)[A_x, \varphi]\chi^2 .$$

Repeating these arguments we can replace (3.11) by

$$\sum_{j,k=0}^{[\mu]} \Lambda^{\mu-j/2} \psi_{\mu+m_1-j/2} R'_{jk} \psi_{\mu+m_1-k/2} \Lambda^{\mu-k/2}$$

where $R'_{jk} \in OPS^\circ(X \times X \times \mathbb{R}, \ \operatorname{Hom}(E))$ and outside of the neighbourhood of $\{\xi = 0\} \cup \{\eta = 0\} \cup \{\tau = 0\}$ the principal symbol of R'_{00} equals

$$(2m_1 + 2\mu + 1)\, |\tau|^{-1} \{a, \varphi\}(x, \xi)\chi^2 ,$$

$$\operatorname{supp} R'_{jk} \subset \operatorname{supp} \chi^{*)} .$$

---

*) This means that $R'_{jk} w = 0$ provided $\operatorname{supp} w \cap \operatorname{supp} \chi = \emptyset$ and $\operatorname{supp} R'_{jk} w \subset \operatorname{supp} \chi$ for arbitrary w.

Repeating these arguments again we obtain that

$$(3.12) \qquad \Psi_{\nu+m_1}\Lambda^{\nu}\mathcal{J}_{\varepsilon} = \sum_{K=0}^{[\nu]} T_{\nu K\varepsilon}\Lambda^{\nu-K}\Psi_{\nu+m_1-K} + \tilde{T}_{\nu\varepsilon} =$$

$$\sum_{K=0}^{[\nu]} T'_{\nu K\varepsilon}\Psi_{\nu+m_1-K}\Lambda^{\nu-K} + \tilde{T}'_{\nu\varepsilon}$$

where $T_{\nu K\varepsilon}$, $T'_{\nu K\varepsilon}$ are operators of order 0 uniformly with respect to ε, $\tilde{T}_{\nu\varepsilon}, \tilde{T}'_{\nu\varepsilon}$ are operators in L_2 bounded uniformly with respect to ε.

Therefore the conjecture of induction implies that modulo terms bounded as ε tends to 0

$$(3.13) \quad i([A_x, \Psi_{2\mu+2m_1+1}]\chi^2\Lambda^{\mu-m_2-1/2}\mathcal{J}_{\varepsilon}v,$$

$$\Lambda^{\mu-m_2-1/2}\mathcal{J}_{\varepsilon}v) \sim (2m_1+2\mu+1)(R\chi\Psi_{\mu+m_1}\mathcal{J}_{\varepsilon}\Lambda^{\mu-m_2}v, \chi\Psi_{\mu+m_1}\mathcal{J}_{\varepsilon}\Lambda^{\mu-m_2}v)$$

where $R = R'_{00}$.

But (3.2), (3.6) imply that if $\mathit{cone\,supp}\, q$ and t_1 are small enough then in the neighbourhood of $WF(v)\cap\{t \leqslant 2t_1\}$ the following inequality holds:

$$\varkappa|\tau|^{-1}\langle\{a,\ell\}(x,\xi)v,v\rangle \geqslant$$

$$c_1|v|^2 - (|\xi|^2+|\eta|^2+|\tau|^2)^{-2}|p(x,\xi,\tau)v|^2$$

where $p(x,\xi,\tau) = -\tau^2 + a(x,\xi)$, $\varkappa = \mathrm{sign}\,\tau^*$, $c_1 > 0$.

Then the sharp Gårding inequality implies that

$$(3.14) \qquad \varkappa\,\mathrm{Re}\,(R\chi\Psi_{\mu+m_1}\mathcal{J}\Lambda^{\mu-m_2}v,$$

$$\chi\Psi_{\mu+m_1}\mathcal{J}_{\varepsilon}\Lambda^{\mu-m_2}v) \geqslant$$

$$c_1 \| \chi \Psi_{\mu+m_1} \mathcal{J}_\varepsilon \Lambda^{\mu-m_2} v \| - C \| \Lambda^{-1/2} \chi \Psi_{\mu+m_1} \mathcal{J}_\varepsilon \Lambda^{\mu-m_2} v \|^2 -$$

$$C \| \omega \chi \Psi_{\mu+m_1} \mathcal{J}_\varepsilon \Lambda^{\mu-m_2} v \|^2 - C \| \Lambda^{-2} P_x \chi \Psi_{\mu+m_1} \mathcal{J}_\varepsilon \Lambda^{\mu-m_2} v \|$$

where $\omega \in OPS^0(X \times X \times \mathbb{R})$, *cone supp* $\omega \subset \{|t| \leq 2t_1\}$, *cone supp* $\omega \cap WF(v) = \emptyset$.

The conjecture of induction and (3.12) imply that the second term in the right side of (3.14) is bounded as ε tends to 0 ; the third term is bounded as ε tends to 0 because of lemma 3.3 and (3.6); lemma 3.3, (3.13) and the conjecture of induction imply that

$$\| [\Lambda^{-2} P_x , \chi \Psi_{\mu+m_1}] \mathcal{J}_\varepsilon \Lambda^{\mu-m_2} v \|$$

is bounded as ε tends to 0 ; repeating the arguments of the first part of the proof one can show that

$$\| \chi \Psi_{\mu+m_1} \mathcal{J}_\varepsilon \Lambda^{\mu-m_2-2} P_x v \|$$

is also bounded as ε tends to 0 ; hence (3.13), (3.14) imply that

$$\text{(3.15)} \quad Re\, \varkappa_i([A_x, \Psi_{2\mu+2m_1+1}]\chi^2 \mathcal{J}_\varepsilon \Lambda^{\mu-m_2-1/2} v, \; \mathcal{J}_\varepsilon \Lambda^{\mu-m_2-1/2} v) \geq$$
$$c_1(2m_1+2\mu+1) \| \chi \Psi_{\mu+m_1} \mathcal{J}_\varepsilon \Lambda^{\mu-m_2} v \|^2 - C$$

where C does not depend on ε, $c_1 > 0$ does not depend on σ.

By the same arguments (but without the sharp Gårding inequality) one can derive that

$$\text{(3.16)} \quad \begin{aligned} &Re\, \varkappa_i([-\mathcal{D}_t^2, \Psi_{2\mu+2m_1+1}]\chi^2 \mathcal{J}_\varepsilon \Lambda^{\mu-m_2-1/2} v) \geq \\ &-2\sigma(2m_1+2\mu+1) \| \chi \Psi_{\mu+m_1} \mathcal{J}_\varepsilon \Lambda^{\mu-m_2} v \|^2 - C \end{aligned}$$

and that

$$\text{(3.17)} \quad \varkappa_i([-\mathcal{D}_t^2, \chi^2] \mathcal{J}_\varepsilon \Lambda^{\mu-m_2-1} v ,$$

$$\psi_{2\mu+2m_1+1}\mathcal{J}_\varepsilon \Lambda^{\mu-m_2-1/2} v) \sim 4(-\chi\chi'\psi_{\mu+m_1+1/2}\mathcal{J}_\varepsilon \Lambda^{\mu-m_2} v,) \geq -C$$
$$\psi_{\mu+m_1+1/2}\mathcal{J}_\varepsilon \Lambda^{\mu-m_2} v) \geq -C$$

because of $\chi\chi' \leq 0$ in the neighbourhood of $WF(v)$.

The boundedness of the left side of (3.9) and inequalities (3.15) -(3.17) imply that $\|\chi\psi_{\mu+m_1}\mathcal{J}_\varepsilon \Lambda^{\mu-m_2} v\|$ is bounded as ε tends to 0 provided σ is small enough; then (3.12) and the conjecture of induction imply that $\|\mathcal{J}_\varepsilon \chi\psi_{\mu+m_1}\Lambda^{\mu-m_2} v\|$ is also bounded as ε tends to 0. Therefore (3.5) holds for $\lambda=\mu$ and the step of induction is made. Thus theorems 3.2, 3.1 have been proved.

3.2. The proof made above fits manifolds with the boundary too, provided $a = a_0(x)\xi_1^2 + a_2(x,\xi')$ at the boundary on which Dirichlet or Neumann condition is given (i.e. $B=\tau$ or $B=i\tau D_1 + B_1\tau$ where $B_1 \in OPS^0(Y, E))$) (see [56]), nevertheless theorems 3.1, 3.2 hold true only after "cutting" of normal waves even in this special case. This proof will fit after certain modifications more general operators $\mathcal{A}$ and more general boundary operators $\mathcal{B}$. But in the most general case we need to use a different method which is based on the same ideas as well as some new ones. Moreover, these additional ideas give us the opportunity to obtain certain results without the cutting of normal waves. In other words we shall prove the normality of the great singularity of $\sigma(t)$ without proving the normality of the great singularity of Γu.

Let us first formulate the main theorems of p.3.3, 3.4. We shall use the notations $\rho^*=(y^*,-\eta^*,\tau^*)\in(T^*Y\setminus 0)\times(\mathbb{R}\setminus 0)$, $\tau^*=\pm 1$, $\hat{\rho}=(y^*,\mp\eta^*)$.

THEOREM 3.4. Assume that we can find $\mu\in\mathbb{R}$ such that at every point $j^{-1}\hat{\rho}\cap\{g(\rho,1)=0\}$ inequalities (0.10) hold.

Then there exists $t_1>0$ such that if $\chi_0\in C_0^\infty(\mathbb{R})$, $q_y = q(y, D_y', D_t)\in OPS^{0'}(X\times\mathbb{R})$ is contained in the small enough conic neighbourhood of ${}^T\rho^*\times\mathbb{R}_t$ then

$$t^s x_1^l \mathcal{D}_t^n \mathcal{D}_x^d \Gamma q_y \chi_0 u_1 \in L_2(X\times[-t_1,t_1],\ hom(E))$$
$$\forall d\in\mathbb{Z}^{+d},\ n, s, k\in\mathbb{Z}^+ \quad \text{such that } |d|+n\leq s+l-m,$$

where $m=m(d)$.

THEOREM 3.5. There exists $t_1>0$ such that if $\chi_0\in C_0^\infty(\mathbb{R})$, $q_y = q(y, D_y', D_t)\in OPS^{0'}(X\times\mathbb{R})$ is contained in the small enough conic neighbourhood of ${}^T\rho^*\times\mathbb{R}_t$, $\zeta\in C^\infty(X)$ then

$$t^{|\alpha|+m+n} D^{\alpha}_{x'} D^{n}_{t} \tilde{\Gamma} \tau q_y \chi_0 u_1 \in L_2(Y\times[-t_1,t_1]),$$

$$t^{|\alpha|+m+n} D^{\alpha}_{x'} D^{n}_{t} \tau \Gamma q_y \chi_0 u_1 \in L_2(Y\times[-t_1,t_1], \, hom\,(E))$$

$$\forall \alpha \in \mathbb{Z}^{+d-1}, \quad n \in \mathbb{Z}^{+}$$

where $m = m(\alpha)$,

$$\tilde{\Gamma}\tilde{v} = \int_0^{\delta} tr(\Gamma v)(\cdot, x_1)\, dx_1 .$$

THEOREM 3.6. Let $\Pi \in OGS^{\circ}(X, E)$. Then the statement of theorem 3.4 holds.

3.3. First we reduce the "wave" equation (2.3) to a first-order system. Let X' be the neighbourhood of Y identified with $[0,\delta)\times Y$. We introduce an operator $\mathcal{J} : C^{\infty}(X',E) \to C^{\infty}(X', E \oplus E)$,

$$\mathcal{J} = \begin{pmatrix} I \\ \varkappa\,(A_0 D_1 + A_1) \end{pmatrix}$$

where $\varkappa = \varkappa(x, D'_x, D_t) \in OPS^{-1'}(X \times \mathbb{R})$ is an elliptic invertible operator with a real principal symbol $\varkappa_{-1}$. Recall that $A = A_0 D_1^2 + 2A_1 D_1 + A_2$, $A_j \in OPS^{2-j'}(X, E)$, $A_0 = a_0(x)$ is invertible. We set

$$U = \mathcal{J}_x \chi_0(t) u_1 \mathcal{J}_y^{\dagger}, \quad U^{\pm} = \mathcal{J}_x \chi_0(t) u_1^{\pm} \mathcal{J}_y^{\dagger}$$

where $\chi_0 \in C_0^{\infty}(\mathbb{R})$, $\chi_0 = 1$ on $[-2t_1, 2t_1]$, $v R^{+} \overset{def}{\equiv} T(R^{*} T v^{\dagger})^{\dagger}$, T is a time reversion operator: $(Tv)(\cdot,t) = v(\cdot,-t)$. Then

$$U = U^{+} + U^{-}, \quad U(y,x,t) = U^{\dagger}(x,y,-t). \tag{3.18}$$

Equation $(2.31)_2$ implies that $u_1 \in C^{\infty}(X\times X, \mathcal{D}'(\mathbb{R}, Hom(E))$; $(2.31)_2$ - (2.32) imply that

$$P_x u_1^{\pm} \equiv \pm\delta(t) K_{\Pi'}(x,y) \quad (mod\ \theta\, C^{\infty}) \tag{3.19}$$

where $\theta = \theta(t)$ is Heavyside function. Proposition 1.7 implies that the right side of (3.19) belongs to

$$C^{\infty}(Y_{y'} \times [0,\delta)^2_{x_1,y_1}, \mathcal{D}'(Y_{x'} \times \mathbb{R}_t, \mathrm{Hom}(E)))$$

and hence

$$u_1^{\pm} \in C^{\infty}(Y_{y'} \times [0,\delta)^2_{x_1,y_1}\ \mathcal{D}'(Y_{x'} \times \mathbb{R}_t, \mathrm{Hom}(E))); \tag{3.20}$$

therefore

$$U^{\pm} \in C^{\infty}(Y_{y'} \times [0,\delta)^2_{x_1,y_1}\ \mathcal{D}'(Y_{x'} \times \mathbb{R}_t, \mathrm{Hom}(E \oplus E)); \tag{3.21}$$

certainly these inclusions remain true with x' and y' permutated. Since

$$P_x \chi_0\, u_1^{\pm} \equiv \pm\delta(t) K_{\Pi'} \quad \text{for} \quad \pm t < 2t_1 \quad (\mathrm{mod}\ \theta\, C^{\infty})$$

then

$$\mathcal{P}_x U^{\pm} \equiv \pm F_x \qquad \text{for} \quad \pm t < 2t_1 \tag{3.22}$$

modulo functions with wave front sets contained in $\{\xi' = \eta' = 0\}$) and

$$\mathcal{P}_x U \equiv 0 \qquad \text{for} \quad |t| < 2t_1\, (\mathrm{mod}\ C^{\infty}) \tag{3.23}$$

where $\mathcal{P} = K D_1 + L$, $\qquad K = \begin{pmatrix} 0 & I \\ I & 0 \end{pmatrix}$, $L \in OPS^{1'}(X \times \mathbb{R}, E \oplus E)$
and the principal symbol of L equals

$$L_1 = \begin{pmatrix} \varkappa_{-1}(a_2 - \tau^2 - a_1 a_0^{-1} a_1) & a_1 a_0^{-1} \\ a_0^{-1} a_1 & -\varkappa_{-1}^{-1}\ a_0^{-1} \end{pmatrix},$$

$$F_x = \begin{pmatrix} \varkappa_x\, \delta(t) K_{\Pi'}(x,y) \\ 0 \end{pmatrix} \mathcal{J}_y^{\dagger};$$

at this point it would be inconvenient for us to write the matrix operators with respect to y to the left of functions.

Similarly,

$$(3.22)^{\dagger} \qquad U^{\pm}\mathcal{P}_y^{\dagger} \equiv \pm F_y \quad \text{for} \quad \pm t < 2t_1$$

(modulo functions with wave front sets contained in $\{\xi' = \eta' = 0\}$)
and

$$(3.23)^{\dagger} \qquad U\mathcal{P}_y^{\dagger} \equiv 0 \quad \text{for } |t| < 2t_1 \ (\mathrm{mod}\ C^{\infty})$$

where $F_y = \mathcal{F}_x(\varkappa_y \delta(t) K_{\Pi'}(x,y) \quad 0)$

Propositions 1.6, 1.7 imply that

$$F_x,\ F_y \in C^{\infty}(Y_{y'} \times [0,\delta)_{x_1,y_1},\ \mathcal{D}'(Y_{x'}) \times \mathbb{R}_t,\ \mathrm{Hom}(E \oplus E));$$

these inclusions remain true for x' and y' permutated.

Boundary conditions (2.33), $(2.33)^{\dagger}$ imply that

$$(2.24) \qquad \mathcal{B}_x U^{\pm} = f_x^{\pm},$$

$$(2.24)^{\dagger} \qquad U^{\pm}\mathcal{B}_y^{\dagger} = f_y^{\pm}$$

where $\mathcal{B} \in OPS^0(Y; E, E \oplus E)\tau$,

$\mathcal{B} = (I \quad 0)\tau \quad$ if $B = \tau$,

$\mathcal{B} = (\varkappa(-iB_1 - A_0^{-1}A_1) \qquad A_0^{-1}\tau)$

if $B = i\tau D_1 + B_1\tau$,

$f_x^{\pm} = -\mathcal{B}_x \mathcal{F}_x\, u_0^{\pm} \mathcal{F}_y^{\dagger},\ f_y^{\pm} = -\mathcal{F}_x\, u_0^{\pm} \mathcal{F}_y^{\dagger} \mathcal{B}_y^{\dagger}.$

Let $\rho^* = (y^*, -\eta^*, \tau^*) \in (T^*Y\setminus 0) \times (\mathbb{R}_\tau \setminus 0)$. Let us consider ξ_1-roots of the polynomial $\det(K\xi_1 + L_1(x,\xi',\tau)) = \det(a(x,\xi') - \tau^2)\ \det a_0^{-1}(x)$. Assume that for $(x,\xi',\tau) = (y^*, -\eta^*, \tau^*)$ these roots are

$$\lambda_1, \ldots, \lambda_N,\ \lambda_{N+1}, \ldots, \lambda_{N+N'},\ \bar{\lambda}_{N+1}, \ldots, \bar{\lambda}_{N+N'}$$

where $\lambda_j \in \mathbb{R}$ for $j = 1, \ldots, N$ and $\lambda_j \in \mathbb{C}_+ \setminus \mathbb{R}$ for $j = N+1, \ldots, N+N'$; all these roots are real and have the multiplicities $s_j\ (j = 1, \ldots, N+N')$; since the characteristic polynomial is real then the multiplicities of λ_j and $\bar{\lambda}_j$ coincide.

Since K, L_1 are Hermitian matrices then there exists in the neighbourhood of ρ^* a positively homogeneous of degree 0 invertible matrix symbol $Q_0(x, \xi', \tau)$ such that

$$Q_0^* K Q = \begin{pmatrix} K_{(1)} & & 0 \\ & \ddots & \\ 0 & & K_{(N+N')} \end{pmatrix} = \tilde{K} ,$$

$$Q_0^* L_1 Q_0 = \begin{pmatrix} L_{1(1)} & & 0 \\ & \ddots & \\ 0 & & L_{1(N+N')} \end{pmatrix} = \tilde{L}_1 ,$$

where $K_{(j)}$ are Hermitian constant matrices, $L_{1(j)}$ are Hermitian matrices. Moreover, for $j=1,\dots,N$ dimension of $K_{(j)}$, $L_{1(j)}$ equals s_j and roots of $\det(K_{(j)}\xi_1 + L_{1(j)})$ are near λ_j (certainly these roots are complex in general); for $j=N+1,\dots,N+N'$

$$K_{(j)} = \begin{pmatrix} 0 & I \\ I & 0 \end{pmatrix}, \quad L_{1(j)} = \begin{pmatrix} 0 & L'_{1(j)} \\ L'^{*}_{1(j)} & 0 \end{pmatrix}$$

where dimension of I and $L'_{1(j)}$ equals s_j and roots of $\det(\xi_1 + L'_{1(j)})$ are near λ_j.

Then there exist operators $Q = Q(x, D'_x, D_t)$, $Q' = Q'(x, D'_x, D_t) \in OPS^{0'}(X\times\mathbb{R}, E\oplus E)$ such that modulo operators with the symbols vanishing in the neighbourhood of ρ^*

$$Q'Q \sim QQ' \sim I$$

$$Q^*(KD_1 + L)Q \sim \tilde{K}D_1 + \tilde{L}$$

where

$$\tilde{L} = \begin{pmatrix} L_{(1)} & & 0 \\ & \ddots & \\ 0 & & L_{(N+N')} \end{pmatrix} \quad \text{and} \quad L_j = \begin{pmatrix} 0 & L'_{(j)} \\ L''_{(j)} & 0 \end{pmatrix}$$

for $j = N+1,\dots,N+N'$; the principal symbols of $L_{(j)}, L'_{(j)}, L''_{(j)}$ are equal to $L_{1(j)}, L'_{1(j)}, L'^{*}_{1(j)}$ respectively.

One can assume without a loss of generality that $\mathit{cone\ supp}\, Q \cup \mathit{cone\ supp}\, Q' \subset \{\xi' \neq 0,\ \tau \neq 0\}$. We introduce

$$\tilde{U}^{\pm} = Q'_x U^{\pm} Q'^{\dagger}_y , \quad \tilde{U} = Q'_x U Q'^{\dagger}_y ,$$

$$\tilde{F}_x = Q_x F_x Q_y^\dagger , \quad \tilde{F}_y = Q_x F_y Q_y^\dagger .$$

We decompose $E \oplus E$ into the sum of subspaces of corresponding dimensions: $E \oplus E = E_1 \oplus \cdots \oplus E_{N+N'}$, $E_j = E_j' \oplus E_j''$ for $j = N+1, \dots, N+N'$ and divide $\tilde{U}, \tilde{U}^\pm, \tilde{F}_x, \tilde{F}_y$ into corresponding components: U_{jk}, $U_{jk}^\pm$, $\tilde{F}_{jkx}$, $\tilde{F}_{jky}$ $(j,k = 1,\dots,N+N')$. If $j \geq N+1$ or (and) $k \geq N+1$ then these components can be devided into two (four) subcomponents.

Then (3.22), $(3.22)^\dagger$, (3.24), $(3.24)^\dagger$ and theorem 2.3 imply that

$$(3.25) \qquad WF_f(\mathcal{P}_{(j)x} U_{jk}^\pm \mp \tilde{F}_{jkx}) \cap (j_x^{-1} \rho^* \times T^*X \cup T^*X \times j_y^{-1}\, {}^T\rho^*) = \emptyset ,$$

$$(3.25)^\dagger \qquad WF_f(U_{jk}^\pm \mathcal{P}_{(k)y}^\dagger \mp \tilde{F}_{jky}) \cap (j_x^{-1} \rho^* \times T^*X \cup T^*X \times j_y^{-1}\, {}^T\rho^*) = \emptyset ,$$

$$(3.26) \qquad WF_f(U_{jk}^\pm) \cap \{\pm t < 0\} = \emptyset ,$$

$$(3.27) \qquad WF_f(\tilde{\mathcal{B}}_x \tilde{U}^\pm - f_x^\pm) \cap (\rho^* \times T^*X \cup T^*Y \times j_y^{-1}\, {}^T\rho^*) = \emptyset ,$$

$$(3.27)^\dagger \qquad WF(\tilde{U}^\pm \tilde{\mathcal{B}}_y^\dagger - f_y^\pm) \cap (j_x^{-1} \rho^* \times T^*Y \cup T^*X \times {}^T\rho^*) = \emptyset$$

where ${}^T\rho^* = (y^*, \eta^{*\prime}, \tau^*)$, $\mathcal{P}_{(j)} = \tilde{K}_{(j)} D_1 + \tilde{L}_{(j)}$, $\tilde{\mathcal{B}} = \mathcal{B} Q$.

Theorem 2.3 implies that if $q_y = q(y, D_y', D_t) \in OPS^{0\prime}(X \times \mathbb{R})$, cone supp q is contained in the small enough conic neighbourhood of ${}^T\rho^*$, $\chi \in C^\infty(\mathbb{R})$, supp $\chi \subset (-2t_1, 2t_1)$ where t_1 is small enough then

$$(3.28) \qquad tr\, \Gamma q_y \chi u_1 = \sum_{jk} \Gamma \mathcal{L}_{jk} U_{jk}$$

where $\mathcal{L}_{jk} \in OPS^{0\prime}(X \times X \times \mathbb{R} ; \mathbb{C}, \ Hom(E_j, E_k))$, cone supp $\mathcal{L}_{jk}$ are contained in the small enough conic neighbourhood of $(y^*, y^*, 0, -\eta^{*\prime}, \eta^{*\prime}, \tau^*)$. This statement remains true if we replace q_y by $q_x = q(x, D_x', D_t)$, cone supp q is contained in the small enough conic neighbourhood of ρ^*.

LEMMA 3.7. Let $j \neq k$ or $j = k > N$ and $\mathcal{L}_{jk} \in OPS^{0'}(X \times X \times \mathbb{R}; \mathbb{C}, Hom(E_j, E_k))$, $cone\ supp\ \mathcal{L}_{jk}$ be contained in the small enough conic neighbourhood of $(y^*, y^*, 0, -\eta^{*'}, \eta^{*'}, \tau^*)$. Then

$$\int_0^\delta \Gamma \mathcal{L}_{jk} U_{jk}\, dx_1 \equiv \Gamma \mathcal{L}'_{jk} r_x r_y U_{jk} \pmod{C^\infty} \tag{3.29}$$

where $\mathcal{L}'_{jk} \in OPS^{-1'}(Y \times Y \times \mathbb{R}; \mathbb{C}, Hom(E_j, E_k))$, $cone\ supp\ \mathcal{L}'_{jk}$ is contained in the small enough conic neighbourhood of $(y^*, y^*, 0, -\eta^{*'}, \eta^{*'}, \tau^*)$.

PROOF. It is easy to show that the principal symbol of operator $R_y^\dagger$ equals the transformed principal symbol of R_x in which (x, ξ, τ) is replaced by $(x, -\xi, \tau)$; the principal symbol of $R_x(R_y^\dagger)$ is written to the left (right) of an element of $Hom(E_x, E_y)$.

Assume first that $j, k \leqslant N$. Then (3.25) - (3.25)' imply that

$$(D_{x_1} + D_{y_1}) U_{jk} \equiv -K_{(j)}^{-1} L_{(j)x} U_{jk} - U_{jk} L_{(k)y}^{\dagger} K_{(k)}^{\dagger -1} = \mathcal{R}_{jk} U_{jk} \tag{3.30}$$

modulo functions which are infinitely smooth in the neighbourhood of $\rho^* \times T^*X \cup T^*X \times {}^T\rho^*$.

If $j \neq k$ then operator $\mathcal{R}_{jk} \in OPS^{1'}(X \times X \times \mathbb{R}, Hom(E_j, E_k))$ is elliptic at the point $(y^*, y^*, -\eta^{*'}, \eta^{*'}, \tau^*)$. Really, its principal symbols maps $V_{jk} \in Hom(E_j, E_k)$ into

$$-K_{(j)}^{-1} L_{1(j)}(x, \xi', \tau) V_{jk} + V_{jk} (K_{(k)}^{-1} L_{1(k)}(y, -\eta', \tau))^T;$$

then the inverse mapping is

$$W_{jk} \longrightarrow \frac{1}{2\pi i} \int_\gamma (z + K_{(j)}^{-1} L_{1(j)}(x, \xi', \tau))^{-1} W_{jk} (z + K_{(k)}^{-1} L_{1(k)}(y, -\eta', \tau))^{-1T} dz$$

where γ is a closed contour with counter-clockwise orientation; all eigenvalues of $-K_{(j)}^{-1} L_{(j)}(x, \xi', \tau)$ are near λ_j and lie inside γ; all eigenvalues of $-K_{(k)}^{-1} L_{(j)}(y, -\eta', \tau)$ are near λ_k and lie outside γ.

Therefore (3.30) implies that there exists $\mathcal{L}''_{jk} \in OPS^{-1'}(X \times X \times \mathbb{R}; \mathbb{C}, Hom(E_j, E_k))$ such that

$$\mathcal{L}_{jk} U_{jk} \equiv -i(D_{x_1}+D_{y_1})\mathcal{L}''_{jk} U_{jk} \ (mod\ C^{\infty});$$

hence

$$\Gamma\mathcal{L}_{jk} U_{jk} \equiv -i\, D_{x_1}\Gamma\mathcal{L}''_{jk} U_{jk} \quad (mod\ C^{\infty})\ ;$$

this equality combined with the theorem 2.4 implies (3.29), $\mathcal{L}'_{jk} = \tau\mathcal{L}''_{jk}$.

Assume now that $k \leqslant N, j > N$; then $U_{jk} = \begin{pmatrix} U''_{jk} \\ U'_{jk} \end{pmatrix}$ (it is convenient for us to use this order of blocks) and (3.25) implies that

$$(D_{x_1} + L'_{(j)x})U'_{jk} \equiv 0, \quad (D_{x_1} + L''_{(j)x})U''_{jk} \equiv 0$$

modulo functions which are infinitely smooth in the neighbourhood of $\rho^* \times T^*X \cup T^*X \cup {}^T\rho^*$. Both these equations are elliptic but the second needs no boundary condition and the first needs the Cauchy condition on Y. Hence $U''_{jk} \equiv 0$.

Similarly, if $k > N$, $j \leqslant N$ then $U_{jk} = (U'_{jk}\ \ U''_{jk})$ and $(3.25)^{\dagger}$ implies that $U'_{jk}(-D_{y_1} + L'^{\dagger}_{(k)y}) \equiv 0$, $U''_{jk}(-D_{y_1} + L''^{\dagger}_{(k)y}) \equiv 0$; hence $U'_{jk} \equiv 0$.

Similarly, if $j > N$, $k > N$ then

$$U_{jk} = \begin{pmatrix} U''\,'_{jk} & U''\,''_{jk} \\ U'\,'_{jk} & U'\,''_{jk} \end{pmatrix}$$

and (3.25) - $(3.25)^{\dagger}$ imply that $U''\,'_{jk} \equiv U''\,''_{jk} \equiv U'\,'_{jk} \equiv 0$ and

$$(D_{x_1} + L'_{(j)x})U'\,''_{jk} \equiv 0, \quad U'\,''_{jk}(-D_{y_1} + L''^{\dagger}_{(k)y}) \equiv 0;$$

in all these cases we can derive (3.29) by means of previous arguments since equality $\lambda_j = \lambda_k$ holds only for $j = k \leqslant N$.

Now consider in detail the structure of systems (3.22), (3.25) and of dual systems (3.22), $(3.22)^{\dagger}$. We assume that

$$-\tau^* æ_{-1}(\rho^*) > 0\ ; \tag{3.31}$$

then (3.22) is t-hyperbolic at $\dot{j}^{-1}\rho^*$ in the sence of [49, 50], i.e.

$$(3.32) \qquad Re < \frac{\partial}{\partial \tau} \mathcal{P}_1(\rho)v, v>>0$$

$$\forall v \in Ker\, \mathcal{P}_1(\rho)\setminus 0 \qquad \forall \rho \in j^{-1}\rho^*$$

where $\mathcal{P}_1$ is the principal symbol of $\mathcal{P}$. Hence $\mathcal{P}_{(j)}$ are also t-hyperbolic:

$$(3.33) \qquad Re < \frac{\partial}{\partial \tau} L_{1(j)}(\varphi^*)v, v>>0$$

$$\forall v \in Ker(K_{(j)}\lambda_j + L_{1(j)}(\rho^*))\setminus 0 \qquad \forall j \leqslant N .$$

Therefore (see [49] or appendix C) for $j \leqslant N$, $K_{(j)}(\rho^*)$ is positive (negative) definite if and only if point $(\hat{\rho}, \lambda_j/\tau^*)$ is positive (negative).

In future we shall prove the normality of the great singularity for functions constructed on the basis of $\mathcal{U}_1^+$; the normality of the great singularity for similar functions constructed on the basis of $\mathcal{U}_1^-$ will follow since $\mathcal{U}_1$ is odd with respect to t.

3.4. Now let us consider boundary value problems (3.24) and (3.27). Since A_B is selfadjoint then

$$(3.34) \qquad <\widetilde{K}v, v> = 0 \qquad \forall v \in Ker\, \widetilde{\mathcal{B}}_0(\rho)$$

where $\widetilde{\mathcal{B}}_0$ is the principal symbol of $\widetilde{\mathcal{B}}$. It is easy to show that $rank\, \widetilde{\mathcal{B}}_0 = D$; then $rank\, \widetilde{\mathcal{B}}_0(\rho^*) = D$ too.

Let

$V = \mathcal{F}_x \chi_0(t) u_0 \mathcal{F}_y^\dagger$, $V^\pm = \mathcal{F}_x \chi_0(t) u_0^\pm \mathcal{F}_y^\dagger \in \mathcal{D}'(X'' \times X'' \times \mathbb{R}$, $Hom(E \oplus E))$ where X'' is the neighbourhood of Y in $\widetilde{X}$ identified with $(-\delta, \delta) \times Y$. Then V satisfies systems (3.23), $(3.23)^\dagger$ and $V^\pm$ satisfy systems (3.22), $(3.22)^\dagger$ with F_x, F_y replaced by

$$F_x^o = \begin{pmatrix} \varkappa_x \; \delta(t) K_{\Pi^o}(x,y) \\ 0 \end{pmatrix} \mathcal{F}_y^\dagger ,$$

$$F_y^o = \mathcal{F}_x (\varkappa_y \delta(t) K_{\Pi^o}(x,y) \qquad o)$$

respectively.

One can transform these systems by means of operators Q, Q' into the systems (3.25), $(3.25)^\dagger$ with $\widetilde{F}_x, \widetilde{F}_y$ replaced by $\widetilde{F}_x^o = Q_x F_x^o Q_y^\dagger$, $\widetilde{F}_y^o = Q_x F_y^o Q_y^\dagger$ respectively.

We shall assume that

(H.5) $\mu_j \in \mathbb{R}$ and inequalities (0.10) hold for $\rho_j = (\hat{\rho}, \lambda_j/\tau^*)$ and $\mu = \mu_j$, $j = 1, \ldots, N$.

We remind the reader that $\mu_j = \lambda_j/\tau^*_{,}$ fit.

PROPOSITION 3.8. Let $q_y = q(y, D_y, D_t) \in OPS^{o'}(X \times \mathbb{R})$, *cone supp* q be contained in the small enough conic neighbourhood of ${}^T\rho^* \times \mathbb{R}$, $\chi \in C_0^\infty(\mathbb{R})$, $supp\, \chi \subset (-2t_1, 2t_1)$, t_1 be small enough. Then

(i) $$D_x^\alpha D_y^\beta D_t^n \hat{\Psi}_{(j)n+|\alpha|+|\beta|+m}\, q_y \chi \overset{+}{V}_{jk} \in L_2(X \times X \times \mathbb{R}, Hom(E_j, E_k))$$

$$\forall \alpha, \beta \in \overset{+d}{\mathbb{Z}}, \; n \in \overset{+}{\mathbb{Z}}, \; j \leqslant N$$

where $m = m(\alpha)$, $\hat{\Psi}_{(j)s}$ correspond to $\hat{\varphi}_{(j)} = \sigma t + \tau^{*-1} \langle \eta^{*'}, y' - x' \rangle + \mu_j (x_1 - y_1)$, $|\sigma|$ is small enough.

(ii) $$D_x^\alpha D_y^\beta D_t^n (x-y)^\gamma t^s q_y \chi \overset{+}{V}_{jk} \in L_2(X \times X \times \mathbb{R}, Hom(E_j, E_k))$$

$\forall \alpha, \beta, \gamma \in \overset{+d}{\mathbb{Z}}$, $n, s \in \overset{+}{\mathbb{Z}}$ such that $|\alpha| + |\beta| + n \leqslant |\gamma| + s - m(\alpha)$, $j > N$.

Moreover one can replace $\hat{\Psi}_{(j)n}$ by $\hat{\Psi}_{(k)n}$ and condition $j \leqslant N$ by condition $k \leqslant N$ in (i) and condition $j > N$ by condition $k > N$ in (ii).

PROOF. It is easy to prove statement (i) by means of the arguments of the section 3.1; statement (ii) follows from the ellipticity of operator $\mathcal{P}_{(j)}$ for $j > N$.

COROLLARY. These statements remain true if we replace $\overset{+}{V}_{jk}$ by f_{kx}, f_{jy} and $X \times X$ by $Y \times X$, $X \times Y$ respectively.

PROPOSITION 3.9. Let $\frac{\partial}{\partial \tau} æ_{-1}(\rho^*) = 0$, $q_y = q(y, D_y', D_t) \in OPS^{o'}(X \times \mathbb{R})$, *cone supp* q be contained in the small enough conic neighbourhood of ${}^T\rho^* \times \mathbb{R}$, $\chi \in C_0^\infty(\mathbb{R})$, $supp\, \chi \subset (-2t_1, 2t_1)$, t_1 be small enough. Then

(i) (3.35) $$D_x^\alpha D_y^\beta \Psi_{(jk)n+|\alpha|+|\beta|+m}\, q_y \chi \overset{+}{U}_{jk} \in$$

$$L_2(X \times X \times \mathbb{R}, Hom(E_j, E_k))$$

$$\forall \alpha, \beta \in \overset{+d}{\mathbb{Z}}, \quad n \in \overset{+}{\mathbb{Z}}, \; j, k \leqslant N$$

where $\Psi_{(jk)s}$ correspond to $\varphi_{(jk)} = \sigma t + \tau^{*-1} \langle \eta^{*'}, y' - x' \rangle - \mu_k y_1 + \mu_j x_1$.

(ii) (3.36) $x_1^l \psi'_{(k)s} D_x^\alpha D_y^\beta D_t^n q_y \chi U^+_{jk} \in L_2(X\times X\times \mathbb{R}, Hom(E_j, E_k))$

$\forall \alpha, \beta \in \mathbb{Z}^{+d}$, $n, l, s \in \mathbb{Z}^+$ such that $|\alpha|+|\beta|+n \leqslant l+s-m(d)$, $k\leqslant N$, $j>N$ where $\psi'_{(k)s}$, $\psi_{(j)s}$ correspond to

$$\varphi'_{(k)} = \sigma t + \tau^{*-1} \langle \eta^{*'}, y' - x' \rangle - \mu_k y_1,$$

$$\varphi_{(j)} = \sigma t + \tau^{*-1} \langle \eta^{*'}, y' - x' \rangle + \mu_j x_1$$

respectively, $|\sigma|$ is small enough.

This statement remains true if we replace condition $k\leqslant N, j>N$ by condition $k>N, j\leqslant N$ and factors x_1^l, $\psi'_{(k)s}$ by factors y_1^l, $\psi_{(j)s}$ respectively.

(iii) $$x_1^l y_1^p \psi''_s D_x^\alpha D_y^\beta D_t^n q_y \chi U^+_{jk} \in L_2(X\times X\times \mathbb{R}, Hom(E_j, E_k))$$

$\forall \alpha, \beta \in \mathbb{Z}^{+d}$, $n, l, p, s \in \mathbb{Z}^+$ such that $|\alpha|+|\beta|+n \leqslant l+s+p-m(d)$, $j, k>N$ where ψ''_s correspond to $\varphi'' = \sigma t + \tau^{*-1} \langle \eta^{*'}, y' - x' \rangle$, $|\sigma|$ is small enough.

PROOF. We shall prove statements (i), (ii); to prove statements (ii), (iii) one need only to replace in this proof U^+_{jk}, $\psi_{(jk)}$ with $k\leqslant N$ by $\tau_y U^+_{jk}$, $\psi_{(j)}$ with $k>N$ respectively, and use proposition 3.8 (ii) instead of proposition 3.8 (i) and finally use the ellipticity of $\mathcal{P}_{(k)}$ with $k>N$.

Theorem 2.3, (3.25), (3.25) and inclusion $\text{cone supp } q \subset \{\tau \neq 0, \eta' \neq 0\}$ imply that we need to prove (3.35), (3.36) only for $\alpha=\beta=0$. Let $\chi_0, \chi, \Lambda^s, \mathcal{J}_\varepsilon$ be the same as in the proof of theorem 3.2, $v_{jk} = q_y \chi_0 U^+_{jk}$. We shall use the induction on $\lambda \in \frac{1}{2}\mathbb{Z}^+$ and prove that

$$\chi \psi_{(jk)\lambda+m_1} \Lambda^{\lambda-m_2} v_{jk} \in L_2(X\times X\times \mathbb{R}, Hom(E_j, E_k)) \quad \forall j\leqslant N, \tag{3.37}$$

$$\chi \psi'_{(k)\lambda+m_1} \Lambda^{\lambda-m_2-1/2} \tau_x v_{jk} \in L_2(Y\times X\times \mathbb{R}, Hom(E_j, E_k)) \quad \forall j. \tag{3.38}$$

Certainly, (3.37) and (3.25) imply (3.38) for $j\leqslant N$.

Assume that for $\lambda \leqslant \nu - \frac{1}{2}$ (3.37), (3.38) are valid. We use the following identity which is similar to (3.9):

$$2\mathrm{Re}\, i(G_j \mathcal{P}_{(j)} w, w) = -\mathrm{Re}(G_j K_{(j)} w, w)_Y - \tag{3.39}$$

$$\mathrm{Re}\, i([\mathcal{P}_{(j)}, G_j]w, w) + \mathrm{Re}\, i((\mathcal{P}_{(j)} - \mathcal{P}^*_{(j)})w, G_j w)$$

provided $G_j^* = G_j$ and set $G_j = \Lambda^{\nu} \chi^2 \psi_{(jк)2\nu+2m_1+1} \Lambda^{\nu}$, $w = \Lambda^{-m_2} \mathcal{J}_\varepsilon v_{jк}$, $j \leq N$.

Repeating the proof of theorem 3.2 one can prove that the left side of (3.39) is bounded as ε tends to 0 and that the second term in the right side of (3.39) equals modulo terms bounded as ε tends to 0 to

$$(2m_1 + 2\nu + 1)\mathrm{Re}(R_{jx} \psi_{(jк)\nu+m_1} \chi \mathcal{J}_\varepsilon \Lambda^{\nu-m_2} v_{jк},$$

$$\psi_{(jк)\nu+m_1} \chi \mathcal{J}_\varepsilon \Lambda^{\nu-m_2} v_{jк}) + \mathrm{Re}(R'_{jx} \psi_{(jк)\nu+m_1+\frac{1}{2}} \chi_1 \mathcal{J}_\varepsilon \Lambda^{\nu-m_2} v_{jк},$$

$$\psi_{(jк)\nu+m_1+\frac{1}{2}} \chi_1 \mathcal{J}_\varepsilon \Lambda^{\nu-m_2} v_{jк})$$

where $\chi_1 = \sqrt{-\chi\chi'}$ for $t > 0$, $\chi_1 \in C_0^\infty(\mathbb{R}^+)$, R_j, $R'_j \in OPS^{0'}(X \times \mathbb{R}, E_j)$ and in the neighbourhood of ρ^* their principal symbols are equal to $-\{\mathcal{P}_{1(j)}, \varphi_{(jк)}\}, \{\mathcal{P}_{1(j)}, t\}$ respectively. Recall that $j \leq N$; then our assumptions combined with lemma C imply that in the neighbourhood of (ρ^*, λ_j)

$$\langle\{\mathcal{P}_{1(j)}, t\}w, w\rangle \geq 4c_0 |w|^2,$$

$$-\langle\{\mathcal{P}_{1(j)}, \varphi_{(jк)}\}w, w\rangle \geq 4c_0 |w|^2$$

$$\forall w \in \mathrm{Ker}\, \mathcal{P}_{1(j)} ; \quad c_0 > 0 .$$

Hence there exists $S \in OPS^{-2}(X \times \mathbb{R})$ such that

$$\langle R_{0j} w, w\rangle + \langle \mathcal{P}^*_{1(j)} S_{-2} \mathcal{P}_{1(j)} w, w\rangle \geq 3c_0 |w|^2 \qquad \forall w$$

in the neighbourhood of $j^{-1}\rho^*$.

Then the sharp Gårding inequality implies that

$$\mathrm{Re}(R_j w^0, w^0)_{\tilde{X}} + \mathrm{Re}(\mathcal{P}^*_{(j)} S \mathcal{P}_{(j)} w^0, w^0)_{\tilde{X}} \geq \tag{3.40}$$

$$3c_0 \|w^0\|^2_{\tilde{X}} - C\|\Lambda^{-1/2} w^0\|^2_X - C\|\omega w^0\|^2_{\tilde{X}}$$

where $w^0 = c_x^+ w$, $\omega \in OPS^{0'}(X \times \mathbb{R})$, $\rho^* \in \mathrm{cone\ supp}\, \omega$.

The inequality (3.40) combined with the equality

$$\mathcal{P}_{(j)} w^0 = -iK_{(j)} \delta(x_1) r_x w + (\mathcal{P}_{(j)} w)^0$$

and proposition 1.2 implies the inequality

$$Re(R_j w, w) \geq$$
$$\geq 2c_0 \|w\|^2 - C\|\omega w\|^2 - C\|\Lambda^{-1/2} w\|^2 - c_1 \|\Lambda^{-1/2} r_x w\|_Y^2 - C\|\Lambda^{-1} \mathcal{P}_{(j)} w\|^2 .$$

Substituting $w = \chi \Psi_{(jk)\nu+m_1} \mathcal{F}_\varepsilon \Lambda^{\nu-m_2} v_{jk}$ and using the first conjecture of the induction ((3.37) with $\lambda \leq \nu - 1/2$) and $(3.25)_j$ we obtain the inequality

(3.41)
$$Re(R_j \chi \Psi_{(jk)\nu+m_1} \mathcal{F}_\varepsilon \Lambda^{\nu-m_2} v_{jk},$$
$$\chi \Psi_{(jk)\nu+m_1} \Lambda^{\nu-m_2} v_{jk}) \geq$$
$$2c_0 \|\chi \Psi_{(jk)\nu+m_1} \mathcal{F}_\varepsilon \Lambda^{\nu-m_2} v_{jk}\|^2 - c_1 \|\chi \Psi_{(jk)\nu+m_1} \mathcal{F}_\varepsilon \Lambda^{\nu-m_2-1/2} v_{jk}\|_Y^2 - C$$

where C does not depend on ε and c_j here and below do not depend on $\nu, m_1, m_2, \varepsilon$ and choice of q, χ.

It is proved in Appendix A that for every $M>0$ there exist $C=C_M$ and operator $\omega = \omega_M \in OPS^{0'}(X \times \mathbb{R})$, *cone supp* $\omega_M \ni \rho^*$ such that the following estimate holds:

$$\|\Lambda^{-1/2} w\|_Y \leq$$

$$M^{-1}\|w\| + C_M \sum_{i=0}^{2s_j} \|D_1^i \Lambda^{-1-i} \mathcal{P}_{(j)} w\| + \|\omega_M w\| + C_M \|\Lambda^{-1/2} w\| \qquad \forall w ;$$

we set $w = \chi \Psi_{(jk)\nu+m_1} \mathcal{F}_\varepsilon \Lambda^{\nu-m_2} v_{jk}$. Then (3.41) implies the inequality

$$Re(R_j \chi \Psi_{(jk)\nu+m_1} \mathcal{F}_\varepsilon \Lambda^{\nu-m_2} v_{jk},$$
$$\chi \Psi_{(jk)\nu+m} \mathcal{F}_\varepsilon \Lambda^{\nu-m_2} v_{jk}) \geq$$
$$c_0 \|\chi \Psi_{(jk)\nu+m_1} \mathcal{F}_\varepsilon \Lambda^{\nu-m_2} v_{jk}\|^2 + (c_0 M - c_1) \|\chi \Psi_{(jk)\nu+m_1} \mathcal{F}_\varepsilon \Lambda^{\nu-m_2-1/2} v_{jk}\|_Y^2 - C$$

provided *cone supp* q is small enough and $t_1 \leq t_1(M)$.

A similar estimate holds for R'_j too. Finally, the first conjecture of the induction implies that the last term in the right side of

(3.40) does not exceed

$$C_2 \| \chi \psi_{(jK)\nu+m_1+\frac{1}{2}} \mathcal{J}_\varepsilon \Lambda^{\nu-m_2} v_{jK} \|^2 + C .$$

Thus the following estimates hold true:

$$2(\nu+m_1)\{C_0 \| \chi \psi_{(jK)\nu+m_1} \mathcal{J}_\varepsilon \Lambda^{\nu-m_2} v_{jK} \|^2 + (C_0 M - C_1) \| \chi \psi'_{(K)\nu+m_1} \mathcal{J}_\varepsilon \Lambda^{\nu-m_2-\frac{1}{2}} v_{jK} \|_Y^2 \} +$$

$$\{C_0 \| \chi_1 \psi_{(jK)\nu+m_1+\frac{1}{2}} \mathcal{J}_\varepsilon \Lambda^{\nu-m_2} v_{jK} \|^2 + (C_0 M - C_1) \| \chi_1 \psi'_{(K)\nu+m_1+\frac{1}{2}} \mathcal{J}_\varepsilon \Lambda^{\nu-m_2-\frac{1}{2}} v_{jK} \|_Y^2 \} \leq$$

$$\mathrm{Re}(K_{(j)} \chi^2 \psi'_{(K)2\nu+2m_1+1} \mathcal{J}_\varepsilon \Lambda^{\nu-m_2} v_{jK}, \ \mathcal{J}_\varepsilon \Lambda^{\nu-m_2} v_{jK})_Y + C \quad (j=1,\dots,N)$$

provided t_1 and $\mathit{cone\ supp}\ q$ are small enough.

We sum these inequalities and obtain the following inequality:

$$(3.42) \qquad \sum_{j=1}^{N} \Big[2(\mu+m_1)\{C_0 \| \chi \psi_{(jK)\nu+m_1} \mathcal{J}_\varepsilon \Lambda^{\nu-m_2} v_{jK} \|^2 +$$

$$(C_0 M - C_1) \| \chi \psi'_{(K)\nu+m_1} \mathcal{J}_\varepsilon \Lambda^{\nu-m_2-\frac{1}{2}} v_{jK} \|_Y^2 \} +$$

$$\{C_0 \| \chi_1 \psi_{(jK)\nu+m_1+\frac{1}{2}} \mathcal{J}_\varepsilon \Lambda^{\nu-m_2} v_{jK} \|^2 +$$

$$(C_0 M - C_1) \| \chi_1 \psi'_{(K)\nu+m_1+\frac{1}{2}} \mathcal{J}_\varepsilon \Lambda^{\nu-m_2-\frac{1}{2}} v_{jK} \|_Y^2 \} \Big] \leq$$

$$\mathrm{Re}(\tilde{K} \chi^2 \psi'_{(K)2\nu+2m_1+1} \mathcal{J}_\varepsilon \Lambda^{\nu-m_2} v_K ,$$

$$\mathcal{J}_\varepsilon \Lambda^{\nu-m_2} v_K)_Y - \sum_{j=N+1}^{N+N'} \mathrm{Re}(K_{(j)} \chi^2 \psi'_{(K)2\nu+2m_1+1} \mathcal{J}_\varepsilon \Lambda^{\nu-m_2} v_{jK} ,$$

$$\mathcal{J}_\varepsilon \Lambda^{\nu-m_2} v_{jK})_Y + C$$

where $v_K = (v_{1K} \dots v_{(N+N')K})^T$.

Recall that for $j \geq N+1$ $v_{jK} = \begin{pmatrix} v''_{jK} \\ v'_{jK} \end{pmatrix}$ and

(3.43)' $$(D_{x_1} + L'_{(j)x})\, v'_{jK} \equiv q_y \chi_o F'_{jKx} ,$$

(3.43)'' $$(D_{x_1} + L''_{(j)x})\, v''_{jK} \equiv q_y \chi_o F''_{jKx} .$$

We know that both these systems are elliptic and the second system needs noboundary condition; hence (3.43)'' combined with proposition 1.7 implies that

(3.44) $$D_x^{\alpha} D_y^{\beta} D_t^{n} t^{s} (x'-y')^{\gamma'} x_1^{\gamma_1} y_1^{\gamma_0} v''_K \in L_2(X\times X\times \mathbb{R}, Hom(E'', E_K))$$

$$\forall \alpha, \beta \in \mathbb{Z}^{+d}, \quad \gamma \in \mathbb{Z}^{+d+1}, n, s \in \mathbb{Z}^{+}$$ such that $|\alpha|+|\beta|+n \leq s+|\gamma|-m(d)$

where $v'_K = (v'_{(N+1)K} \dots v'_{(N+N')K})^T$,

$v''_K = (v'_{(N+1)K} \dots v'_{(N+N')K})^T$, $E' = E'_{N+1} \oplus \dots \oplus E'_{N+N'}$, $E'' = E''_{N+1} \oplus \dots \oplus E''_{N+N'}$;

hence if we replace v_K by v''_K then (3.38) would be valid for every $\lambda \in \frac{1}{2}\mathbb{Z}^+$.

Since $K_{(j)} = \begin{pmatrix} 0 & I \\ I & 0 \end{pmatrix}$ for $j \geq N+1$ then the second term (sum) in the right side of (3.42) is bounded as ε tends to 0.

Consider now the first term in the right side of (3.42); using this term we want to obtain in the left side the terms which we lack.

Equality (3.34) implies that in the neighbourhood of ρ^*

$$\tilde{K} = \beta_0 \tilde{\mathcal{B}}_0 + (\beta_0 \tilde{\mathcal{B}}_0)^*$$

where β_0 is the principal symbol of operator $\beta \in OPS^0(Y\times \mathbb{R}, E \oplus E, E)$; hence the second conjecture of the induction ((3.38) with $\lambda \leq \nu - \frac{1}{2}$) combined with proposition 3.8, its corollary and lemma 3.3 imply that modulo terms bounded as ε tends to 0

(3.45) $$Re(\tilde{K} \chi^2 \psi'_{(K)2\nu+2m_1+1} \mathcal{J}_\varepsilon \Lambda^{\nu-m_2} v_K,$$
$$\mathcal{J}_\varepsilon \Lambda^{\nu-m_2} v_K)_Y \sim Re(\beta [\tilde{\mathcal{B}}, \chi^2 \psi'_{(K)2\nu+2m_1+1}] \mathcal{J}_\varepsilon \Lambda^{\nu-m_2} v_K ,$$

$$\mathcal{F}_\varepsilon \Lambda^{\nu-m_2} v_K)_Y \sim -(2\nu+2m_1+1) \times$$

$$Re(R'' \chi \psi'_{(K)\nu+m_1} \mathcal{F}_\varepsilon \Lambda^{\nu-m_2-\frac{1}{2}} v_K,$$

$$\chi \psi'_{(K)\nu+m_1} \mathcal{F}_\varepsilon \Lambda^{\nu-m_2-\frac{1}{2}} v_K)_Y -$$

$$Re(R''' \chi_1 \psi'_{(K)\nu+m_1+\frac{1}{2}} \mathcal{F}_\varepsilon \Lambda^{\nu-m_2-\frac{1}{2}} v_K,$$

$$\chi_1 \psi'_{(K)\nu+m_1+\frac{1}{2}} \mathcal{F}_\varepsilon \Lambda^{\nu-m_2-\frac{1}{2}} v_K)_Y$$

where $R'', R''' \in OPS^0(Y \times \mathbb{R}, E \oplus E)$ and in the neighbourhood of ρ^* their principal symbols are equal to $i|\tau|\beta_0\{\tilde{\mathcal{B}}_0, \varphi''\}$, $-i|\tau|\beta_0 \frac{\partial}{\partial\tau}\tilde{\mathcal{B}}_0$ respectively.

We shall prove below that in the neighbourhood of ρ^*

$$(3.46) \quad |\tau| \; Re < i\beta_0\{\tilde{\mathcal{B}}_0, \varphi''\} w, w > \geq 3c_0 |w|^2,$$

$$(3.46)' \quad -|\tau| \; Re < i\beta_0 \frac{\partial}{\partial\tau} \tilde{\mathcal{B}}_0 w, w > \geq 3c_0 |w|^2$$

$$\forall w \in Ker\, \tilde{\mathcal{B}}_0 \cap Ran\, e'_0, \; c_0 > 0$$

where e'_0 is the coordinate projector to E'; hence in the neighbourhood of ρ^*

$$Re < R''_0 w, w > + c_3 |\tilde{\mathcal{B}}_0 w|^2 + c_3 |(I - e'_0) w|^2 \geq 2c_0 |w|^2 \quad \forall w;$$

then the sharp Garding inequality implies that

$$(3.47) \quad Re(Rw, w)_Y + c_3 \|\tilde{\mathcal{B}} w\|_Y^2 + c_3 \|(I - e'_0) w\|_Y^2 \geq$$

$$2c_0 \|w\|_Y^2 - C \|\Lambda^{-1/2} w\|_Y^2 - \|\omega w\|_Y^2$$

where $\omega \in OPS^0(Y \times \mathbb{R})$, $cone\ supp\ \omega \ni \rho^*$. We set $w = \chi \psi'_{(K)\nu+m_1} \mathcal{F}_\varepsilon \Lambda^{\nu-m_2-1/2} v_K$. Then the second conjecture of the induction combined with proposition 3.8, its corollary and lemma 3.3, implies that the second term in the left side and the second and the third terms in the right side are bounded as ε tends to 0 and that the third term in the left side does not exceed

$$c_4 \sum_{j=1}^{N} \|\chi \psi_{(K)\nu+m_1} \mathcal{F}_\varepsilon \Lambda^{\nu-m_2-\frac{1}{2}} v_{jK}\|_Y^2 +$$

$$c_4 \| \chi \psi'_{(k)\nu+m_1} \mathcal{J}_\varepsilon \Lambda^{\nu-m_2-\frac{1}{2}} v''_k \|^2_Y + C \ ;$$

but the second term in this expression is bounded as ε tends to 0 because of (3.38) holds for v''_{jk} , $j > N$ for all χ ; hence (3.47) implies that

$$\mathrm{Re}\,(R'' \chi \psi'_{(k)\nu+m_1} \mathcal{J}_\varepsilon \Lambda^{\nu-m_2-\frac{1}{2}} v_k ,$$

$$\chi \psi'_{(k)\nu+m_1} \mathcal{J}_\varepsilon \Lambda^{\nu-m_2-\frac{1}{2}} v_k)_Y \geqslant$$

$$c_0 \| \chi \psi_{(k)\nu+m_1} \mathcal{J}_\varepsilon \Lambda^{\nu-m_2-\frac{1}{2}} v'_k \|^2_Y - c_5 \sum_{j=1}^{N} \| \chi \psi'_{(k)\nu+m_1} \mathcal{J}_\varepsilon \Lambda^{\nu-m_2-\frac{1}{2}} v_{jk} \|^2_Y - C.$$

A similar inequality holds for R''' too:

$$\mathrm{Re}\,(R''' \chi_1 \psi'_{(k)\nu+m_1+\frac{1}{2}} \mathcal{J}_\varepsilon \Lambda^{\nu-m_2-\frac{1}{2}} v_k ,$$

$$\chi_1 \psi'_{(k)\nu+m_1+\frac{1}{2}} \mathcal{J}_\varepsilon \Lambda^{\nu-m_2-\frac{1}{2}} v_k)_Y \geqslant$$

$$c_0 \| \chi_1 \psi_{(k)\nu+m_1+\frac{1}{2}} \mathcal{J}_\varepsilon \Lambda^{\nu-m_2-\frac{1}{2}} v'_k \|^2_Y - c_5 \sum_{j=1}^{N} \| \chi_1 \psi'_{(k)\nu+m_1+\frac{1}{2}} \mathcal{J}_\varepsilon \Lambda^{\nu-m_2-\frac{1}{2}} v_{jk} \|^2_Y - C.$$

Then (3.42), (3.45) combined with these two inequalities imply the estimate

$$2(\nu+m_1) \sum_{j=1}^{N} \{ c_0 \| \chi \psi_{(jk)\nu+m_1} \mathcal{J}_\varepsilon \Lambda^{\nu-m_2} v_{jk} \|^2 +$$

$$(c_0 M - c_6) \| \chi \psi'_{(k)\nu+m_1} \mathcal{J}_\varepsilon \Lambda^{\nu-m_2-\frac{1}{2}} v_{jk} \|^2_Y \} +$$

$$\sum_{j=1}^{N} \{ c_0 \| \chi_1 \psi_{(jk)\nu+m_1+\frac{1}{2}} \mathcal{J}_\varepsilon \Lambda^{\nu-m_2} v_{jk} \|^2 +$$

$$(c_0 M - c_6)\|\chi_1 \psi'_{(k)\nu+m_1+\frac{1}{2}} \mathcal{J}_\varepsilon \Lambda^{\nu-m_2-\frac{1}{2}} v_{jk}\|_Y^2 \} +$$

$$2c_0(\nu+m_1)\|\chi\psi_{(k)\nu+m_1} \mathcal{J}_\varepsilon \Lambda^{\nu-m_2-\frac{1}{2}} v'_k\|_Y^2 +$$

$$c_0 \|\chi_1 \psi'_{(k)\nu+m_1+\frac{1}{2}} \mathcal{J}_\varepsilon \Lambda^{\nu-m_2-\frac{1}{2}} v'_k\|_Y^2 \leq C.$$

Now we choose the large enough M (and therefore we choose t_1 and $\mathrm{cone\ supp}\ q$) and obtain the inclusions

$$\chi\psi_{(jk)\nu+m_1} \Lambda^{\nu-m_2} v_{jk} \in L_2(X\times X\times \mathbb{R}, \mathrm{Hom}(E_j, E_k)),$$

$$\chi\psi'_{(k)\nu+m_1} \Lambda^{\nu-m_2-\frac{1}{2}} r_x v_k \in L_2(Y\times X\times \mathbb{R}, \mathrm{Hom}(E\oplus E, E_k)).$$

Thus we proved (3.37), (3.38) for $\lambda=\nu$; so the step of the induction is made.

Note that (3.38) combined with the elliptic system (3.43)' implies (3.36) for U^+_{jk} replaced by U'^+_{jk} ; remind that (3.36) for U^+_{jk} replaced by U''^+_{jk} was proved before.

It remains to prove inequalities (3.46) and (3.46)'. Note that by Euler identity $\{\tilde{\mathcal{B}}_0, \mathcal{Y}''\}(\rho^*) = -(1-\sigma)\frac{\partial}{\partial\tau}\tilde{\mathcal{B}}_0(\rho^*)$; hence we need to prove (3.46)' only. Since (3.31) and $\frac{\partial}{\partial\tau}\varkappa_{-1}(\rho^*)=0$ hold then $\frac{\partial}{\partial\tau}\mathcal{B}_0(\rho^*)=0$ and $\frac{\partial}{\partial\tau}L_1(\rho^*)$ is nonnegative definite with $\mathrm{Ker}\,\frac{\partial}{\partial\tau}L_1(\rho^*)=\binom{0}{E}$. We have for $w\in \mathrm{Ker}\,\tilde{\mathcal{B}}_0 \cap \mathrm{Ran}\,e'_0$ the following chain of equalities:

$$-i\langle \beta_0 \frac{\partial}{\partial\tau}\tilde{\mathcal{B}}_0 w, w\rangle = -i\langle \beta_0 \tilde{\mathcal{B}}_0 \frac{\partial Q_0}{\partial\tau} w, w\rangle =$$

$$-i\langle \beta_0 \tilde{\mathcal{B}}_0 Q_0^{-1} \frac{\partial Q_0}{\partial\tau} w, w\rangle = -i\langle \tilde{K} Q_0^{-1} \frac{\partial Q_0}{\partial\tau} w, w\rangle +$$

$$i\langle (\beta_0 \tilde{\mathcal{B}}_0)^* Q_0^{-1} \frac{\partial Q_0}{\partial\tau} w, w\rangle = -i\langle \tilde{K} Q_0^{-1} \frac{\partial Q_0}{\partial\tau} w, w\rangle =$$

$$-i\langle Q_0 K \frac{\partial Q_0}{\partial\tau} e'_0 w, w\rangle.$$

Note that $e'(\tau) = Q_0(\tau) e'_0 Q_0^{-1}(\tau)$ is the projector to invariant subspace of $-K^{-1}L_1(\tau)$ corresponding to eigenvalues lying in the half-plane $\{ \operatorname{Im} z > \varepsilon \}$ where $\varepsilon > 0$ is small enough and that $e'^{*} K e' = 0$; then the final expression equals

$$-i \langle e'^{*} K \frac{\partial e'}{\partial \tau} Q_0 w, Q_0 w \rangle .$$

Lemma B and the nonnegativeness of $\partial L_1 / \partial \tau$ imply that $i e'^{*} K \partial e' / \partial \tau$ is a Hermitian nonnegative definite matrix and $\operatorname{Ker} i e^{*} K \frac{\partial e'}{\partial \tau} \cap \operatorname{Ran} e' = \{ w \in \operatorname{Ran} e' : \exp(isK^{-1}L_1) w \in \operatorname{Ker} \partial L_1 / \partial \tau \quad \forall s \in \mathbb{R}^{+} \}$; it is easy to check up that the last set equals $\{0\}$. Thus $i e'^{*} K \partial e' / \partial \tau$ is positive definite on Ran e' and hence (3.46)' holds.

Proposition 3.9 has been proved. Theorems 3.4, 3.5 follow from proposition 3.9; theorem 3.6 follow from the statement:

PROPOSITION 3.10. Let $\Pi \in OGS^{0}(X, E)$, $\frac{\partial}{\partial \tau} \varkappa_{-1}(\rho^{*}) = 0$, $q_y = q(y, D'_y, D_t) \in OPS^{0'}(X \times \mathbb{R})$, *cone supp* q be contained in the small enough conic neighbourhood of $T\rho^{*} \times \mathbb{R}$, $\chi \in C_0^{\infty}(\mathbb{R})$, *supp* $\chi \subset (-2t_1, 2t_1)$, t_1 be small enough. Then

(i) If either $K_{(j)}$ or $K_{(k)}$ is definite then

$$D_x^{\alpha} D_y^{\beta} D_t^{n} t^{s} (x' - y')^{\gamma'} x_1^{\gamma_1} y_0^{\gamma_0} q_y \chi U_{jk}^{+} \in$$

$$L_2(X \times X \times \mathbb{R}, \operatorname{Hom}(E_j, E_k))$$

$\forall \alpha, \beta \in \mathbb{Z}^{+d}, \gamma \in \mathbb{Z}^{+d+1}, s, n \in \mathbb{Z}^{+}$ such that $|\alpha| + |\beta| + n \leq |\gamma| + s - m(d)$.

(ii) In a general case

$$x_1^{\gamma_1} y_1^{\gamma_0} \psi_s'' D_x^{\alpha} D_y^{\beta} D_t^{n} q_y \chi U_{jk}^{+} \in L_2(X \times X \times \mathbb{R}, \operatorname{Hom}(E_j, E_k)) \tag{3.48}$$

$\forall \alpha, \beta \in \mathbb{Z}^{+d}, \gamma \in \mathbb{Z}^{+2}, n, s \in \mathbb{Z}^{+}$ such that

$|\alpha| + |\beta| + n \leq |\gamma| + s - m(d)$;

remind that ψ_s'' correspond to $\varphi'' = \sigma t + \langle \eta^{*\prime}, y' - x' \rangle$, $|\sigma|$ is small enough.

PROOF. Statement (i) can be proved by a considerably simplified method of the proof of proposition 3.9; one needs only to use system

$(3.25)_j$ or $(3.25)^+_k$ without boundary conditions and replace $\chi(t)$, $\varphi_{(jk)}$ by $\chi_2(-\ell x_1+t^2)$, $\varphi'''=\ell(x_1+y_1)+<\zeta, x'-y'>+\sigma t$ respectively where ℓ is large enough positive number, $|\zeta|+|\sigma|\leq 1$, $\chi_2\in C_0^\infty(\mathbb{R})$, $supp\,\chi_2\subset(-2s_1, 2s_1)$, $\chi_2=1$ on $[-s_1, s_1]$, $\chi_2'(s)\leq 0$ for $s>0$, s_1 is small enough; remind the reader that in this case $WF_f(u^\pm)\subset\{x_1=y_1=0\}\cup\{x_1=\xi'=\eta=0\}\cup\{y_1=\xi'=\eta=0\}\cup\{\xi=\eta=0\}$ and this inclusion remains true for $U^\pm$.

Note that proposition 1.7 implies that (3.48) for $s=0$ holds; combining (3.48) with $s=0$ and (3.35), (3.36) one obtains (3.48) for arbitrary s.

3.5. It remains for us to examine the zone of normal (and almost normal) rays, i.e. the neighbourhood of $\{\eta'=0\}$.

THEOREM 3.11. There exists $t_1>0$ such that if $\chi_0\in C_0^\infty(\mathbb{R})$, $q_y=q(y, D_y', D_t)\in OPS^{0'}(X\times\mathbb{R})$, $cone\,supp\,q$ is contained in the small enough conic neighbourhood of the set $\{y_1=\eta'=0\}$ then

$$t^{|\alpha|+m+n}D_{x'}^{\alpha}D_t^{n}\tilde{\Gamma}q_y\chi_0u_1\in L_2(Y\times[-t_1, t_1]),$$

$$t^{|\alpha|+m+n}D_{x'}^{\alpha}D_t^{n}\tau\Gamma q_y\chi_0u_1\in L_2(Y\times[-t_1, t_1], hom(E))$$

$$\forall\alpha\in\mathbb{Z}^{+d-1}, n\in\mathbb{Z}^+$$

where $m=m(\alpha)$.

In the proof we use the scheme which was used for the proof of theorems 3.4 - 3.6 with essential simplifications and certain complications. These complications are due to the fact that the smoothness with respect to every separable variable x', x, y', y, t was equal before to the smoothness with respect to all these variables and now it does not remain true and now needs to take into account that B_1 is a pseudo-differential operator with respect to x' but not (x',t); in particular it may occur that

$$WF(u_1)\cap\{t=\xi'=\eta'=0\}=\sum_i Y_i\times Y_i\times 0\times(\mathbb{R}_\tau\setminus 0)$$

where Y_i are connected components of Y; moreover if $v|_Y=0 \not\Rightarrow B_1v|_{Y_i}=0$ that it may occur that

$$WF(u_1)\cap\{t=\xi'=\eta'=0\}=Y\times Y\times 0\times(\mathbb{R}_\tau\setminus 0).$$

Therefore the points $(y^*, \mp\eta^{*\prime}, \tau^*)$, $(y^*, y^*, -\eta^{*\prime}, \eta^{*\prime}, \tau^*)$ should be replaced by the sets $Y\times 0\times\tau^*$, $Y\times Y\times 0\times 0\times\tau^*$

respectively.

In the neighbourhood of $Y\times 0\times\tau^*$, the refined decomposition $E\oplus E=E_1\oplus\ldots\oplus E_{N+N'}$, $E_j = E_j'\oplus E_j''$ for $j=N+1,\ldots N+N'$, by means of pseudo-differential operators is in general impossible; but it is possible more rough decomposition $E\oplus E = E_I\oplus E_{II}\oplus E'\oplus E''$ where $\dim E_I=\dim E_{II}=D_+, \dim E'=\dim E''=D_-, D_\pm$ are dimensions of positive and negative invariant subspaces of a_0 (in general $D_\pm$ depend on a connected component of Y), K_I and K_{II} are positive and negative definite, L is elliptic; all previous notations remain.

Obviously, analogy of lemma 3.7 is valid.

We shall examine different components of U in the neighbourhood Ω of

$$j_x^{-1}(Y\times 0)\times T^*X\times\tau^* U\, T^*X\times j_y^{-1}(Y\times 0)\times\tau^*.$$

Obviously, as before

$$U^{'',}\equiv U^{,'}\equiv 0 \tag{3.49}$$

modulo functions which are infinitely smooth in Ω

LEMMA 3.12. $U^{I,I}$ and $U^{II,II}$ are infinitely smooth in the neighbourhood of Ω.

PROOF. Note that

$$\mathcal{P}_{Ix}U^{-I}\equiv -F_x^I \pmod{C^\infty(\Omega)}$$

where $WF_f(F_x^I)\subset\{t=0\}$, $F_x^I\in C^\infty(X\times X\setminus diag(Y\times Y), H^{-2}(\mathbb{R}))$

and $WF_f(U^{-I,})\subset\{t\leqslant 0\}$; then the theory of the propagation of the singularities for symmetric hyperbolic systems implies that

$$U^{-I,}\in C^\infty(X\times X\setminus diag(Y\times Y), H^{-2}(\mathbb{R}))\oplus C^\infty(\Omega), WF_f(U^{-I,})\cap\Omega\subset\{t=0\}.$$

One can prove these inclusions also for $U^{+II,}$, $U^{-II,}$, $U^{+,I}$; for $U^{-,II}$ and $U^{+,I}$ one should use the dual systems. Hence these inclusions hold for $U^{I,I}$ an $U^{II,II}$ too. But $U^{I,I}$ satisfies the system

$$\mathcal{P}_{I,x} U^{I,I} \equiv 0 \; (\mathrm{mod}\; C^{\infty}(\Omega))$$

and then the above-mentioned theory of the propagation of the singularities combined with the inclusion $WF_{\ell}(U^{I,I}) \cap \Omega \subset \{t=0\}$ implies that $U^{I,I} \in C^{\infty}(\Omega)$; $U^{II,II}$ can be considered by the same method.

Thus all components of U different from $U^{I,II}$, $U^{II,I}$, $U^{\prime,I}$, $U^{\prime,II}$, $U^{I,\prime\prime}$, $U^{II,\prime\prime}$, $U^{\prime,\prime\prime}$ belong to $C^{\infty}(\Omega)$.

Let $\tilde{q}_y = \tilde{q}(y, D_y, D_t) \in OPS^0(\tilde{X} \times \mathbb{R})$, $\mathrm{cone\; supp}\; q$ is contained in the small enough conic neighbourhood of the set $\mathcal{M} = \{\rho \in N^* Y \times \tau,\ g(\rho) = 0\}$ and $\mathrm{cone\; supp}(I - \tilde{q}) \cap \mathcal{M} = \emptyset$. Then $\tilde{q} = \tilde{q}^+ + \tilde{q}^-$, $\tilde{q}_y^{\pm} = \tilde{q}^{\pm}(y, D_y, D_t) \in OPS^0(\tilde{X} \times \mathbb{R})$, $\mathrm{cone\; supp}\; \tilde{q}^{\pm}$ is contained in the small enough conic neighbourhood of the set

$$\mathcal{M}^{\pm} = \{\rho = (y, \eta_1, 0, \tau^*) \in N^* Y \times \tau^*,$$
$g(\rho) = 0$ and $(y, \pm \eta_1/\tau^*)$ is a negative point$\}$,

$$\mathrm{cone\; supp}(I - \tilde{q}^{\pm}) \cap \mathcal{M}^{\pm} = \emptyset .$$

The one can apply theorem 3.2 to $\tilde{q}_y^{\pm} \chi u_0^{\pm}$ and $\tilde{q}_y^{\pm} \chi u_0^{\mp}$ with $\varphi = \sigma t + \nu(x_1 - y_1)$ where ν is a large enough number with appropriate sign, $|\sigma| \leqslant 1$ and prove inclusions

$$(3.50) \quad \tau_x \tau_y t^{|\alpha|+|\beta|+n+m} D_x^{\alpha} D_y^{\beta} D_t^{n} \tilde{q}_y^{\pm} \chi u_0^{\pm} \in L_2(Y \times Y \times \mathbb{R}, \mathrm{Hom}(E))$$

$$\forall \alpha, \beta \in \mathbb{Z}_+^d,\ n \in \mathbb{Z}^+;\ m = m(d)$$

and the similar inclusions for $\tilde{q}_y^{\pm} \chi u_0^{\mp}$; hence the same inclusions hold for $\tilde{q}_y \chi u_0$.

Let $q_y = q(y, D_y', D_t) \in OPS^{0'}(X \times \mathbb{R})$, $\mathrm{cone\; supp}\; q$ is contained in the small enough conic neighbourhood of $Y \times 0 \times \tau^*$; then $q_y \chi u_0 \equiv q_y \tilde{q}_y \chi u_0 \; (\mathrm{mod}\; C^{\infty})$; therefore (3.50) holds also for $q_y \chi u_0$. Hence the following statement holds true:

LEMMA 3.13. Let $q_y = q(y, D_y', D_t) \in OPS^0(Y \times \mathbb{R})$, $\mathrm{cone\; supp}\; q$ be contained in the small enough conic neighbourhood of $Y \times 0 \times \tau^*$, $\chi \in C_0^{\infty}(\mathbb{R})$, $\mathrm{supp}\, \chi \subset (-2t_1, 2t_1)$, $0 < t_1$ is small enough; then

$$D_{x'}^{\alpha} D_{y'}^{\beta} D_t^{n} t^{|\alpha|+|\beta|+n+m} q_y \chi \bar{f} \in L_2(Y \times Y \times \mathbb{R}, \ldots)$$

$$\forall \alpha, \beta \in \mathbb{Z}^{+\, d-1}, \quad n \in \mathbb{Z}^{+}$$

where $\bar{f} = r_y f_x, r_x f_y, m = m(d)$.

LEMMA 3.14. Let $W \in \mathcal{D}'(Y \times \mathbb{R}, E \oplus E)$ satisfy under-determined system

$$\tilde{\mathcal{B}} W \equiv 0$$

modulo functions which are infinitely smooth in the neighbourhood of $Y \times 0 \times \mathcal{T}^*$; then modulo the same functions

$$W^{I} \equiv \beta^{I,II} W^{II} + \beta^{I,''} W'' + \beta^{I} f,$$

$$W^{II} \equiv \beta^{II,I} W^{I} + \beta^{II,''} W'' + \beta^{II} f,$$

$$W' \equiv \beta^{',II} W^{II} + \beta_{II}^{','' } W'' + \beta_{II}^{'} f,$$

$$W' \equiv \beta^{',I} W^{I} + \beta_{I}^{',''} W'' + \beta_{I}^{'} f,$$

where every operator B (we omit the indices) is a pseudo-differential operator of order 0 on $Y \times \mathbb{R}$ and

(i) β is a classical pseudo-differential operator provided B_1 is a differential operator.

(ii) In a general case (i.e. if B_1 is a pseudo-differential operator), $\beta = \bar{\beta} + \tilde{\beta}$ where β is a classical pseudo-differential operator and $\tilde{\beta}$ is a pseudo-differential operator with the symbol $\tilde{\beta}(x', \xi', \tau)$ satisfying the estimates

$$(3.51) \quad |D_{x'}^{\alpha} D_{\xi'}^{\gamma} D_{\tau}^{n} \tilde{\beta}(x', \xi', \tau)| \leq C_{\alpha\gamma n} (1 + |\tau| + |\xi'|)^{-n-1} (1 + |\xi'|)^{1-|\gamma|}$$

$$\forall \alpha, \beta, \gamma ;$$

$\tilde{\beta}$ can be expanded into an asymptotic sum $\tilde{\beta} \sim \sum_{k=0}^{\infty} \tilde{\beta}_k$ where $\tilde{\beta}_k$ are positively homogeneous of degree $-k$ and satisfy the estimates

$$(3.51)' \quad |D_{x'}^{\alpha} D_{\xi}^{\gamma} D_{\tau}^{n} \tilde{\beta}(x', \xi', \tau)| \leq C_{\alpha\gamma n} (|\tau| + |\xi'|)^{-n-1} |\xi'|^{1-k-|\gamma|}$$

$$\forall \alpha, \gamma, n, k ,$$

in the sense that for $|\xi'| \geq 1$

$$(3.51)'' \quad |D_{x'}^{\alpha} D_{\xi'}^{\gamma} D_{\tau}^{n} (\tilde{\beta} - \sum_{K=0}^{N-1} \tilde{\beta}_K)| \leq C_{\alpha\gamma n N} (|\tau| + |\xi'|)^{-n-1} |\xi'|^{1-N-|\gamma|}$$

$$\forall \alpha, \gamma, n, N ;$$

neglible operators are operators β', such that for every $n \in \mathbb{Z}^+$ operator $(Ad\, t)^n \beta'$ acts from $\mathcal{D}'(Y, H^s(\mathbb{R}))$ into $C^\infty(Y, H^{s+n+1}(\mathbb{R}))$ $\forall s \in \mathbb{R}$; all these operators commute with D_t.

PROOF. We set $\xi' = 0$; then elements of E_I, E_{II}, E', E'' are respectively

$$W^I = (w^I \;\; \varkappa_{-1} \tau a_0^{1/2} w^I)^T,$$
$$W^{II} = (w^{II} \;\; \varkappa_{-1} \tau a_0^{1/2} w^{II})^T,$$
$$W' = (w' \;\; \varkappa_{-1} \tau a_0^{1/2} w')^T,$$
$$W'' = (w'' \;\; \varkappa_{-1} \tau a_0^{1/2} w'')^T$$

where $w^I, w^{II} \in E_+$, $w', w'' \in E_-$, $E_\pm$ are positive and negative invariant subspaces of a_0, $a_0^{1/2}$, $(-a_0)^{1/2}$ are positive roots on E_+, E_- respectively. Recall that $\varkappa_{-1} \tau < 0$. Note that either $\mathcal{B} = (I \;\; 0)$ or $\mathcal{B}_0 = (0 \;\; I)$; hence $\mathcal{B}_0 W = 0$ implies either $w^I = -w^{II}$, $w' = -w''$ or $w^I = w^{II}$, $w' = w''$ respectively. Hence lemma 3.14 holds.

REMARK. Certainly, the dual statement holds true (concerning system $W \tilde{\mathcal{B}}_y^+ \equiv f$).

Lemmas 3.12 - 3.14 and this remark imply that

$$(3.52) \quad D_{x'}^{\alpha} D_{y'}^{\beta} D_t^{n} t^{|\alpha|+|\beta|+n+m} r_x r_y q_y \chi U \in L_2(Y \times Y \times \mathbb{R},\ Hom(E \oplus E))$$

$$\forall \alpha, \beta \in \mathbb{Z}^{+\, d-1}, \quad n \in \mathbb{Z}^+ .$$

These inclusions combined with lemma 3.7 imply theorem 3.11.

THEOREM 3.15. Let V be every component of U different from $U^{II,I}$ and $U^{I,II}$, $q_y = q(y, D_y', D_t) \in OPS^{0'}(X \times \mathbb{R})$, *cone supp* q be contained in the small enough conic neighbourhood of $Y \times 0 \times \tau^*$, $\chi \in C_0^\infty(\mathbb{R})$, *supp* $\chi \subset (-2t_1, 2t_1)$, t_1 be small enough. Then

$$x_1^K t^s \Gamma D_x^{\alpha} D_y^{\beta} D_t^{n} q_y \chi V \in L_2(X \times \mathbb{R}, \ldots) \quad \forall \alpha, \beta \in \mathbb{Z}^{+d}, \; s, k, n \in \mathbb{Z}^+$$

such that $|\alpha| + |\beta| + n \leq s + k - m$, $m = m(d)$.

PROOF. We need to consider components $V = U^{\prime,\mathrm{I}}, U^{\prime,\mathrm{II}}, U^{\mathrm{I},\prime\prime}, U^{\mathrm{II},\prime\prime}, U^{\prime,\prime\prime}$. We consider $V = U^{\prime,\mathrm{I}}$ (the other components can be considered by the same method). Then V satisfies systems

$$(K' D_{x_1} + L'_x) V \equiv 0,$$

$$V(K_{\mathrm{I}} D_{y_1} + L_{\mathrm{I},y})^{\dagger} \equiv 0$$

modulo functions which are infinitely smooth in the neighbourhood of $\{\xi' = \eta' = 0\}$. Hence $W = V|_{x_1 = y_1}$ satisfies system

$$D_{x_1} W \equiv -K'^{-1} L'_x W + W L^{\dagger}_{\mathrm{I},y} K^{\dagger -1}_{\mathrm{I}} = \mathcal{U} W$$

modulo the same functions. Note that

$$\mathcal{U} \in OPS^{1'}([0,\delta) \times Y \times Y, Hom(E', E_{\mathrm{I}}));$$

its principal symbol is

$$\mathcal{U}_1 = -K'^{-1} L'_1(x_1, x', \xi', \tau) \otimes Id + Id \otimes L^{T}_{\mathrm{I},1}(x_1, y', \eta', \tau) K^{T-1}_{\mathrm{I}} \quad ;$$

eigenvalues of $\mathcal{U}$ are $\mu_{ij} = -\mu'_i + \mu_{\mathrm{I},j}$ where μ'_i, $\mu_{\mathrm{I},j}$ are eigenvalues of $K'^{-1} L'_1(x, \xi', \tau)$ and $K^{-1}_{\mathrm{I}} L_{\mathrm{I},1}(x_1, y', \eta', \tau)$ respectively. Then $\mathrm{Im}\, \mu_{ij} > 0$ because of $\mathrm{Im}\, \mu'_i < 0$, $\mathrm{Im}\, \mu_{\mathrm{I},j} = 0$; therefore

$$W \equiv h \tau_x W = h \tau_x \tau_y V \tag{3.53}$$

modulo functions which are infinitely smooth in the neighbourhood of $\{\xi' = \eta' = 0\}$, where $h \in OP^0([0,\delta) \times Y \times Y \times \mathbb{R}, Hom(E', E_{\mathrm{I}}))$; moreover (3.53) holds modulo functions which are infinitely smooths in the neighbourhood of $\{\xi' = 0\} \cup \{\eta' = 0\}$. Inclusions (3.52) combined with (3.53) and proposition 1.7 imply the theorem.

It is easy to prove the following statements by means of methods used in the proof of proposition 3.9.

PROPOSITION 3.16 (i) Let $q_y \in OPS^{0'}(X \times \mathbb{R})$, *cone supp* q_y be contained in the small enough conic neighbourhood of the point $(y^*, \eta^*, \tau^*) \in (T^*Y \setminus 0) \times (\mathbb{R}_\tau \setminus 0)$, $\chi \in C_0^\infty(\mathbb{R})$, $\chi \subset (-2t_1, 2t_1)$, t_1 be small enough; then

$$D^{\alpha}_x D^{\beta}_y D^{n}_t \Psi^{\pm}_{|\alpha|+|\beta|+n+m} q_y \chi U^{\pm} \in L_2(X \times X \times \mathbb{R}, Hom(E \oplus E))$$

$$\forall \alpha, \beta \in \mathbb{Z}^{+d}, \quad n \in \mathbb{Z}^{+}$$

where $m = m(\alpha)$, $\psi_n^{\pm}$ correspond to $\varphi^{\pm} = \mp t + \sigma(x_1 + y_1)$, $|\sigma|$ is small enough.

(ii) Let $q_y \in OPS^{0'}(X \times \mathbb{R})$, $cone\ supp\ q_y$ is contained in the small enough conic neighbourhood of $Y \times 0 \times \tau^*$, $\chi \in C_0^{\infty}(\mathbb{R})$, $supp\ \chi \subset (-2t_1, 2t_1)$ is small enough. Then

$$D_x^{\alpha} D_y^{\beta} D_t^{n} \psi^{+}_{|\alpha|+|\beta|+m+n} q_y \chi U^{I,II} \in L_2(X \times X \times \mathbb{R}, Hom(E_{I,II}))$$

$$\forall \alpha, \beta \in \mathbb{Z}^{+d}, \quad n \in \mathbb{Z}^{+},$$

$$y_1^{l} \psi_s^{+} D_x^{\alpha} D_y^{\beta} D_t^{n} q_y \chi U^{I,''} \in L_2(X \times X \times \mathbb{R}, Hom(E_I, E''))$$

$\forall \alpha, \beta \in \mathbb{Z}^{+d}$, $n, l, s \in \mathbb{Z}^{+}$ such that

$$|\alpha| + |\beta| + n \leq l + s - m,$$

where $m = m(\alpha)$, $\psi_s^{\pm}$ are as in (i).

The similar inclusions with ψ_s^{+} replaced by ψ_s^{-} hold for $U^{II,I}$, $U^{II,''}$; all these inclusions remain true if we replace y_1^{l} by x_1^{l} and $U^{I,''}$, $U^{II,''}$ by $U^{',II}$, $U^{',I}$ respectively.

(iii) Let the conditions of (ii) be fulfilled. Then the statements of (ii) remain true if we replace $\psi_s^{\pm}$ by $\hat{\psi}_s$ corresponding to $\hat{\varphi} = -(x_1 + y_1) + \sigma t$, σ is small enough.

COROLLARY 3.17. Let $\mathcal{L} \in OPS^{0'}(X \times X \times \mathbb{R}, \quad Hom(E \oplus E))$, $\chi_0 \in C_0^{\infty}(\mathbb{R})$; then

$$(x_1 + y_1)^{|\alpha|+|\beta|+n+m} D_x^{\alpha} D_y^{\beta} D_t^{n} \mathcal{L} \chi_0 U \in L_2(\{|t| \leq t_1, x_1 + y_1 \geq 2\nu|t|\}, Hom(E \oplus E))$$

and

$$x_1^{|\alpha|+n+m} D_x^{\alpha} D_t^{n} \Gamma \mathcal{L} \chi_0 U \in L_2(\{|t| \leq t_1,\ x_1 \geq \nu|t|\}, hom(E \oplus E))$$

$$\forall \alpha, \beta \in \mathbb{Z}^{+d}, \quad n \in \mathbb{Z}^{+}; \ m = m(\alpha),$$

provided t_1 is small enough, ν is large enough.

COROLLARY 3.18. Let $\mathcal{L} \in OPS^{0'}(X \times X \times \mathbb{R}, Hom(E_I, E_{II}))$, $cone\ supp\,\mathcal{L}$ be contained in the small enough conic neighbourhood of $\{\xi' = \eta' = 0\}$, $\chi_0 \in C_0^\infty(\mathbb{R})$. Then $(x_1+y_1)^{|\alpha|+n+m} D_x^\alpha D_t^n \Gamma \mathcal{L} \chi_0 U^{I,II} \in L_2(\{-t_1 \leq t \leq 0\} \cup \{0 \leq t \leq t_1, x_1 \leq \nu^{-1} t\}, hom(E_I, E_{II}))\ \forall \alpha \in \mathbb{Z}^{+d}, n \in \mathbb{Z}^+; m = m(\alpha)$, provided t_1 is small enough, ν is large enough. The similar inclusions with t replaced by $-t$ hold for $U^{II,I}$.

We shall not use statements 3.16 - 3.18 in what follows but they seem to be interesting by themselves.

§4. Calculation of the "great" singularity

4.0. Here we shall calculate $\Gamma u_0, \Gamma u_1, \sigma_0 = \int_X \Gamma u_0\, dx$ and $\sigma_1 = \int_X \Gamma u_1\, dx$ modulo more smooth in the neighbourhood of $t=0$ functions. We shall again consider three different zones: outside of the boundary, near the boundary for $\eta' \neq 0$ and near the boundary for $\eta' = 0$. In every case we shall apply the formal method of successive approximation for u_0 or u_1 and derive the formal series; the terms in these series have increasing orders of singularity and contain powers of t increasing faster. Then we shall derive the formal series for $\Gamma u_0, \Gamma u_1$, σ_0 and σ_1; the terms in these series have decreasing orders of singularity and hence we shall obtain the formal asymptotics for $\Gamma u_0, \Gamma u_1, \sigma_0, \sigma_1$ with respect to the smoothness. These asymptotics will be justified by the results of the previous paragraph.

In reality in the normal rays zone $\{\eta' = 0\}$ this method can be applied only partially because there is no normality of the great singularity of Γu_1 and because B_1 is not a classical pseudo-differential operator with respect to (x', t). But the procedure describe fits the calculation $\tau_x \tau_y D_{x_1}^\mu D_{y_1}^\nu u_0$; applying lemma 3.14 we shall derive the asymptotics for $\tau_x \tau_y D_{x_1}^\mu D_{y_1}^\nu u_1$; this asymptotics is complete provided B_1 is a differential operator and incomplete *) in a general case. Using this asymptotics and applying lemma 3.7 we shall derive asymptotics for σ_1; we shall also construct Γu_1 in the form of a sum of oscillatory integrals; both these

*) I.e. contains only a finite number of terms and the remainder term is only finitely smooth.

asymptotics for σ_1, Γu_1 will be complete provided B is a differential operator and incomplete in a general case.

4.1. First let X be a closed manifold. First of all we construct parametrices with the proper propagation of singularities for $P=-D_t^2+A$.

We assume that 0 is not an eigenvalue of A.

Let us consider at $\mathrm{Ran}\,\Pi^+$ the abstract Cauchy problem

$$(4.1)\qquad Pv=f,\quad v|_{t\ll 0}=0 .$$

This problem is an abstract hyperbolic problem and has the parametrix

$$(4.2)\qquad G_h=F^{-1}_{\tau\to t}(-(\tau-i0)^{-1}+A)^{-1}F_{t\to\tau}$$

defined for functions $f\in\mathcal{S}'_{(+)}(\mathbb{R},\mathrm{Ran}\,\Pi^+)$ where $\mathcal{S}'_\pm(\mathbb{R})=\{f\in\mathcal{S}'(\mathbb{R}),$ $\operatorname{supp} f$ is semibounded from below (above)$\}$.

Namely:

$$(4.3)\qquad f\in\mathcal{S}'(\mathbb{R},\mathrm{Ran}\,\Pi^+),\ \operatorname{supp} f\subset\{t\geq t_1\}\Rightarrow$$

$$PG_h f=f,\qquad \operatorname{supp} G_h f\subset\{t\geq t_1\} .$$

Inversely

$$(4.4)\qquad v\in\mathcal{S}'_{(+)}(\mathbb{R},\mathrm{Ran}\,\Pi^+),\ Pv=f\Rightarrow v=G_h f .$$

Moreover, $G_h f$ at $(-\infty,t)$ depends only on f at $(-\infty,t)$ and G_h maps $\mathcal{S}'_{(+)}(\mathbb{R},\mathrm{Ran}\,\Pi^+)\cap(H^{s;0}_{x;t})_{loc}$ into $\mathcal{S}'_{(+)}(\mathbb{R},\mathrm{Ran}\,\Pi^+)\cap(H^{s+1;1}_{x;t})_{loc}$.

Let $\mathcal{S}'^{\pm}(\mathbb{R})=\{f\in\mathcal{S}'(\mathbb{R}),\ \operatorname{supp} f\subset\overline{\mathbb{R}}^{\pm}\}$. Then

$$(4.5)\qquad f\in\mathcal{S}'^{\pm}(\mathbb{R},\mathrm{Ran}\,\Pi^+)\cap t^{p}(H^{s;0}_{x;t})_{loc}\Longrightarrow$$

$$G_h f\in\mathcal{S}'^{\pm}(\mathbb{R},\mathrm{Ran}\,\Pi^+)\cap t^{p+1}(H^{s+1;0}_{x;t})_{loc}\ ,$$

$$D_t f\in\mathcal{S}'^{+}(\mathbb{R},\mathrm{Ran}\,\Pi^+)\cap t^{p+1}(H^{s;0}_{x;t})_{loc}\qquad \forall s\in\mathbb{R},\ p\in\mathbb{Z}^+.$$

Really, the energy estimate

$$(4.6)\qquad |G_h f(t)|_{s+1} + |D_t G_h f(t)| \leqslant c \int_{-\infty}^{t} |f(\tau)|_s \, d\tau$$

where $|\cdot|_s$ contains no integration with respect to t combined with the Hölder inequality implies that for

$$f \in \mathscr{S}'^{+}(\mathbb{R}, Ran\, \Pi^{+}) \cap t^{\rho}(H^{s;0}_{x;t})_{loc}$$

$$\int_0^T t^{-2\rho-2}(|G_h f(t)|^2_{s+1} + |D_t G_h f(t)|^2_s)\, dt \leqslant$$

$$c_\nu(T) \int_0^T \int_0^t \tau^{-2\nu} t^{2\nu-2\rho-1} |f(\tau)|^2_s \, d\tau\, dt \leqslant$$

$$c'_\nu(T) \int_0^T \tau^{-2\rho} |f(\tau)|^2_s \, d\tau < \infty$$

provided $-\frac{1}{2} < \nu < \rho$.

Let $\Phi^{\mp}_s(\mathbb{R}) = \{f \in \mathscr{S}'^{\mp}(\mathbb{R}) \cap \mathcal{E}'(\mathbb{R}) : \forall \rho \in \mathbb{Z}^+ \;\; t^{\rho} f \in H^{s+\rho;0}_{x;t}\}$. Then

$$(4.7)\qquad f \in \Phi^{-}_s(\mathbb{R}, Ran\,\Pi^{+}) \Rightarrow \theta(-t) D^j_t G_h f \in \Phi^{-}_{s+2-j}(\mathbb{R}, Ran\,\Pi^{+})$$

$$\forall s \in \mathbb{R}, \quad j = 0, 1, 2;$$

where θ is the Heaviside function.

Really, (4.6) combined with the Hölder inequality implies that

$$\int_{-T}^{0} |t|^{2\rho}(|G_h f(t)|^2_{s+2} + |D_t G_h f(t)|^2_{s+1})\, dt \leqslant$$

$$c_\nu(T) \int_{-T}^{0} \int_{-T}^{t} |\tau|^{-2\nu} |t|^{2\nu+2\rho+1} |f(\tau)|^2_{s+1} \, d\tau\, dt \leqslant$$

$$c'_\nu(T) \int_{-T}^{0} |\tau|^{2\rho+2} |f(\tau)|^2_{s+1} \, d\tau < \infty$$

provided $-\frac{1}{2} > \nu > -(p+1)$; hence we obtain (4.7) for $j = 0, 1$.
Equality $D_t^2 G_h f = -f + A G_h f$ implies (4.7) for $j=2$.
(4.7) implies that for $f \in \Phi_s^-(\mathbb{R}, Ran\, \Pi^+)$

$$D_t^j G_h f|_{t=0} \in H_x^{s+3/2-j} \qquad (j = 0, 1);$$

then for appropriate $v \in H_{x;t}^{s+2;\infty}$

$$D_t^j (G_h f - v)|_{t=0} = 0 \qquad j = 0, 1;$$

since $P(G_h f - v) = -Pv \in H_{x;t}^{s;\infty}$ for $t \geqslant 0$ then

$$\theta(t) G_h f \in (H_{x;t}^{s+2;0} \oplus t H_{x;t}^{s+1;0})_{loc}$$

and

$$\theta(t) D_t G_h f \in (H_{x;t}^{s+1;0} \oplus t H_{x;t}^{s;0})_{loc}.$$

Therefore

(4.8) $$f \in \Phi_s^-(\mathbb{R}, Ran\, \Pi^+) \Rightarrow G_h f \in \Phi_{s+2}^-(\mathbb{R}, Ran\, \Pi^+) \oplus$$

$$\mathcal{S}'^+(\mathbb{R}, Ran\, \Pi^+) \cap (H_{x;t}^{s+2;0} \oplus t\, H_{x;t}^{s+1;0})_{loc}.$$

Finally, the energy estimate implies that

(4.9) $$f = \delta(t) v_0 + \delta'(t) v_1, \quad v_j \in H_x^{s+j} \cap Ran\, \Pi^+ \Rightarrow G_h f \in C(\overline{\mathbb{R}}^+, H_x^{s+1}) \cap$$

$$C^1(\overline{\mathbb{R}}^+, H_x^s) \subset \mathcal{S}'^+(\mathbb{R}, Ran\, \Pi^+) \cap (H_{x;t}^{s+1;0})_{loc} \quad \forall s \in \mathbb{R}.$$

Consider now at $Ran\, \Pi^-$ an abstract elliptic equation $Pu = f$; this equation has the parametrix G_e expressed modulo infinitely smoothing operator by formula

(4.10) $$G_e \equiv F_{\tau \to t}^{-1} (-\tau^2 + A)^{-1} F_{t \to \tau};$$

the right side operator is defined for functions $f \in \mathcal{E}'(\mathbb{R}, \mathrm{Ran}\,\Pi^-)$; G_e is properly supported.

Namely, $PG_e f \equiv f \pmod{C^\infty(X \times \mathbb{R})}$ for every $f \in \mathcal{D}'(\mathbb{R}, \mathrm{Ran}\,\Pi^-)$. Operator G_e maps $\mathcal{S}'(\mathbb{R}, \mathrm{Ran}\,\Pi^-) \cap (H^{s;0}_{x;t})_{loc}$ into $\mathcal{S}'(\mathbb{R}, \mathrm{Ran}\,\Pi^-) \cap (H^{s+2;2}_{x;t})_{loc}$; moreover if $Pv \equiv f \pmod{H^{s;0}_{x;t}}$ at interval (t_1, t_2), $v, f \in \mathcal{S}'(\mathbb{R}, \mathrm{Ran}\,\Pi^-)$ then $v \equiv G_e f \pmod{H^{s+2;2}_{x;t}}$ at any compact subinterval.

One can prove all these statements by means of standart elliptic techniques.

Finally,

$$f \in \mathcal{S}'^{+}(\mathbb{R}, \mathrm{Ran}\,\Pi^-) \cap t^{p}(H^{s;0}_{x;t})_{loc} \Rightarrow D^{j}_{t} G_e f \in \tag{4.11}$$

$$\sum_{k=0}^{p} t^{p-k} (H^{s+2+k-j;0}_{x;t})_{loc} \cap \mathcal{S}'^{+}(\mathbb{R}, \mathrm{Ran}\,\Pi^-) \oplus \Phi^{-}_{s+p+2-j}(\mathbb{R}, \mathrm{Ran}\,\Pi^-)$$

$$\forall j = 0, 1, 2, \quad s \in \mathbb{R}, \ p \in \mathbb{Z}^+,$$

$$f \in \Phi^{\mp}_{s}(\mathbb{R}, \mathrm{Ran}\,\Pi^-) \Rightarrow G_e f \in \Phi^{+}_{s+2}(\mathbb{R}, \mathrm{Ran}\,\Pi^-) \oplus \Phi^{-}_{s+2}(\mathbb{R}, \mathrm{Ran}\,\Pi^-) \tag{4.12}$$

$$\forall s \in \mathbb{R}.$$

One can prove these statements by means of formula

$$[G_e, t] \equiv -G_e [P, t] G_e. \tag{4.13}$$

Finally

$$f = \delta(t) v_0 + \delta'(t) v_1, \ v_j \in H^{s+j}_{x} \cap \mathrm{Ran}\,\Pi^- \Rightarrow \tag{4.14}$$

$$G_e f \in \Phi^{+}_{s+1}(\mathbb{R}, \mathrm{Ran}\,\Pi^-) \oplus \Phi^{-}_{s+1}(\mathbb{R}, \mathrm{Ran}\,\Pi^-) \quad \forall s \in \mathbb{R}.$$

Let us now introduce an operator $G = G_h \Pi^+ + G_e \Pi^-$; this operator can be expressed modulo infinitely smoothing operator by formula

$$G \equiv F^{-1}_{t \to \tau} (-(\tau - i0)^2 + A)^{-1} F_{t \to \tau}; \tag{4.15}$$

the right side operator is defined for functions $f \in \mathcal{E}'(\mathbb{R})$.

Summarizing the properties of G_h and G_e we obtain the proposition:

PROPOSITION 4.1. Operator G has properties:

(4.16)
$$f\in \mathcal{S}'^{+}(\mathbb{R})\cap t^{p}(H^{s;0}_{x;t})_{loc} \Rightarrow$$
$$Gf\in \mathcal{S}'^{+}(\mathbb{R})\cap \sum_{k=0}^{p+1} t^{p-k+1}(H^{s+k+1;0}_{x;t})_{loc}\oplus \Phi^{-}_{s+p+2}(\mathbb{R})$$
$$\forall s\in\mathbb{R},\ p\in\mathbb{Z}^{+};$$

(4.17)
$$f\in\Phi^{-}_{s}(\mathbb{R})\Rightarrow Gf\in\Phi^{-}_{s+2}(\mathbb{R})\oplus$$
$$\mathcal{S}'^{+}(\mathbb{R})\cap(H^{s+2;0}_{x;t}\oplus tH^{s+1;0}_{x;0})_{loc}\quad \forall s\in\mathbb{R};$$

(4.18)
$$f=\delta(t)v_0+\delta'(t)v_1,\ v_j\in H^{s+j}_{x}\Rightarrow$$
$$Gf\in\mathcal{S}'^{+}(\mathbb{R})\cap(H^{s+1;0}_{x;t})_{loc}\oplus\Phi^{-}_{s+1}(\mathbb{R})\quad\forall s\in\mathbb{R};$$

(4.19)
$$PGf\equiv f\ (\mathrm{mod}\ C^{\infty}(X\times\mathbb{R}));$$
$$f\equiv 0(\mathrm{mod}\ H^{s;0}_{x;t}(X\times\mathbb{R}^{-}_{t_1})_{loc})\Rightarrow Gf\equiv 0(\mathrm{mod}\ H^{s+1;1}_{x;t}(X\times\mathbb{R}^{-}_{t_1})_{loc})$$
$$\forall s,t_1\in\mathbb{R};$$

(4.20)
$$Pv\equiv f\ (\mathrm{mod}\ H^{s;0}_{x;t}(X\times\mathbb{R}^{-}_{t_1})_{loc}),$$
$$v\equiv 0\ (\mathrm{mod}\ H^{s+1;\nu}_{x;t}(X\times\mathbb{R}^{-}_{t_2})_{loc}\Rightarrow$$
$$v\equiv Gf\ (\mathrm{mod}\ H^{s+1;\nu}_{x;t}(X\times\mathbb{R}^{-}_{t_1})_{loc})$$
$$\forall s,t_1,t_2\in\mathbb{R},\quad t_2<t_1,\ \nu=0,1.$$

REMARK. We have constructed the parametrix $G=G^{+}$ which propagates singularities in the direction of increasing t. Similarly one can can construct the parametrix G^{-} which propagates singularities in the direction of decreasing t. All our statements remain true with permutated $\mathbb{R}^+$ and $\mathbb{R}^-$, $\mathcal{S}'^+$ and $\mathcal{S}'^-$, $\mathcal{S}'_{(+)}$ and $\mathcal{S}'_{(-)}$, Φ^+ and Φ^- and with $\tau - i0$ replaced by $\tau + i0$, in (2.20) inequality $t_2 < t_1$ replaced by $t_2 > t_1$. Note that here only the hyperbolic part of parametrix (G_h) changes; the elliptic part parametrix (G_e) does not change.

Let $X = \mathbb{R}^d$ and $\bar{A} = a(D_x)$ be an operator with a symbol which does not depend on x and is positively homogeneous of the degree 2; then $\bar{G}^{\pm}$ can be expressed modulo infinitely smoothing operators by formula

$$(4.21) \qquad \bar{G}^{\pm} \equiv F^{-1}_{\xi,\tau\to x,t}(-(\tau \mp i0)^2 + a(\xi))^{-1}(1-\zeta(\xi))F_{x,t\to\xi,\tau}$$

where $\zeta \in C_0^\infty(\mathbb{R}^d)$, $\zeta = 0$ for $|\xi| > 1/2$ and $\zeta = 1$ for $|\xi| < 1/4$.

4.2. Let $X \supset \Omega \supset\supset \Omega' \supset\supset \Omega''$ be small enough open sets; then there exists a coordinate system in Ω and $E|_\Omega$ is trivial. Let $\omega \in C_0^\infty(\Omega)$, $\omega = 1$ in Ω'; let $y \in \Omega''$. We assume for the sake of convènience that $dx = dx_1 \dots dx_d$ at y in our coordinate system. Then theorem 2.2 implies that

$$(4.22) \qquad P_x(\omega(x)u^{\pm}) \equiv \pm\delta(t)\omega(x)K_\Pi(x,y) \quad (\mathrm{mod}\ \mathcal{S}'^{\pm}(\mathbb{R}) \cap H^{\infty;0}_{x,y;t}) \quad \text{for } t < 2t_2 \quad \text{provided}$$

$0 < t_2$ is small enough.

Let $\bar{A} = a(y, D_x)$ be the principal part of A frozen at y; then $A = \bar{A} - R$ in Ω where

$$(4.23) \qquad R = \sum_{0<|\alpha|+K<M} (x-y)^\alpha R_{\alpha K}(y, D_x) + \tilde{R}_M ,$$

$$(4.23)' \qquad \tilde{R}_M = \sum_{|\alpha|+K=M} (x-y)^\alpha \tilde{R}_{\alpha K}(x, y, D_x),$$

$R_{\alpha K}$ and $\tilde{R}_{\alpha K}$ are operators of order $2-K$, the symbols of $R_{\alpha K}$ are positively homogeneous of degree $2-K$ for $|\xi| > 1/2$ and are equal to

$$-\frac{1}{\alpha!}\left(\frac{\partial}{\partial y}\right)^\alpha A_{2-K}(y,\xi)$$

there.

Let $\overline{P} = -D_t^2 + \overline{A}$; then

$$\overline{P}(\omega u^{\pm}) \equiv R(\omega u^{\pm}) \pm \delta(t)\omega K_{\Pi} \tag{4.24}$$

$(\mathrm{mod}\ \mathcal{S}'^{\pm}(\mathbb{R}) \cap H^{\infty;0}_{x,y;t})$ for $\pm t < 2t$ and (4.20) implies that $\omega u^{\pm} \equiv \overline{G}^{\pm}(\omega u^{\pm}) \pm \overline{G}^{\pm}(\delta(t)\omega K_{\Pi}) (\mathrm{mod}\ H^{\infty;0}_{x,y,t})$ for $\pm t < 2t_2$ where $\overline{G}^{\pm}$ are given by (4.21) with $a(\xi) = a(y,\xi)$. Iterating this equality we obtain that for an arbitrary M

$$\omega u^{\pm} \equiv \pm \sum_{K=0}^{M-1} (\overline{G}^{\pm} R)^{K} \overline{G}^{\pm}(\delta(t)\omega K_{\Pi}) + (\overline{G}^{\pm} R)^{M} (\omega u^{\pm})\ (\mathrm{mod}\ H^{\infty;0}_{x,y,t}) \tag{4.25}$$

for $\pm t < 2t_2$.

Recall that

$$u^{\pm} = \pm G^{\pm} \delta(t) K_{\Pi}. \tag{4.26}$$

Formula (4.25) is strict. If we set formally $M=\infty$ then we obtain the formal series of successive approximation

$$\omega u^{\pm} \sim \pm \sum_{K=0}^{\infty} (\overline{G}^{\pm} R)^{K} \overline{G}^{\pm}(\delta(t)\omega K_{\Pi}). \tag{4.25'}$$

The orders singularities of the terms in these series increase because $\mathrm{ord}\, R = 2$ and $\overline{G}^{\pm}$ are operators of order -1 in the Sobolev spaces. These series are not interesting by themselves.

We assign the weight w to every operator: $w(b(x,y,D_x)) = -\mathrm{ord}\, b$, $w(G^{\pm}) = w(\overline{G}^{\pm}) = 2$, $w((x-y)^{\alpha}) = |\alpha|$; the weight of the product equals the sum of weights of factors. We decompose Π into the sum

$$\Pi = \sum_{0 \leq |\alpha|+K < M} (x-y)^{\alpha} \Pi_{\alpha K}(y, D_x) + \tilde{\Pi}_M, \tag{4.27}$$

$$\Pi = \sum_{|\alpha|+K=M} (x-y)^{\alpha} \tilde{\Pi}_{\alpha K}(x, y, D_x) \tag{4.27'}$$

where $\Pi_{\alpha k}, \tilde{\Pi}_{\alpha k} \in OPS^{-k}(X, E)$ and symbols of $\Pi_{\alpha k}$ are positively homogeneous of degree $-k$ for $|\xi| > 1/2$.

Applying relation $K_q(x,y)(x_j - y_j) = K_{[x_j;q]}(x,y)$ we then obtain equality

$$(4.28) \qquad K_\Pi(x,y) = \sum_{j=0}^{M-1} K_{(j)}(x,y) + \tilde{K}_{(M)}(x,y) ,$$

$K_{(j)}(x,y)$, $\tilde{K}_{(M)}(x,y)$ are Schwartz kernels of operators $\Pi_{(j)} \in OPS^{-j}(X, E)$, $\tilde{\Pi}_{(M)} \in OPS^{-M}(X,E)$ respectively; symbols of $\Pi_{(j)} = \Pi_{(j)}(y, D_x)$ are equal to

$$\sum_{|\alpha|+k=j} \frac{i^{|\alpha|}}{\alpha!} \left(\frac{\partial}{\partial y} \right)^\alpha \left(\frac{\partial}{\partial \xi} \right)^\alpha \pi_{-k}(y,\xi)$$

for $|\xi| > 1/2$ where π_{-k} are terms in decomposition of the symbol of Π positively homogeneous of degree $-k$; $\Pi_{(j)}(y,\xi)$ are positively homogeneous of degree $-j$ for $|\xi| > \frac{1}{2}$ and vanish for $|\xi| < 1/4$. In particular

$$(4.29) \qquad \Pi_{(0)}(y,\xi) = \pi(y,\xi) ,$$

$$(4.29)' \qquad \Pi_{(1)}(y,\xi) = \pi_{-1}(y,\xi) + i \sum_{j=1}^{d} \frac{\partial^2}{\partial y_j \partial \xi_j} \pi(y,\xi)$$

for $|\xi| > 1/2$.

We substitute (4.23), (4.23)', (4.28), (4.26) into (4.25), remove the parentheses and group together all terms with the same weights;

$$(4.30) \qquad \omega u^{\pm} \equiv \sum_{w=0}^{M-1} u^{\pm}_{(w)} + \tilde{u}^{\pm}_{(M)}$$

$$(\mathrm{mod}\, H^{\infty;0}_{x,y;t}) \qquad \text{for } \pm t < 2t_2 ;$$

$\tilde{u}^{\pm}_{(M)}$ contains all terms with the weights greater than, or equal to, M ; we set $w(K_{(j)}) = j$, $w(\delta(t)) = -2$ and hence the least weight in (4.30) equals 0.

It is easy to calculate the first and the second terms in (4.30):

$$(4.31) \qquad u^{\pm}_{(0)} = \pm \bar{G}^{\pm}(\delta(t)K_{(0)}),$$

$$(4.31)' \qquad u^{\pm}_{(1)} = \pm\{\bar{G}^{\pm}\bar{R}\,\bar{G}^{\pm}\delta(t)K_{(0)} + \bar{G}^{\pm}\delta(t)K_{(1)}\}$$

where $\bar{R} = \sum_{j=1}^{d}(x_j - y_j)(\frac{\partial}{\partial y_j}a)(y,D_x) + A_1(y,D_x)$.

These terms as well as all the other terms with the weights less than M, contain only parametrices $\bar{G}^{\pm}$ and pseudo-differential operators $b(y,\mathcal{D}_x)$ with the symbols positively homogeneous for $|\xi|>1/2$; $\tilde{u}^{\pm}(M)$ contains operators with inhomogeneous symbols and parametrices $G^{\pm}$ as well as $\bar{G}^{\pm}$.

First consider terms with the weights less than M. We stage that

$$(4.32) \qquad F_{x,t\to\xi,\tau}u^{\pm}_{(w)} = \pm F_w(y,\xi,\tau\mp i0)e^{-i\langle y,\xi\rangle}$$

where $F_w(y,\xi,\tau)$ is even with respect to τ, positively homogeneous of degree $(-w-2)$ with respect to (ξ,τ) for $|\xi|>1/2$, vanishes for $|\xi|<1/4$, is meromorphic function of $\tau\in\mathbb{C}$; its poles coincide with the zeros of $g(y,\xi,\tau) = det(-\tau^2 + a(y,\xi))$ and lie on the segments $\{\tau\in\mathbb{R}\cup i\mathbb{R},\ C^{-1}|\xi|\leqslant|\tau|\leqslant C|\xi|\}$ of the real and imaginary axes; moreover $F_W(y,\xi,\tau)$ satisfies the estimate

$$(4.33) \qquad |F_W(y,\xi,\tau)| \leqslant C|Im\,\tau|^{-1-2w}|\xi|^{w+1}$$

for $|Im\,\tau|<\varepsilon|\xi|$ $(\varepsilon>0)$, i.e. $F_w(y,\xi,\tau)$ has singularities of degree - type on $\mathbb{R}$ and hence $F_w(y,\xi,\tau\mp i0)$ are distributions [31, 85].

Really, consider the action of the Fourier transform $F_{x,t\to\xi,\tau}$; $\delta(t)K_{(w)}(x,y)$ is transformed into $(2\pi)^{-d-1}\Pi_{(w)}(y,\xi)\exp(-i\langle y,\xi\rangle)$ because of the choice of the coordinate system, the operators $b(y,D_x)$ are transformed into the operators of multiplication by $b(y,\xi)$, the parametrices $\bar{G}^{\pm}$ are transformed into the operators of multiplication by $(-(\tau\mp i0)^2 + a(y,\xi))^{-1}(1-\zeta(\xi))$; finally operators of multiplication by $(x_j - y_j)$ are transformed into $(i\frac{\partial}{\partial\xi_j} - y_j)$, i.e. into operators $i\frac{\partial}{\partial\xi_j}$ acting only on $F_w(y,\xi,\tau)$.

Then the estimate

$$|(-\tau^2 + a(y,\xi))^{-1}| \leqslant C|Im\,\tau|^{-1}|Re\,\tau|^{-1}$$

implies all our statements; it is easy to show that the number of factors $(-\tau^2+a(y,\xi))^{-1}$ in F_w does not exceed $(2w+1)$ because $\frac{\partial}{\partial\xi_j}$ can give only one additional factor.

In particular

(4.34) $$F_0(y,\xi,\tau)=(2\pi)^{-d-1}(-\tau^2+a(y,\xi))^{-1}\,\Pi(y,\xi)\ ,$$

(4.34)' $$F_1(y,\xi,\tau)=(2\pi)^{-d-1}(-\tau^2+a(y,\xi))^{-1}\cdot$$

$$\Big\{\sum_{j=1}^{d}-i\,\frac{\partial}{\partial\xi_j}\Big(\frac{\partial a(y,\xi)}{\partial y_j}(-\tau^2+a(y,\xi))^{-1}\,\Pi(y,\xi)\Big)-$$

$$A_1(y,\xi)(-\tau^2+a(y,\xi))^{-1}\,\Pi(y,\xi)+$$

$$\Pi_{-1}(y,\xi)+\sum_{j=1}^{d} i\,\frac{\partial^2\Pi(y,\xi)}{\partial y_j\,\partial\xi_j}\Big\}\ .$$

We set $u_{(w)}=u^+_{(w)}+u^-_{(w)}$; then

(4.35) $$F_{x,t\to\xi,\tau}\,u_{(w)}=\mathcal{F}_w(y,\xi,\tau)e^{-i\langle y,\xi\rangle}\ ,$$

$$\mathcal{F}_w(y,\xi,\tau)=F_w(y,\xi,\tau-i0)-F_w(y,\xi,\tau+i0)\ .$$

Then $\mathcal{F}_w$ is positively homogeneous of degree $(-2-w)$ with respect to (ξ,τ) for $|\xi|>1/2$, $\operatorname{supp}\mathcal{F}_w\subset\{|\xi|\geq\frac14\ ,\ g(y,\xi,\tau)=0\}\subset\{|\xi|\geq\frac14\ ,\ C^{-1}|\xi|\leq|\tau|\leq C|\xi|\}$. Since F_w is even with respect to τ then replacing τ by $-\tau$ is equivalent to the permutation $\tau+i0$ and $\tau-i0$; hence $\mathcal{F}_w$ is odd with respect to τ . We apply the inverse Fourier transform $F^{-1}_{\xi\to x}$ and set $x=y$; we obtain the equality

$$F_{t\to\tau}\,\Gamma u_{(w)}=\int\mathcal{F}_w(y,\xi,\tau)d\xi\,;$$

note that the integrand has a compact support for every τ ; hence the integral on the right side converges and defines a function which is odd and positively homogeneous of degree $(d-w-2)$ with respect to τ for $|\tau|>\frac12 C$, i.e.

(4.36) $$F_{t\to\tau}\,\Gamma u_{(\omega)} = c_\omega(y)\,|\tau|^{d-\omega-2}\,\mathrm{sign}\,\tau$$

for $|\tau| > \frac{1}{2}C$ where

(4.37) $$c_\omega(y) = \int \mathcal{F}_\omega^{hom}(y,\xi,1)\,d\xi\ ,$$

$\mathcal{F}_\omega^{hom}$ is a function which is with respect to (ξ,τ) and coincides with $\mathcal{F}_\omega$ for $|\xi| > \frac{1}{2}$.

We state that

(4.38) $$\mathcal{F}_\omega \in C^\infty(\Omega''_y,\ \mathcal{S}'(\mathbb{R}^{d+1}_{\xi,\tau}))\ ,$$

(4.38)' $$\mathcal{F}_\omega^{hom}(\cdot,1) \in C^\infty(\Omega'',\mathcal{E}'(\mathbb{R}^d_\xi\setminus 0)).$$

Really, introduce the spherical coordinate system $(r,\sigma) \in \mathbb{R}^+\times \mathbf{S}^{d-1}$ in $\mathbb{R}^d_\xi\setminus 0$; then all the symbols positively homogeneous of degree k with respect to ξ can be rewritten in the form $r^k b(y,\tau)$ and hence they are holomorphic with respect to r ; since all the factors in F_ω which have singularities are $(-\tau^2 + r^2 a(y,\tau))^{-1}$ then replacing $\tau\in\mathbb{R}\setminus 0$ by $\tau(1\pm i0)$ is equivalent to replacing r by $r(1\mp i0)$ and hence (4.38) - (4.38)' hold true.

Therefore

$$F_{t\to\tau}\,\Gamma u_{(\omega)} \in C^\infty(\Omega'',\mathcal{S}'(\mathbb{R}_\tau,\mathrm{hom}(E))),\ c_\omega\in C^\infty(\Omega'',\mathrm{hom}(E))\ ,$$

(4.39) $$\Gamma u_{(\omega)} \equiv c_\omega(y)\,\Phi_{d-\omega-2}(t)\ (\mathrm{mod}\ C^\infty)$$

where $\Phi_k(t) = F^{-1}_{\tau\to t}\ |\tau|^k\ \mathrm{sign}\ \tau$; for $k\leqslant -1$ one needs to replace $|\tau|^k\,\mathrm{sign}\,\tau$ by its certain regularization at $t=0$ [31].

Thus we obtain the formal asymptotics with respect to smoothness

(4.40) $$\Gamma u \sim \sum_{\omega=0}^{\infty} c_\omega(y)\,\Phi_{d-\omega-2}(t),$$

$c_w \in C^\infty(X, hom(E))$. Note that all the terms in this asymptotics have the property of the normality of singularity:

(4.41) $$\Phi_K \in C^\infty(\mathbb{R}\setminus 0),$$
$$(tD_t)^n \Phi \in H_{loc}^{-1/2-K-\varepsilon}(\mathbb{R}) \qquad \forall n \in \mathbb{Z}^+, \varepsilon > 0.$$

In Appendix D we shall calculate $c_0(y)$ and $c_1(y)$; now we shall justify asymptotics (4.40).

THEOREM 4.2. There exists $t_3 > 0$ such that

$$\Gamma u - \sum_{w=0}^{N-1} c_w(y)\Phi_{d-w-2}(t) \in C^\infty(X, H^{-d+N+\frac{3}{2}-\varepsilon}([-t_3, t_3], hom(E)))$$

$$\forall N \in \mathbb{Z}^+, \varepsilon > 0.$$

These inclusions remain true if one applies an operator $(tD_t)^n$ with an arbitrary $n \in \mathbb{Z}^+$ to their left parts.

COROLLARY 4.3. If $\chi \in C_0^\infty(\mathbb{R})$, $supp\,\chi \subset (-t_3, t_3)$, $\chi = 1$ in the neighbourhood of $t=0$, then the following asymptotics holds:

(4.42) $$F_{t\to\tau}\chi\Gamma u \sim \sum_{w=0}^{\infty} c_w(y)\,|\tau|^{d-w-2}\, sign\,\tau \qquad \tau \to \pm\infty$$

REMARK. Since all the previous arguments as well as the following proofs fit not only u but also $q\chi_0 u$ where $\chi_0 \in C_0^\infty(\mathbb{R})$, $q = q(x,y,D_x,D_y,D_t) \in OPS^0(X\times X\times\mathbb{R}, Hom(E))$, or $q = q(x, D_x, D_t) \in OPS^0(X\times\mathbb{R}, E)$ or $q = q(x, D_x) \in OPS^0(X, E)$ etc then the statements of the theorem and the corollary remain true for these functions, perhaps with different coefficients c_w ; in particular, in the case of the manifolds with the boundary these statements hold true for u_0 and also with the same coefficients c_w.

PROOF OF THEOREM 4.2. Consider some term with the weight M. We state that it can be transformed into the sum of terms with the same weights which do not contain factors $(x-y)^\alpha$. Note that if

$v \equiv 0 \pmod{H^{\infty;0}_{x,y,t}}$ for $\pm t < 0$ then

$G^\pm v \equiv 0 \pmod{H^{\infty;2}_{x,y,t}}$ for $\pm t < 0$ and

$$P(x_j - y_j)G^\pm v = (x_j - y_j)PG^\pm v + [P, x_j]G^\pm v \equiv$$

$$\equiv (x_j - y_j)v + [P, x_j] G^{\pm} v \quad (\mathrm{mod}\ H^{\infty;0}_{x,y;t})$$

and (4.20) implies that

$$(4.43)\quad (x_j-y_j)G^{\pm} v \equiv G^{\pm}(x_j-y_j)v + G^{\pm}[P,x_j]G^{\pm} v \quad (\mathrm{mod}\ H^{\infty;0}_{x,y;t}) .$$

The weights of all tree terms in both sides of this equality are equal to $w(v)+3$ . . This equality remains true if one replaces $G^{\pm}, P$ by $\overline{G}^{\pm}, \overline{P}$ respectively. Using this equality we can transfer all powers of $(x-y)$ to the right, to $\delta(t) K_{\Pi_{(s)}}(x,y)$; a some part of these powers will be lost. On the other hand

$$(x-y)^{\beta} K_{\Pi_{(s)}}(x,y) = K_{\Pi'_{(s+|\beta|)}}(x,y)$$

where $\Pi'_{(s+|\beta|)} \in OPS^{-s-|\beta|}(X,E)$; this transformation does not change weight too.

Now we can apply (4.16) - (4.18) and proposition 1.7 and obtain inclusion

$$\tilde{u}^{\pm}_{(M)} \in \mathcal{S}'^{\pm}(\mathbb{R}) \cap \sum_{j=0}^{M_1} t^j (H^{-m-j+M;0}_{x,y;t})_{loc} \oplus \Phi^{\mp}_{-m+1}(\mathbb{R})$$

where $m = m(d)$, $M_1 = M_1(M)$ and the smoothness with respect to x and y is total.*)

Therefore

$$\tilde{u}_{(M)} = \tilde{u}^{+}_{(M)} + \tilde{u}^{-}_{(M)} \in \sum_{j=0}^{M_1} t^j (H^{-m-j+M;0}_{x,y,t})_{loc}$$

and the smoothness with respect to x and y is total.

Let $q_y = q(y, D_y, D_t) \in OPS^0(X \times \mathbb{R})$, cone supp $q \subset \{\tau \neq 0, \eta \neq 0\}$, cone supp $(I-q) \cap \{g(y,\eta,\tau)=0\} = \phi$; then for $\chi_0 \in C_0^{\infty}(\mathbb{R})$

*) $v \in H^{s}_{x,y}$ and the smoothness with respect to x and y is total means that $v \in H^{s;\infty}_{x,t} \cap H^{s;\infty}_{y;x}$; $v \in H^{s}_{x,y,z}$ and the smoothness with respect to x, y and z is total means that $v \in H^{s;\infty}_{x;y,z} \cap H^{s;\infty}_{y;x,z} \cap H^{s;\infty}_{z;x,y}$ etc.

$$q_y \chi_0 \tilde{u}_{(M)} = q_y \sum_{j=0}^{M_1} t^j v_j = \sum_{j=0}^{M_1} t^j v_j'$$

where $v_j \in H^{-m-j+M;0}_{x,y;t}$; $v_j' \in H^{-m-j+M}_{x,y,t}$ and the smoothness with respect to x, y, t is total. Hence

(4.44) $$\Gamma q_y \chi_0 \tilde{u}_{(M)} = \sum_{j=0}^{M_1} t^j v_j'' ,$$

$v_j'' \in H^{-m-j+M;\infty}_{t\;;\;y}$ (perhaps with greater $m = m(d)$).

On the other hand, for $w < M$

(4.45) $$D_x^{\alpha} D_t^{n} t^{m+n} \Gamma u_{(w)} \in L_2(X\times \mathbb{R}, hom(E))$$
$$\forall \alpha \in \mathbb{Z}^{+d},\ n \in \mathbb{Z}^{+}$$

and $$(1-q_y)\chi_0 u_{(w)} \in C^{\infty}(X \times \mathbb{R}^d \times \mathbb{R},\ Hom(E))$$

because $cone\ supp\,(1-q) \cap supp\ \mathcal{F}_w = \emptyset$; theorem 3.1 states *) that

$$D_y^{\alpha} D_t^{n} t^{m+n+|\alpha|} \Gamma u \in L_2(X\times[-t_1,t_1], hom(E))$$
$$\forall \alpha \in \mathbb{Z}^{+d},\ n \in \mathbb{Z}^{+}$$

and we know that

$$(I-q_y)\chi_0 u \in C_0^{\infty}(X\times X\times \mathbb{R}, Hom(E))$$

because $cone\ supp\ q \cap WF(u) = \emptyset$; therefore

(4.46) $$D_y^{\alpha} D_t^{n} t^{m+n+|\alpha|} \Gamma q_y \chi_0 \tilde{u}_{(M)} \in L_2(X\times[-2t_3, 2t_3], hom(E))$$
$$\forall \alpha \in \mathbb{Z}^{+d},\quad n \in \mathbb{Z}^{+};\ \ t_3 = \tfrac{1}{2}\min(t_1, t_2).$$

We set $V = \chi \Gamma q_y \chi_0 \tilde{u}_{(M)}$, $\chi \in C_0^{\infty}(\mathbb{R})$, $supp\,\chi \subset (-2t_3, 2t_3)$ and consider

$$\|\mathcal{F}_{\varepsilon} D_y^{\alpha} D_t^{n} V\|^2 = \sum_{j=0}^{M_1} (\mathcal{F}_{\varepsilon}^2 D_y^{2\alpha} D_t^{2n} V, t^j \chi v_j'') =$$

---

*) Here and only here we use theorem 3.1.

$$\sum_{j=0}^{M_1} (\Lambda^{m+j-M} t^j \mathcal{J}_\varepsilon^2 D_y^{2\alpha} D_t^{2n} V, \Lambda^{-m-j+M} v_j'')$$

where Λ^s, $\mathcal{J}_\varepsilon$ were introduced in section 3.1. (4.44) implies that the second factor belongs to $L_2(X\times\mathbb{R}, hom(E))$; (4.46) implies that the L_2-norm of the first factor is bounded as ε tends to 0 provided $2|\alpha|+2n+m+j-M\leqslant j-m$; i.e. provided $|\alpha|+n\leqslant M/2-m$. Hence

$$\Gamma q_y \chi_0 \tilde{u}_{(M)} \in H^{M/2-m}_{y,t}(X\times[-t_3,t_3], hom(E))$$

and hence

$$\Gamma u - \sum_{\omega=0}^{M-1} c_\omega(y)\Phi_{d-\omega-2}(t) \in H^{M/2-m}_{y,t}(X\times[-t_3,t_3], hom(E))$$

$$\forall M\in\mathbb{Z}^+ .$$

This sequence of inclusions combined with (4.45) implies theorem 4.2.

4.3. To derive the asymptotics for $_\pm\Gamma u_1$, σ_1 we need to derive the "asymptotics" for $f_x^\pm = -B_x u_0^\pm$; therefore we need to derive the "asymptotics" for $u_0^\pm$.

Recall that u_0 was constructed in section 2.4 by means of $\tilde{u}$-solution of the problem (2.3) - (2.5) on extended manifold $\tilde{X}$ with initial conditions $\tilde{u}|_{t=0}=0$, $\tilde{u}_t|_{t=0}=K_{\tilde{\Pi}^+}(x,y)$.

Arguments of the previous section imply the following "asymptotics"

$$\text{(4.47)}\qquad u_0^\pm \equiv \sum_{\omega=0}^{M-1} u^\pm_{(\omega)} (mod\, \mathcal{S}^\pm(\mathbb{R})\cap \sum_{j=0}^{M_1} t^j H^{M-j-m;0}_{x,y;t} \oplus \Phi^\mp_{M-m}(\mathbb{R})),$$

$$F_{x,t\to\xi,\tau}\, u^\pm_{(\omega)} = \pm F_\omega(y,\xi,\tau\mp i0)e^{-i\langle y,\xi\rangle},$$

F_ω are functions of the same type as before and formulas (4.34), (4.34)' remain valid; $M_1=M_1(M)$.

Since the smoothness with respect to x and y is total in every term in (4.47) and in the remainder term too then we can apply operator

Obviously

$$(4.48)\qquad F_{x',t\to\xi',\tau}\, \tau_x D^{\mu}_{x_1} u^{\pm}_{(\omega)} = \pm vrai \int_{-\infty}^{\infty} \xi_1^{\mu} F_{\omega}(y,\xi,\tau) e^{-i\langle y,\xi\rangle} d\xi_1 .$$

Note that $F_{\omega}(y,\xi,\tau)$ is infinitely smooth in zone $\{ |\xi_1| \geq C(|\tau|+|\xi'|+1)\}$ and in this zone

$$F_{\omega}(y,-\xi,-\tau) - F_{\omega}(y,\xi,\tau)(-1)^{\omega}$$

vanishes together with all its derivatives at $y\in Y$, $\xi'=\tau=0$ because F_{ω} was constructed by means of homogeneous symbols of the differential operator A by means of symbols of operators $\widetilde{\Pi}^{+}$, θ with the transmission property; recall that F_{ω} is positively homogeneous of degree $-\omega-2$; hence $F_{\omega} = F'_{\omega} + F''_{\omega}$ where F'_{ω} is the symbol of a pseudo-differential operator with the transmission property and F''_{ω} vanishes for $|\xi_1| \geq C(|\tau|+|\xi'|+2)$; hence $\tau_x D^{\mu}_{x_1} u^{\pm}_{(\omega)}$ is infinitely smooth with respect to y_1 (with corresponding decreasing of smoothness with respect to other variables). We shall examine zone

$$\{C_1^{-1}|\xi'| \leq |\tau| \leq C_1|\xi'|,\ C_1^{-1}|\eta'| \leq |\tau| \leq C_1|\eta'|\},$$

C_1 is arbitrary.

Modulo functions which are infinitely smooth in this zone

$$(4.49)\qquad D^{\nu}_{y_1} \tau_x D^{\mu}_{x_1} u^{\pm}_{\omega} \in \mathcal{S}'^{\pm}(\mathbb{R}) \cap \sum_{j=0}^{M_\omega} t^{j} (\mathcal{H}^{\omega-m-j-\nu;0}_{x',y';y_1,t})_{loc} \oplus \Phi^{\mp}_{\omega-\nu-m}(\mathbb{R})$$

$$\forall \nu \in \mathbb{Z}^{+}$$

and the smoothness with respect to x' and y' is total.

Using equality

$$\widetilde{G}^{\pm} v \equiv D_t^2 \widetilde{A}^{-1} \widetilde{G}^{\pm} v - \widetilde{A}^{-1} v \pmod{C^{\infty}}$$

for $v \in \mathcal{S}'_{(\pm)}(\mathbb{R})$ one can easily show that the same inclusions hold for $\widetilde{u}^{\pm}_{(M)}$ too for $\nu \leq M-m$.

Finally, note that we could freeze the operators at point $(0,y')$ instead of y and decompose the operators in powers of $(x_1, x'-y')$

instead of $(x-y)$. Using formulas (4.43) we could transfer powers of x_1 to the left to τ_x instead of transfering these powers to the right; powers of $(x'-y')$ are transferred to the right as before. Since the smoothness with respect to x and y was total then x_1 attaining τ_x would annul the term; thus in (4.48) we can assume without a loss of generality that F_ω depends only on y' instead of y.

Summarizing we obtain:

PROPOSITION 4.4. Modulo functions which are infinitely smooth in zone

$$\{C_1^{-1}|\xi'|\leq|\tau|\leq C_1|\xi'|,\ C_1^{-1}|\eta'|\leq|\tau|\leq C_1|\eta'|\}$$

(4.50)
$$D_{y_1}^{\nu} f^{\pm} = -D_{y_1}^{\nu} B_x u_0^{\pm} \equiv$$
$$D_{y_1}^{\nu} \sum_{\omega=0}^{M-1} f_{(\omega)}^{\pm} \ (\mathrm{mod}\ \mathcal{S}'^{\pm}(\mathbb{R}) \cap$$
$$\sum_{j=0}^{M_1} t^j (\mathcal{H}^{M-j-m-\nu;0}_{x',y';y_1,t})_{loc} \oplus \Phi^{\mp}_{M-m-\nu}(\mathbb{R})) \qquad \forall \nu \leq M-m$$

where $m=m(d)$ and the smoothness with respect to x' and y' is total,

(4.51)
$$F_{x',t'\to\xi',\tau} f^{\pm}_{(\omega)} =$$
$$\pm\, \mathrm{vrai} \int_{-\infty}^{\infty} F'_\omega(y',\eta_1,\xi',\tau) e^{-iy_1\eta_1 - i\langle y',\xi'\rangle} d\eta_1 (1-\zeta(\xi')),$$

$F'_\omega(y',\eta_1,\xi',\tau)$ is positively homogeneous of degree $(-\omega-2+\mu)$ with respect to (η_1,ξ',τ) *) and F'_ω is either a product in some order of factors of the following types or a sum of such products:

(i) F is homogeneous with respect to (η_1,ξ',τ), meromorphic with respect to (η_1,τ); its poles coincide with the zeros of $g(y',\eta_1,\xi',\tau)$;

(ii) $F=F(y',\eta_1,\xi')$ is a positively homogeneous symbol of a pseudo-differential operator with the transmission property;

(iii) $F=F(y',\xi')$ is a positively homogeneous symbol

*) Recall that $\mu = \mathrm{ord}\, B$

of a pseudo-differential operator;

$$\zeta \in C_0^\infty(\mathbb{R}^{d-1}),\ \zeta = 0 \text{ for } |\xi'| > \frac{1}{2} \text{ and } \zeta = 1 \text{ for } |\xi'| < 1/4.$$

In particular

$$F_0'(y', \eta_1, \xi', \tau) = -(2\pi)^{-d} b(y', \eta_1, \xi') \cdot (-\tau^2 + a(y', \eta_1, \xi'))^{-1} \Pi(y', \eta_1, \xi') \tag{4.52}$$

where $b = I$ for $\mu = 0$, $b = i\eta_1 + b_1(y', \xi')_+$ for $\mu = 1$.

Thus the "asymptotics" for $f^\pm = B_x u_1^\pm$ are derived.

4.4. We need to construct parametrices with the proper propagation of singularities for the problem $Pv = f$, $Bv = \varphi$. The construction of section 4.1 fits $\varphi = 0$ but we need to revise the scales of spaces. The matter is that certain statements of section 4.1 hold only for $H^s = \mathcal{D}(A^{s/2})$ at least for the hyperbolic part of parametrix these statements do not remain true for the usual spaces $H^s (s \in \mathbb{Z}^+)$: for the smoothness of $G_h^\pm f$ with respect to x one needs not only the smoothness of f with respect to but also the agreement conditions. Spaces $H^s = \mathcal{D}(A^{s/2})$ do not fit the successive approximation. It is convenient to have the smoothness with respect to t equal 0 because of the energy estimates and the weight factors t^p. Moreover to restrict to Y we need a certain smoothness with respect to x_1.

Fortunately there is the "convenient" variable y' in the zone under consideration: all operators contain y' only as a parameter; we shall not consider the variable y_1 in this section at all.

Note first that theorem 2.3 (iii) implies that

$$WF_f^{',1}(G_h^\pm) \cap \{\pm t < t_0\} \subset \{\pm t \geq 0, |x - y| + |\xi' - \eta'|(|\xi'| + |\tau|)^{-1} + |\xi' - \eta'|(|\eta'| + |\tau|)^{-1} \leq \pm c_0 t\} \cup \{|\xi'| + |\eta'| \leq \pm c_0 t |\tau|\}. \tag{4.53}$$

It is easy to show that

$$\text{(4.54)} \qquad WF_f^{\prime,1}(G_e) \subset \{t=0,\ x=y,\ \xi=\eta\} \cup$$

$$\{t=0,\ x=y\in Y,\ \xi'=\eta'\} \cup \{t=0,\ x,y\in Y,\ \xi'=\eta'=0\}.$$

Therefore (4.53) holds for $WF^1_{,f}(G^{\pm})$ and $WF_f^{\prime,1}(\bar{G}^{\pm})$ too and

$$\text{(4.55)} \qquad f\in\mathcal{E}',\ WF_f^{\prime,n}(f)\cap\{\pm t<\bar{t}\}\subset\{\pm t\geq 0,\ C_1^{-1}|\tau|\leq|\xi'|\leq C_1|\tau|,$$

$$C_1^{-1}|\tau|\leq|\eta'|\leq C_1|\tau|\} \Rightarrow WF_f^{\prime,n+2}(G^{\pm}f)\cap\{\pm t<\bar{t}\}\subset$$

$$\{\pm t\geq 0,\ C_2^{-1}|\tau|\leq|\xi'|\leq C_2|\tau|,\ \ C_2^{-1}|\tau|\leq|\eta'|\leq C_2|\tau|\}$$

where $\bar{t}\leq\bar{t}(C_1)>0$, $C_2=C_2(C_1,t)\to C_1$ as $t\to 0$.

Let us

$\Phi_s^{\mp}(\mathbb{R})=\{u\in\mathcal{E}'(\mathbb{R})\cap\mathcal{S}'^{\mp}(\mathbb{R}):\forall p,j\in\mathbb{Z}^+\ \ t^{p+j}D_t^j u\in\mathcal{H}^{s+p;0}_{x',y';t,x_1}$ and the smoothness with respect to x' and y' is total $\}$,

$$\Phi_s^{\mp\prime}(\mathbb{R})=\{u\in\mathcal{E}'(\mathbb{R})\cap\mathcal{S}'^{\mp}(\mathbb{R})\ :\ \forall p,j\in\mathbb{Z}^+\quad t^{p+j}D_t^j u\in\mathcal{H}^{s+p;0}_{y';x,t}\},$$

$$\Phi_s^{\mp\prime\prime}(\mathbb{R})=\{u\in\mathcal{E}'(\mathbb{R})\cap\mathcal{S}'^{\mp}(\mathbb{R})\ :\ \forall\alpha\in\mathbb{Z}^{+d},\quad n,j\in\mathbb{Z}^+$$

$$t^{|\alpha|+j}D_x^{\alpha}D_t^{n}u\in\mathcal{H}^{s;0}_{y;x,t}\}.$$

Repeating the arguments of section 4.1 one can prove that

$$f\in\mathcal{S}'^{\pm}(\mathbb{R},\operatorname{Ran}\Pi^+)\cap t^{p}(\mathcal{H}^{s;0}_{y';x,t})_{loc}\Rightarrow D_x^{\alpha}D_t^{k}G_h^{\pm}f\in\mathcal{S}'^{\pm}(\mathbb{R},\operatorname{Ran}\Pi^+)\cap$$

$$t^{p+1}(\mathcal{H}^{s;0}_{y';x,t})_{loc}\qquad\forall\alpha,k:\ |\alpha|+k=1$$

---

*) It is easy to prove that $WF_f^{\prime,1}(\bar{G}^{\pm})\subset\{|x-y|\leq\pm c_0 t|\tau|,\ \xi'=\eta'\}\cup\{\xi'=\eta'=0\}$.

and

$$f \in \Phi_s^{\mp'}(\mathbb{R}, \mathrm{Ran}\, \Pi^+) \Rightarrow$$

$$\theta(\mp t) D_x^\alpha D_t^k G_h^\pm f \in \Phi^{\mp'}_{s+2-|\alpha|-k}(\mathbb{R}, \mathrm{Ran}\, \Pi^+)$$

$$\forall \alpha, k : |\alpha| + k \leq 2,$$

$$D_x^\alpha D_t^k G_h^\pm f \in \Phi^{\mp'}_{s+1}(\mathbb{R}, \mathrm{Ran}\, \Pi^+) \oplus$$

$$\mathcal{S}'^{\pm}(\mathbb{R}, \mathrm{Ran}\, \Pi^+) \cap (\mathcal{H}^{s+1;0}_{y';x,t} \oplus t\, \mathcal{H}^{s;0}_{y';x,t})_{loc} \qquad \forall \alpha, k : |\alpha| + k \leq 1 .$$

Assume that $WF_f^{',n}(f) \cap \{\pm t \leq \bar{t}\} \subset \{\pm t \geq 0, C_1^{-1}|\xi'| \leq |\tau| \leq C_1|\xi'|, C_1^{-1}|\eta'| \leq |\tau| \leq C_1|\eta'|\}$

where $\bar{t} = \bar{t}(C_1)$; then (4.55) implies that

$$WF_f^{',n+2}(G_h^\pm f - q\chi_0\, G_h^\pm f) \cap \{\pm t < \bar{t}\} = \emptyset$$

for appropriate $q \in OPS^{0'}(X \times X \times \mathbb{R})$, *cone supp* $q \subset \{\xi' \neq 0, \eta' \neq 0, \tau \neq 0\}$, $\chi_0 \in C_0^\infty(\mathbb{R})$ and hence

$$f \in \mathcal{S}'^{\pm}(\mathbb{R}, \mathrm{Ran}\, \Pi^+) \cap \sum_{j=0}^{p} t^{p-j} (\mathcal{H}^{s+j;0}_{x',y';x_1,t})_{loc} \oplus \Phi^{\mp}_{s+p}(\mathbb{R}, \mathrm{Ran}\, \Pi^+), \tag{4.56}$$

$$WF_f^{',n}(f) \cap \{\pm t < \bar{t}\} \subset \{\pm t \geq 0, C_1^{-1}|\xi'| \leq |\tau| \leq C_1|\xi'|, C_1^{-1}|\eta'| \leq |\tau| \leq C_1|\eta'|\} \Rightarrow$$

$$\chi D_x^\alpha D_t^k G_h^\pm f \in \mathcal{S}'^{\pm}(\mathbb{R}, \mathrm{Ran}\, \Pi^+) \cap \sum_{j=0}^{p+1} t^{p-j+1} \mathcal{H}^{s+1+j-|\alpha|-k;0}_{x',y';x_1,t} \oplus$$

$$\Phi^{\mp}_{s+p+2-|\alpha|-k}(\mathbb{R}, \mathrm{Ran}\, \Pi^+) \qquad \forall \alpha, k : |\alpha| + k \leq 1$$

provided $\chi \in C_0^\infty(\mathbb{R})$, $\mathrm{supp}\, \chi \subset \{\pm t < \bar{t}\}$, $\bar{t} \leq \bar{t}(C_1)$; here all the smoothnesses with respect to x' and y' are total.

One can show by means of standart elliptic techniques that

$$f \in \mathcal{S}'^{\pm}(\mathbb{R}, \mathrm{Ran}\, \Pi^-) \cap t^p (\mathcal{H}^{s;0}_{y';x,t})_{loc} \Rightarrow$$

$$D_x^\alpha D_t^k G_e f \in \mathcal{S}'^{\pm}(\mathbb{R}, \mathrm{Ran}\,\Pi^-) \cap \sum_{j=0}^{p} t^{p-j} \Lambda^{-j} (\mathcal{H}^{s;0}_{y';x,t})_{loc} \oplus$$

$$\Lambda^{-p} \Phi_s^{\mp''}(\mathbb{R}, \mathrm{Ran}\,\Pi^-) \qquad \forall \alpha, k: |\alpha|+k \leq 2 ,$$

$$f \in \Phi_s^{\mp''}(\mathbb{R}, \mathrm{Ran}\,\Pi^-) \Rightarrow D_x^\alpha D_t^k G_e f \in \Phi_s^{+''}(\mathbb{R}, \mathrm{Ran}\,\Pi^-) \oplus$$

$$\Phi_s^-(\mathbb{R}, \mathrm{Ran}\,\Pi^-) \qquad \forall \alpha, k: |\alpha|+k \leq 2$$

where $\Lambda^\nu = (I + |D_x'|)^\nu$; certainly the inclusions in the right sides can be improved.

Hence

(4.56)' $$f \in \mathcal{S}'^{\pm}(\mathbb{R}, \mathrm{Ran}\,\Pi^-) \cap$$

$$\sum_{j=0}^{p} t^{p-j} (\mathcal{H}^{s+j;0}_{x',y';x_1,t})_{loc} \oplus \Phi^{\mp}_{s+p}(\mathbb{R}, \mathrm{Ran}\,\Pi^-),$$

$$WF_f'^{n}(f) \cap \{\pm t < \bar{t}\} \subset \{C_1^{-1}|\xi'| \leq |\tau| \leq C_1|\xi'| , C_1^{-1}|\eta'| \leq |\tau| \leq C_1|\eta'|\} \Rightarrow$$

$$\chi D_x^\alpha D_t^k G_e f \in \mathcal{S}'^{\pm}(\mathbb{R}, \mathrm{Ran}\,\Pi^-) \cap \sum_{j=0}^{p} t^{p-j} (\mathcal{H}^{s+j+2-|\alpha|-k;0}_{x',y';x_1,t})_{loc} \oplus$$

$$\Phi^{\mp}_{s+p+2-|\alpha|-k}(\mathbb{R}, \mathrm{Ran}\,\Pi^-) \quad \forall \alpha, k: |\alpha|+k \leq 2$$

provided $\mathrm{supp}\,\chi \subset \{\pm t < \bar{t}\}$ here the smoothness with respect to x' and y' is total.

Combining (4.56), (4.56)' we obtain

(4.56)'' $$f \in \mathcal{S}'^{\pm}(\mathbb{R}) \cap \sum_{j=0}^{p} t^{p-j} (\mathcal{H}^{s+j;0}_{x',y';x_1,t})_{loc} \oplus \Phi^{\mp}_{s+p}(\mathbb{R}) ,$$

$$WF_f'^{n}(f) \cap \{\pm t < \bar{t}\} \subset \{\pm t \geq 0 \quad ,$$

$$C_1^{-1}|\xi'|\leq|\tau|\leq C_1|\xi'|,\ C_1^{-1}|\eta'|\leq|\tau|\leq C_1|\eta'|\}\ \Rightarrow$$

$$\chi D_x^{\alpha} D_t^{k} G^{\pm} f \in \mathcal{S}'^{\pm}(\mathbb{R}) \cap \sum_{j=0}^{p} t^{p-j+1} \mathcal{H}^{s+1+j-|\alpha|-k;0}_{x',y'\,;\,x_1,t} \oplus$$

$$\Phi^{\mp}_{s+p+2-|\alpha|-k}(\mathbb{R}) \qquad \forall \alpha,k: |\alpha|+k\leq 2$$

provided $\chi\in C_0^{\infty}(\mathbb{R})$, $\operatorname{supp}\chi\subset\{\pm t<\bar t\}$, $\bar t\leq \bar t\,(C_1)$; here the smoothnesses with respect to x' and y' are total.

The simpler properties of $G^{\pm}$ are:

(4.57)
$$PG^{\pm}f\equiv f \quad (\operatorname{mod} C^{\infty}),$$
$$BG^{\pm}f=0;$$
$$f\equiv 0 \text{ for } \pm t<t' \Rightarrow G^{\pm}f\equiv 0 \text{ for } \pm t<t' \ (\operatorname{mod} C^{\infty}),$$

(4.58)
$$Pv\equiv f,\ Bv=0 \text{ for } \pm t<t';\ v\equiv 0 \text{ for } \pm t<t''(t''<t')\Rightarrow$$
$$v\equiv G^{\pm}f \text{ for } \pm t<t' \ (\operatorname{mod} C^{\infty});$$

here f does not depend on y.

Hence there exist operators $G^{\pm\prime}$: $\mathcal{S}'_{(\pm)}(Y\times\mathbb{R}) \to \mathcal{S}'_{(\pm)}(X\times\mathbb{R})$ such that

(4.57)'
$$PG^{\pm\prime}\varphi\equiv 0 \quad (\operatorname{mod} C^{\infty}),$$
$$BG^{\pm\prime}\varphi=\varphi,$$
$$\varphi\equiv 0 \text{ for } \pm t<t' \Rightarrow G^{\pm\prime}\varphi\equiv 0 \text{ for } \pm t<t' \ (\operatorname{mod} C^{\infty}),$$

(4.58)'
$$Pv\equiv 0,\ Bv\equiv\varphi \text{ for } \pm t<t',\ v\equiv 0$$
$$\pm t<t''\ (t''<t')\Rightarrow v\equiv G^{\pm\prime}\varphi \text{ for } \pm t<t' (\operatorname{mod} C^{\infty}).$$

Moreover

(4.56)'''
$$\varphi\in\mathcal{S}'^{\pm}(\mathbb{R})\cap\sum_{j=0}^{p} t^{p-j}\,(H^{s+j+\frac{3}{2}-\mu;0}_{x',y';t})_{loc}\oplus\hat{\Phi}^{\mp}_{s+p+\frac{3}{2}-\mu}(\mathbb{R}),$$
$$WF(\varphi)\cap\{\pm t\leq\bar t\}\subset\{\pm t\geq 0,$$

$$C_1^{-1}|\xi'|\leq|\tau|\leq C_1|\xi'|,\ C_1^{-1}|\eta'|\leq|\tau|\leq C_1|\eta'|\}\Rightarrow \chi D_x^{\alpha} D_t^{k} G^{\pm\prime} g\in \mathcal{S}'^{\pm}(\mathbb{R})\cap$$

$$\sum_{j=0}^{p} t^{p-j+1} \mathcal{H}^{s+1+j-k-|\alpha|;\,0}_{x',y'\ ;\ x_1,t} \oplus \Phi^{\mp}_{s+p+2-|\alpha|-k} \qquad \forall \alpha\in \mathbb{Z}^{+d},\ k\in \mathbb{Z}^{+}$$

provided $\chi\in C_0^{\infty}(\mathbb{R})$, $\operatorname{supp}\chi\subset\{\pm t<\bar t\}$, $\bar t=\bar t(C_1)$; here $\mu=\operatorname{ord} B$,

$\hat\Phi_s^{\pm}(\mathbb{R})=\{g\in\mathcal{E}'(\mathbb{R})\cap\mathcal{S}'^{\pm}(\mathbb{R});\ \forall p\in\mathbb{Z}^{+}\quad t^{p} g\in H^{s+p\,;\,p}_{x',y';\,t}$
and the smoothness with respect to x' and y' is total$\}$, all the smoothnesses with respect to x' and y' are total.

(4.57), (4.57)', (4.58), (4.58)' imply

(4.58)''
$$Pv\equiv f,\quad Bv\equiv g \quad \text{for } \pm t<t',$$
$$v\equiv 0 \qquad \pm t<t'' \quad (t''<t')\ \Rightarrow$$
$$v\equiv G^{\pm} f + G^{\pm\prime} g \ \text{ for } \pm t<t' \ (\operatorname{mod} C^{\infty}).$$

Recall that

(4.53)'
$$WF'^{1}_{f}(G^{\pm})\cap\{\pm t<t_0\}\subset$$
$$\{\pm t\geq 0,\ |x-y|+|\xi'-\eta'|(|\xi'|+|\tau|)^{-1}+|\xi'-\eta'|(|\xi'|+|\tau|)^{-1}\leq\pm c_0 t\}\cup$$
$$\{\ |\xi'|+|\eta'|\leq\pm c_0\, t|\tau|\},$$

(4.53)''
$$WF_{f}(G^{\pm\prime})\cap\{\pm t<t_0\}\subset\{\pm t\geq 0,\ |x_1|+|x'-y'|+$$
$$|\xi'-\eta'|(|\xi'|+|\tau|)^{-1}+|\xi'-\eta'|(|\eta'|+|\tau|)^{-1}\leq\pm c_0 t\}\cup$$
$$\{\ |\xi'|+|\eta'|\leq\pm c_0\, t\,|\tau|\ \}.$$

Finally we shall give the explicit construction for the parametrices $G^{\pm}$, $G'^{\pm}$ in the case $X=\bar{\mathbb{R}}^{+}\times\mathbb{R}^{d-1}$, $A=a(D_x)$, $B=\tau$ or $B=i\tau D_1+b_1(D')\tau$, $a(\xi)$, $b_1(\xi')$ are positively homogeneous of degrees 2,1 respectively. We shall examine zone $\{|\xi'|\geq C_1^{-1}|\tau|\}$ for an arbitrary $C_1>0$.

We set

(4.59)
$$\bar G^{\pm\prime}=F^{-1}_{\xi',\tau\to x',t}\,\mathcal{G}'(x_1,\xi',\tau\mp i0)\ (1-\varsigma(\xi'))F_{x',t\to\xi',\tau}$$

where $\mathcal{G}'(x_1, \xi', \tau)$ will be defined below. We know that for $\tau \in \mathbb{R} \cup i\mathbb{R}$ the polynomial $g(\xi_1, \xi', \tau)$ has no real root. Let γ be a contour with counter-clock-wise orientation; γ is contained in $\mathbb{C}_+$ and contains inside all roots of $g(\xi_1, \xi', \tau)$ lying in $\mathbb{C}_+$. We set

$$\mathcal{G}'(x_1, \xi', \tau) = \int_\gamma (-\tau^2 + a(\xi_1, \xi'))^{-1} e^{ix_1\xi_1} d\xi_1 \, S^{-1}(\xi', \tau) \ , \tag{4.60}$$

$$S(\xi', \tau) = \int_\gamma (-\tau^2 + a(\xi_1, \xi'))^{-1} d\xi_1 \tag{4.61}$$

provided $B = \tau$ and

(4.60)'
$$\mathcal{G}'(x_1, \xi', \tau) =$$

$$\int_\gamma (-\tau^2 + a(\xi_1, \xi'))^{-1} (-i\xi_1 + b_1^*(\xi')) e^{ix_1\xi_1} d\xi_1 \, S'^{-1}(\xi', \tau) \ ,$$

(4.61)'
$$S'(\xi', \tau) =$$

$$\int_\gamma (i\xi_1 + b_1(\xi'))(-\tau^2 + a(\xi_1, \xi'))^{-1} (-i\xi_1 + b_1^*(\xi')) \, d\xi_1$$

provided $B = i\tau D_1 + b_1(D')$.

Integrals (4.60), (4.61) can be replaced by integrals along the real axis. Integral (4.60)' can be replaced by the principal value of the integral along the real axis. Moreover, integral (4.60) can be replaced by an integral along the real axis too; really

$$(i\xi_1 + b_1(\xi'))(-\tau^2 + a(\xi))^{-1} (-i\xi_1 + b_1^*(\xi')) =$$

$$a_0^{-1} + i\xi_1^{-1} a_0^{-1} \{ b_1^*(\xi') a_0 - a_0 b_1(\xi') - 2i a_1(\xi') \} a_0^{-1} + O(\xi_1^{-2}) \quad \text{as} \quad \xi_1 \to \infty ;$$

recall that $a(\xi) = a_0 \xi_1^2 + 2a_1(\xi')\xi_1 + a_2(\xi')$.

Note that $\mathcal{A}_B$ is symmetric if and only if the coefficient at ξ_1^{-1} vanishes; hence

$$S'(\xi',\tau) = \int_{-\infty}^{\infty} ((i\xi_1 + b_1(\xi'))(-\tau^2 + a(\xi))^{-1}(-i\xi_1 + b_1^*(\xi')) - a_0^{-1}) d\xi' .$$

We need to prove that $S(\xi',\tau)$ is invertible ($S'(\xi',\tau)$ can be considered in the same way). Obviously

$$i(S^* - S)(\xi',\tau) = \mathrm{Im}\,\tau^2 \int_{-\infty}^{\infty} (-\bar{\tau}^2 + a(\xi))^{-1}(-\tau^2 + a(\xi))^{-1} d\xi_1$$

is a definite matrix for $\tau \overline{\in} \mathbb{R} \cup i\mathbb{R}$; hence $S(\xi',\tau)$ is invertible for $\tau \overline{\in} \mathbb{R} \cup i\mathbb{R}$ and $\|S^{-1}(\xi',\tau)\| \leq C\, |\mathrm{Im}\,\tau|^{-1} |\mathrm{Re}\,\tau|^{-1}$.

Note that $\mathcal{G}'(x_1, \xi', \tau)$ is the solution of the problem

$$(-\tau^2 + a(D_1, \xi'))\mathcal{G}' = 0 \ ,$$
$$rb(D_1, \xi')\mathcal{G}' = Id ,$$
$$\mathcal{G}' = o(1) \quad \text{as} \quad x_1 \to +\infty .$$

Since $\{\mathcal{A}, B\}$ is elliptic then $\{-D_t^2 + \mathcal{A}, B\}$ is elliptic in zone $\{ |\tau| \leq C^{-1} |\xi'| \}$ and hence there is another formula for $\mathcal{G}'$ in this zone [4]; this formula implies that $\mathcal{G}'$ is holomorphic for $|\tau| \leq C^{-1} |\xi'|$ even if $\tau \in \mathbb{R} \cup i\mathbb{R}$. Conversely $\mathcal{G}'(x_1, \xi', \tau)$ is holomorphic for $|\tau| \geq C|\xi'|$, $\tau \in \mathbb{R}$ ($\tau \in i\mathbb{R}$) if and only if a_0 is negative (positive) definite.

Using the parametrices in the whole space and $\bar{G}^{\pm\prime}$ we can construct parametrices $\bar{G}^{\pm}$ too. Exactly, let $\mathcal{G}'(\xi',\tau)$ be a one-dimensional (on $\bar{\mathbb{R}}^+$) integral operator with the Schwartz kernel

(4.62) $$K_{\mathcal{G}}(x_1, y_1, \xi', \tau) =$$

$$\frac{1}{2\pi} \int_{-\infty}^{\infty} \{ e^{i(x_1 - y_1)\xi_1} - \mathcal{G}(x_1, \xi', \tau) b(\xi) e^{-iy_1\xi_1} \} (-\tau^2 + a(\xi))^{-1} d\xi_1 , \ \tau \in \mathbb{C} \setminus (\mathbb{R} \cup i\mathbb{R}) ;$$

then

(4.63) $$\bar{G}^{\pm} = F^{-1}_{\xi',\tau \to x',t}\, \mathcal{G}'(\xi', \tau \mp i0)(1 - \zeta(\xi))\, F_{x',t \to \xi',\tau}$$

are appropriate parametrices.

In particular, for $v \in \mathcal{S}'(\mathbb{R}^{d-1})$, $\eta_1 \in \mathbb{R}$

(4.64) $$\bar{G}^{\pm}(e^{ix_1\eta_1} v) = F^{-1}_{\xi',\tau \to x',t} \{ e^{ix_1\eta_1} - \mathcal{G}'(x_1, \xi', \tau \mp i0)$$

$$b(\eta_1, \xi')\} (-(\tau \mp i0)^2 + a(\eta_1, \xi'))^{-1} (1 - \chi(\xi')) F_{x',t \to \xi',\tau} v \, .$$

4.5. Now we can use the method of successive approximation. Let $\bar{\mathcal{A}} = a(0, y', D_x)$, $\bar{B} = b(y', D_x)$; we freeze the principal parts of $\mathcal{A}, B$ at the point $(0, y')$; then $\mathcal{A} = \bar{\mathcal{A}} - R$, $P = \bar{P} - R$, $B = \bar{B} - L$ where

(4.65) $$R = \sum_{0 < |\alpha| + k < M} x_1^{\alpha_1} (x' - y')^{\alpha'} R_{\alpha k}(y', D_x) + \tilde{R}_{M'}$$

(4.65)' $$\tilde{R}_M = \sum_{|\alpha| + k = M} x_1^{\alpha_1} (x' - y')^{\alpha'} \tilde{R}_{\alpha k}(x, y', D_x) \, ,$$

$R_{\alpha k}$, $\tilde{R}_{\alpha k}$ are differential operators of order $2-k$ $(k = 0, 1, 2)$ with homogeneous symbols,

(4.66) $$L = \sum_{0 < |\alpha'| + k < M} (x' - y')^{\alpha'} L_{\alpha' k}(y', D'_x) + \tilde{L}_M \, ,$$

(4.6)' $$\tilde{L}_M = \sum_{|\alpha'| + k = M} (x' - y')^{\alpha'} \tilde{L}_{\alpha' k}(x', y', D'_x) \, ,$$

$L_{\alpha' k}$, $\tilde{L}_{\alpha' k}$ are pseudo-differential operators of order $1-k$; the symbols of $L_{\alpha' k}$ are positively homogeneous for $|\xi'| > 1/2$ and vanish for $|\xi'| < 1/4$; we consider the most complicated case $B = i\tau D_1 + B_1(x', D'_x)\tau$; $L = 0$ provided $B = \tau$.

Let $\omega(x) = \omega'(x')\omega_1(x_1)$ be a function of the same type as before, $\omega_1 = 0$ for $x_1 > 2\delta$, $\omega_1 = 1$ for $x_1 < \delta$, $\chi_0 \in C_0^\infty(\mathbb{R})$, $\chi_0 = 1$ for $|t| < 2t_2$, $q'_y = q'(y', D'_y, D_t) \in OPS^0(Y \times \mathbb{R})$, *cone supp* $q' \subset \{\tau \neq 0, \eta' \neq 0\}$, $q_y = \omega_1(y_1) q'_y$.

Then (2.31)-(2.33) imply that

(4.67) $$P_x(\omega(x) q_y \chi_0(t) u_1^{\pm}) \equiv \pm q_y \omega(x) \delta(t) K_{\Pi'}(x,y) \pmod{C^\infty}$$

for $\pm t < 2t_2$,

(4.67)' $$B_x \omega(x) q_y \chi_0 u_1^{\pm} \equiv q_y \omega(x) \chi_0 f^{\pm} \pmod{C^\infty}$$

where $\Pi' \in OGS^0(X, E)$, $f^{\pm} = -B_x u_0^{\pm}$,

(4.67)'' $$\omega(x) q_y \chi_0 u_1^{\pm} \equiv 0 \pmod{C^\infty} \qquad \pm t < 0 .$$

Hence

(4.68) $$\overline{P}(\omega q_y \chi_0 u_1^{\pm}) \equiv R \omega q_y \chi_0 u_1^{\pm} \pm q_y \omega \delta(t) K_{\Pi'}(x,y) \pmod{C^\infty}$$

for $\pm t < 2t_2$,

(4.68)' $$\overline{B}(\omega q_y \chi_0 u_1^{\pm}) \equiv L \tau_x \omega q_y \chi_0 u_1^{\pm} + q_y \omega \chi_0 f^{\pm} \pmod{C^\infty};$$

then (4.58)'' implies that

(4.69) $$\omega q_y \chi_0 u_1^{\pm} \equiv (\overline{G}^{\pm} R + \overline{G}^{\pm\prime} L \tau_x) \omega q_y \chi_0 u_1^{\pm} \pm \overline{G}^{\pm}(q_y \omega \delta(t) K_{\Pi'}(x,y)) + \overline{G}^{\pm\prime} q_y \omega \chi_0 f^{\pm} \pmod{C^\infty} \quad \text{for } \pm t < 2t_2 .$$

Iterating this equality we obtain that for an arbitrary M

(4.70) $$\omega q_y \chi_0 u_1^{\pm} \equiv \sum_{k=0}^{M-1} (\overline{G}^{\pm} R + \overline{G}^{\pm\prime} L \tau_x)^k (\pm \overline{G}^{\pm} q_y \omega \delta(t) K_{\Pi'}(x,y) + \overline{G}^{\pm\prime} q_y \omega \chi_0 f^{\pm}) + (\overline{G}^{\pm} R + \overline{G}^{\pm\prime} L \tau_x)^M \omega q_y \chi_0 u_1^{\pm} \pmod{C^\infty} \quad \text{for } \pm t < 2t_2 .$$

Moreover

$$(4.71)\quad \omega q_y \chi_0 u_1^{\pm} \equiv \pm G^{\pm}(q_y \omega \delta(t) K_{\Pi'}(x,y)) + G^{\pm'}(q_y \omega \chi_0 f^{\pm})$$

$$(\mathrm{mod}\ C^{\infty}) \quad \text{for} \quad \pm t < 2t_2 .$$

Note that $q_y(\omega\delta(t)K_{\Pi'}(x,y))$ is the Schwartz kernel of operator $Q \in OGS^0(X\times\mathbb{R}, E)$ *). Repeating with obvious modifications the arguments of section 4.2 one can decompose $K_Q = q_y(\omega\delta(t)K_{\Pi'}(x,y))$ into the sum

$$(4.72)\quad K_Q(x,y,t) = \sum_{w=0}^{M-1} K'_{(w)}(x,y,t) + \widetilde{K}'_{(M)}(x,y,t)$$

where $K'_{(w)}$, $\widetilde{K}'_{(M)}$ are Schwartz kernels of operators $Q'_{(w)} \in OGS^{-w}(X\times\mathbb{R}, E)$, $\widetilde{Q}'_{(M)} \in OGS^{-M}(X\times\mathbb{R}, E)$ respectively and the symbols of $Q'_{(w)}$ are either the products $F_w(y', \eta_1, \xi')\, h_w(y', \xi', \tau')$ or the sum of such products where F_w are the symbols of singular Green operators positively homogeneous of degree $(-w-1)$, h_w are the pseudo-differential symbols positively homogeneous of degree 0 for $|\xi'| > 1/2$, $\mathrm{supp}\, h_w \subset \{\, |\xi'| \geq 1/4 ,\ C^{-1}|\xi'| \leq |\tau| \leq C|\xi'|\,\}$. In particular

$$Q'_{(0)}(y', \xi_1, \eta_1, \xi', \tau) = \Pi'_0(y', \xi_1, \eta_1, \xi')\, q'_0(y', -\xi', \tau) \qquad |\xi'| > 1/2$$

where Π'_0 is the principal symbol of Π'

Proposition 4.4 implies that

$$(4.73)\quad D^{\nu}_{y_1} q_y \chi_0 f^{\pm} \equiv D^{\nu}_{y_1} \sum_{w=0}^{M-1} \tilde{f}^{\pm}_{(w)}\ (\mathrm{mod}\ \mathcal{S}'^{\pm}(\mathbb{R}) \cap \sum_{j=0}^{M_1} t^j \mathcal{H}^{M-j-m-\nu;0}_{x',y';y_1,t} \oplus$$

$$\Phi^{\mp}_{M-\nu}(\mathbb{R})) \qquad \forall \nu < M-m \quad \text{for} \quad \pm t < 2t_2$$

where $\tilde{f}^{\pm}_{(w)}$ are expressed by oscillatory integrals (4.51) with $F'_w(y', \eta_1, \xi', \tau)$ replaced by either products $F'_w(y', \eta_1, \xi', \tau)\, h_w(y', \xi', \tau)$ or the sums of such products, h_w are as above and all the smoothnesses with respect to x' and y' are total. In particular, $h_0(y', \xi', \tau) = q_0(y', -\xi', \tau)$ for $|\xi'| > 1/2$.

*) All operators Q: commute with D_t and hence t denotes the difference of the corresponding arguments.

We assign the weight w to every operator and function:

$$w(b(x,y,D_x))=-\operatorname{ord} b,\quad w(G^{\pm})=w(\bar{G}^{\pm})=2,\quad w(G^{\pm\prime})=w(\bar{G}^{\pm\prime})=$$

$$1+\mu,\quad w(\tau)=-1,\quad w(x_1^{\alpha_1}(x'-y')^{\alpha'})=|\alpha|,\quad w(K'_{(j)})=j-2,\quad w(\tilde{f}^{\pm}_{(j)})=j-\mu\ ;$$

the weight of the product equals the sum of the weights of factors.

Repeating the corresponding arguments of section 4.2 we obtain that

(4.74) $$\omega q_y \chi_0 u_1^{\pm} \equiv \sum_{w=0}^{M-1} v^{\pm}_{(w)} + \tilde{v}^{\pm}_{(M)} \qquad (\operatorname{mod} C^{\infty}) \quad \text{for} \quad \pm t < 2t_2\ ;$$

in particular

(4.74)' $$v^{\pm}_{(0)} = \pm \bar{G}^{\pm} K'_{(0)} + \bar{G}^{\pm\prime} \tilde{f}^{\pm}_{(0)}\ .$$

Consider first the terms with the weight less than M. We apply the Fourier transform $F_{x',t\to\xi',\tau}$.

PROPOSITION 4.5. $F_{x',t\to\xi',\tau}\, v^{\pm}_{(w)}$ equals the sum of the following oscillatory integrals:

(4.75) $$\pm v\tau a i \int_{-\infty}^{\infty}\int_{-\infty}^{\infty} F_w(y',\xi_1,\eta_1,\xi',\tau\mp i0) \exp i(x_1\xi_1 - y_1\eta_1 - \langle y',\xi'\rangle)\, d\xi_1\, d\eta_1 \cdot$$
$$h_w(y',\xi',\tau)\,\omega_w(y_1)$$

(we omit the index of summation) where $F_w(y',\xi_1,\eta_1,\xi',\tau)$ is positively homogeneous of degree $(-w-3)$ with respect to (ξ_1,η_1,ξ',τ) and is the product of the functions of the following types:

(i) $F(y',\xi_1,\xi',\tau)$ is homogeneous with respect to (ξ_1,ξ',τ) meromorphic with respect to (ξ_1,τ), its poles coincide with the zeros of $g(y',\xi_1,\xi',\tau)$ and F tends to 0 as ξ_1 tends to ∞;

(ii) $F(y',\eta_1,\xi',\tau)$ is homogeneous with respect to (η_1,ξ',τ) meromorphic with respect to (η_1,τ), its poles coincide with the zeros of $g(y',\eta_1,\xi',\tau)$ and F tends to 0 as η_1 tends to ∞;

(iii) $F(y',\xi',\tau)$ is positively homogeneous with respect to (ξ',τ) and holomorphic with respect to $\tau\in\mathbb{C}\setminus(\mathbb{R}\cup i\mathbb{R})\cap\{|\tau|\geq C^{-1}|\xi'|\}$.

(iv) $F(y',\eta_1,\xi',\tau)$ is a pseudo-differential symbol with the transmission property, positively homogeneous of a nonpositive degree with respect to (η_1, ξ', τ).

(v) $F(y', \xi_1, \eta_1, \xi')$ is the symbol of a singular Green operator positively homogeneous of a negative degree with respect to (ξ_1, η_1, ξ').

This product necessarily contains the factor of the type (i) and it contains the factor of the type (v) if and only if it contains no factor of the type (ii) $\cup$ (iii) $\cup$ (iv).

If $\Pi \in OGS^o(X, E)$ then this product contains no factors of the type (ii).

$h_\omega(y', \xi', \tau)$ is a pseudo-differential symbol positively homogeneous of degree 0 for $|\xi'| > 1/2$, $\operatorname{supp} h_\omega \subset \{ |\xi'| \geq 1/4,\ C_1^{-1} |\xi'| \leq |\tau| \leq C_1 |\xi'| \}$, $\omega_\omega \in C^\infty(\bar{\mathbb{R}}^+)$, $\operatorname{supp} \omega_\omega \subset [0, 2\delta]$ and for $y_1 \leq \delta$ either $\omega_\omega = 1$ or $\omega_\omega = 0$.

One can assume without a loss of generality that all factors except of h_ω are scalar: the general product can be decomposed into the sum of such products.

PROOF. Obviously $\pm F_{x', t \to \xi', \tau} K'_{(\omega)}(x, y, t)$ equals the sum of oscillatory integrals (4.75) in which F_ω are functions of the type (v). We know that $F_{x', t \to \xi', \tau} \tilde{f}^{\pm}_{(\omega)}$ are the sums of oscillatory integrals

(4.76) $$\pm vrai \int_{-\infty}^{\infty} F'_\omega(y', \eta_1, \xi', \tau \mp i0) \cdot$$

$$\exp i(-y_1 \eta_1 - \langle y', \xi' \rangle) \, d\eta_1 \, h_\omega(y', \xi', \tau) \omega_\omega(y_1)$$

where $F'_\omega(y', \eta_1, \xi', \tau)$ are the products of factors of the types (ii), (iv). In what follows we shall consider the oscillatory integrals (4.76) in which F'_ω are the products of factors of the types (ii), (iii), (iv).

(4.59) - (4.61)' imply that operator $\bar{G}^{\pm\prime}$ transforms an oscillatory integral (4.76) into the oscillatory integral (4.75) with

$$F(y', \xi_1, \eta_1, \xi', \tau) = (-\tau^2 + a(\xi_1, \xi'))^{-1} \cdot$$

$$(-i\xi_1 + b_1^*(\xi')) S'^{-1}(y', \xi', \tau) F'(y', \eta_1, \xi', \tau)$$

provided $B = i\tau D_1 + B_1 \tau$ and with $F(y', \xi_1, \eta_1, \xi', \tau) = (-\tau^2 + a(\xi_1, \xi_2'))^{-1} S^{-1}(y', \xi', \tau) F'(y', \eta_1, \xi', \tau)$

provided $B=\tau$; hence the new factors of types (i), (iii) appear and $\deg F = \deg F' - 1 - \mu$ where $\deg F$ denotes the degree of positive homogeneity.

Inversely, operator τ_x transforms an oscillatory integral (4.75) into the oscillatory integral (4.76) with

$$F'(y', \eta_1, \xi', \tau) = v.p.\, i \int_{-\infty}^{\infty} F(y', \xi_1, \eta_1, \xi', \tau)\, d\xi_1 \;; \tag{4.77}$$

obviously $\deg F = \deg F' + 1$.

If F is the product of factors of the types (i) - (iv) then F' is the product of factors of the types (ii) - (iv).

Let now F be the product of factors of the types (i), (v). Without a loss of generality we can assume that factor of the type (i) equals $g^{-n}(y', \xi_1, \xi', \tau)\, \xi_1^s$ with $s < 2Dn$ since all factors of the type (i) which have appeared and will appear can be decomposed into sums of such ratios. Thus

$$F(y', \xi_1, \eta_1, \xi', \tau) = g^{-n}(y', \xi_1, \xi', \tau)\, \xi_1^s\, F''(y', \xi_1, \eta_1, \xi')$$

where F'' is a factor of the type (v). By Malgrange division theorem the right side can be decomposed into the sum

$$\sum_{k=0}^{2n-1} g(y', \xi_1, \xi', \tau)^{-n}\, \xi_1^k\, F_k(y', \eta_1, \xi', \tau) + F'''(y', \xi_1, \eta_1, \xi', \tau)$$

where F_k are factors of the type (iv), F''' is the symbol of a singular Green operator on $X \times \mathbb{R}$. Substituting this decomposition in the right side of (4.77) we obtain that F' is the sum of products of factors of the types (ii).

(4.64) implies that operator $\overline{G}^{\pm}$ transforms an oscillatory integral (4.75) into the oscillatory integral (4.75) again and $\deg(F_{trans.}) = \deg(F_{orig.}) - 2$. Moreover, if $F_{orig.}$ contains factors of the types (i) - (iv) then $F_{trans.}$ contains factors of these types too; if $F_{orig.}$ contains factors of the types (i), (v) then repeating the preceding arguments one can show that

$$F_{trans.} - (-\tau^2 + a(\xi_1, \xi'))^{-1} F_{orig.}$$

is the sum of products containing factors (i), (v) too.

The Fourier transform $F_{x', t \to \xi', \tau}$ maps (pseudo) differential operators $c(y',D'_x)$ into operators of multiplication by $c(y',\xi')$ and operators of multiplication by $(x_j - y_j)$ into operators $(i\frac{\partial}{\partial \xi_j} - y_j)$ $(j = 2,\dots,d)$; all these operators transform oscillatory integrals (4.75), (4.76) into oscillatory integrals of the same type. Operator D_{x_1} transforms oscillatory integrals (4.75) into oscillatory integrals (4,75) again because all the factors depending on ξ_1 have the transmission property. Since $x_1 e^{ix_1\xi_1} = -i\frac{\partial}{\partial \xi_1} e^{ix_1\xi_1}$ then integrating by parts we obtain that operator of multiplication by x_1 transforms oscillatory integrals (4.75) into oscillatory integrals of the same type. Moreover every considered operator changes the degree of positive homogeneity in agreement with its assigned weight.

Proposition 4.5 has been proved.

In particular

$$F_0'(y',\xi_1,\eta_1,\xi',\tau) = -(2\pi)^{-d-1}(-\tau^2 + a(y',\xi_1,\xi'))^{-1} b^*(y',\xi_1,\xi')\cdot \tag{4.78}$$

$$S'^{-1}(y',\xi',\tau)\,b(y',\eta_1,\xi')(-\tau^2 + a(y',\eta_1,\xi'))^{-1}\,\Pi(y',\eta_1,\xi') +$$

$$(2\pi)^{-d}\{Id - (-\tau^2 + a(y',\xi_1,\xi'))^{-1} b^*(y',\xi_1,\xi') S'^{-1}(y,\xi',\tau)\}(-\tau^2 + a(y',\xi_1,\xi'))^{-1}\Pi_0'(y',\xi_1,\eta_1,\xi')$$

provided $B = i\tau D_1 + B_1\tau$; this formula with $b(y',\xi_1,\xi') = i\xi_1 + b_1(y',\xi')$, $S'(y',\xi',\tau)$ replaced by Id, $S(y',\xi',\tau)$ respectively is valid for $B=\tau$ too.

Moreover $h_0(y',\xi',\tau) = q_0(y',-\xi',\tau)$ for $|\xi'|>1/2$, $\omega_0 = 1$ for $y_1 < \delta$.

Let $v_{(w)} = v^+_{(w)} + v^-_{(w)}$. We apply the inverse Fourier transform $F^{-1}_{\xi'\to x'}$ and set $x=y$; proposition 4.5 implies that

$$F_{t\to\tau}\Gamma v_{(w)} = \omega_w(y_1) Q_w(y,\tau)^{*)} \tag{4.79}$$

where

$$Q_w = v\tau a i\int_{-\infty}^{\infty} \mathcal{F}''_w(y',\eta_1,\tau) e^{iy_1\eta_1} d\eta_1 , \tag{4.80}$$

---

*) We omit the sign and the index of summation.

$$\mathcal{F}''_w = \int_{\mathbb{R}^d_\xi} \{ F_w(y', \xi_1, \xi_1 - \eta_1, \xi', \tau - i0) - \tag{4.81}$$
$$F_w(y', \xi_1 - \eta_1, \xi', \tau + i0)\} h_w(y', \xi', \tau) d\xi .$$

Note that for every (η_1, τ) the integrand in (4.81) has a compact support and hence integral (4.81) converges and defines the function $\mathcal{F}''_w \in C^\infty(Y \times \mathbb{R}_{\eta_1}, \mathcal{S}'(\mathbb{R}_\tau))$ which is positively homogeneous of degree $(d-w-3)$ with respect to (η_1, τ) for $|\tau| > C_2$ and vanishes for $|\tau| < C_2^{-1}$.

PROPOSITION 4.6 (i). If F_w contains no factors of the type (ii) then $\mathcal{F}''_w \in C^\infty(Y \times \mathbb{R}^2_{\eta_1, \tau})$.
(ii) Let F_w contain a factor of the type (ii), $(y^*, \eta^{*\prime}, \tau^*) \in (T^*Y \setminus 0) \times (\mathbb{R} \setminus 0)$. Assume that there exists $\mu \in \mathbb{R}$ such that inequalities (0.10) hold at every point $\rho \in j^{-1}(y^*, -\eta^*/\tau^*) \cap \{ g(\rho, 1) = 0 \}$. Then $\mathcal{F}''_w \in C^\infty(Y \times \mathbb{R}^2_{\eta_1, \tau})$ provided cone supp q_y is contained in the small enough conic neighbourhood of $(y^*, \eta^{*\prime}, \tau^*)$.

PROOF. (i) We introduce the spherical coordinate system $(r, \tau) \in \mathbb{R}^+ \times S^{d-2}$ in $\mathbb{R}^{d-1}_{\xi'} \setminus 0$.

Note that factors of the types (i), (iii) are meromorphic with respect to r, ξ_1, τ and that the replacement of $\tau \in \mathbb{R} \setminus 0$ by $\tau(1 \mp i0)$ in these factors is equivalent to the replacement of r, ξ_1 by $r(1 \pm i0)$, $\xi_1(1 \pm i0)$ respectively in the same factors; hence $F_w(\cdot, \tau \mp i0) \in C^\infty(Y \times \mathbb{R}_{\eta_1} \times \mathbb{R}_\tau, \mathcal{S}'(\mathbb{R}^d_\xi))$. Hence $\mathcal{F}''_w \in C^\infty(Y \times \mathbb{R}^2_{\eta_1, \tau})$.
(ii) Let now F_w contain factors of the types (i) - (iv) and cone supp is contained in the small enough conic neighbourhood of $(y^*, \eta^{*\prime}, \tau^*) \in (T^*Y \setminus 0) \times (\mathbb{R} \setminus 0)$; then cone supp $h_w(y, -\xi', \tau)$ is contained in this neighbourhood too. Factors of the types (i), (ii) are holomorphic outside of the zeros of $g(y', \xi_1, \xi', \tau)$ and $g(y', \xi_1 - \eta_1, \xi', \tau)$ respectively; recall that we replaced η_1 by $\xi_1 - \eta_1$. Hence one can prove by the preceding arguments that sing supp $\mathcal{F}''_w$ is contained in the small enough conic neighbourhood of the set $\mathcal{M} = \{ (y^*, \lambda_k - \lambda_j, \tau^*), \quad j, k = 1, \ldots, s \}$ where λ_k are the roots of the polynomial $g(y^*, \xi_1, -\eta^{*\prime})$.

Therefore $F''_w \subset \{ |\eta_1| \leq C_3 |\tau| \}$ provided C_3 is large enough.

We state that $F''_w \subset \{ |\tau| \leq C_4 |\eta_1| \}$. Really, if $|\tau| \geq C_4 |\eta_1|$ then the replacement of $\tau \in \mathbb{R} \setminus 0$ by $\tau(1 \mp i0)$ in factors of the types (i) - (iii) is equivalent to the replacement of r, ξ_1

by $\tau(1\pm i0)$, $\xi_1(1\pm i0)$ respectively in these factors; we do not need to replace η_1 by $\eta_1(1\pm i0)$ because $|\eta_1|/|\tau|$ is small enough; hence $F_\omega(\cdot,\tau\mp i0)\in C^\infty(Y\times \mathbb{R}^2_{\eta_1,\tau}\cap\{|\eta_1|<C_4^{-1}|\tau|\}, \mathcal{S}'(\mathbb{R}^d_\xi))$ and hence $\mathcal{F}''_\omega\in C^\infty(Y\times\mathbb{R}^2_{\eta_1,\tau}\cap\{|\eta_1|<C_4^{-1}|\tau|\})$.

Thus (4.82) $\quad \operatorname{sing\,supp}\mathcal{F}''_\omega\subset\{C_4^{-1}|\tau|\leqslant|\eta_1|\leqslant C_4|\tau|\}$.

(iii) Let now conditions of the second statement be fulfilled, then matrices

$$\left(\xi'\frac{\partial}{\partial\xi'}+\mu\frac{\partial}{\partial\xi_1}\right)a(y^*,\lambda_j,-\eta^{*\prime})$$

are positive definite at $\operatorname{Ker}(-\tau^{*2}+a(y^*,\lambda_j,-\eta^{*\prime}))$, $\quad j=1,\ldots,s$. Hence the replacement of $\tau\in\mathbb{R}\setminus 0$ by $\tau(1\mp i0)$ in factors of the types (i) - (iii) is equivalent to the replacement of τ,ξ_1 by $\tau(1\pm i0)$, $\xi_1\pm i\mu 0$ respectively in these factors; hence $F_\omega(\cdot,\tau\mp i0)\in C^\infty(Y\times\mathbb{R}^2_{\eta_1,\tau},\mathcal{S}'(\mathbb{R}^d_\xi))$ in the neighbourhood of $(y^*,-\eta^{*\prime},\tau^*)\times\mathbb{R}^2_{\xi_1,\eta_1}$ and hence $\mathcal{F}''_\omega\in C^\infty(Y\times\mathbb{R}^2_{\eta_1,\tau})$.

Consider now $\mathcal{F}''_\omega(y',\eta_1,\tau)$ for $|\eta_1|>C_4|\tau|$. We know that $\mathcal{F}''_\omega$ is infinitely smooth in this zone. Since factors of the types (i), (ii), (iv), (v) can be decomposed into asymptotic series in negative powers of ξ_1, $\xi_1-\eta_1$ as $\xi_1\to\infty$, $\xi_1-\eta_1\to\infty$ (together or separately) then one can show that $\mathcal{F}''_\omega$ can be decomposed into asymptotic series in negative power of η_1 as $\eta_1\to\infty$; hence for $|\tau|>C_1$, $|\eta_1|>C_4|\tau|$ $\mathcal{F}''_\omega(y',\eta_1,\tau)\,\tau^{-d+\omega+3}$ is a symbol with the transmission property positively homogeneous of degree 0 with respect to (η_1,τ).

Assume first that $\mathcal{F}''_\omega\in C^\infty(Y\times\mathbb{R}^2_{\eta_1,\tau})$.

Then $W_\omega=F^{-1}_{\tau\to t}Q_{\omega,}(y,\tau)$ is the Schwartz kernel of an operator $H\in OP^{-d-\omega-3}(\overline{\mathbb{R}}^+\times\mathbb{R})$; $y'\in Y$ is considered as a parameter. Then proposition 1.7 implies that

$$(4.83)\qquad y_1^j t^k W_\omega\in C^\infty(Y,\mathcal{H}_{t\,;\,y_1}^{\omega+j+k+2-d-\varepsilon;\infty}(\overline{\mathbb{R}}^+\times\mathbb{R},\hom(E))_{loc})$$

$$\forall j,k\in\mathbb{Z}^+,\ \varepsilon>0.$$

Moreover

$$(4.84)\qquad F_{t\to\tau}W_\omega=|\tau|^{d-\omega-2}\,Q^\pm_\omega(y',|\tau|y_1)\quad\text{for}\quad \pm\tau>C_1$$

where

$$Q^{\pm}_{w}(y',s) = \mathrm{vrai}\int_{-\infty}^{\infty} \mathcal{F}^{\prime\prime\,hom}_{w}(y',\eta_1,\pm 1)e^{i\eta_1 s}\,d\eta_1 , \tag{4.85}$$

$\mathcal{F}^{\prime\prime\,hom}_{w}$ is a positively homogeneous function which coincides with $\mathcal{F}^{\prime\prime}_{w}$ for large τ.

Then

$$D^{\alpha'}_{y'} D^{k}_{s} Q^{\pm}_{w} = O(s^{-\infty}) \quad \text{as} \quad s \to +\infty \tag{4.86}$$

uniformly with respect to $y' \in Y$ $\forall \alpha', k$ i.e. $F_{t\to\tau} W_w$ is the term of boundary - layer type.

Thus if either $\Pi \in OGS^{\circ}(X,E)$ or $cone\ supp\, q_y$ is contained in the small enough conic neighbourhood of $(y^*, \eta^{*\prime}, \tau^*) \in (T^*Y\setminus 0)\times(\mathbb{R}\setminus 0)$ and there exists $\mu \in \mathbb{R}$ such that for every point $(y^*, \lambda_j, -\eta^{*\prime}, \tau^*)$, $j=1,\dots,s$ inequalities (0.10) hold then we derived the formal asymptotics with respect to smoothness

$$\Gamma q_y \chi_0 u_1 \sim \sum_{w=0}^{\infty} W_w \tag{4.87}$$

and formal asymptotics

$$F_{t\to\tau}\, \Gamma q_y \chi_0 u_1 \sim \tag{4.88}$$

$$\sum_{w=0}^{\infty} |\tau|^{d-w-2} Q^{\pm}_{w}(y', |\tau| y_1) \quad \text{as} \quad \tau \to \pm\infty .$$

Consider now the general case; we shall derive the formal asymptotics for $\tilde{\Gamma} q_y \chi_0 u_1$. Recall that $\tilde{\Gamma} v = \int_0^{3\delta} \mathrm{tr}\, \Gamma v\, dy_1$.

Note that $\mathcal{F}^{\prime\prime}_{w} \in C^{\infty}(Y\times\mathbb{R}_{\tau}, \mathcal{S}'(\mathbb{R}_{\eta_1}))$ because the replacement of $\tau \in \mathbb{R}\setminus 0$ by $\tau(1\mp i0)$ in factors of the types (i) - (iii) is equivalent to the replacement of τ, ξ_1, η_1 by $\tau(1\pm i0)$, $\xi_1(1\pm i0)$, $\eta_1(1\pm i0)$ respectively in these factors.

Let $\psi \in C^{\infty}(\bar{\mathbb{R}}^+)$, $\psi = 0$ for $y_1 < \delta_1$, $\psi = c = const$ for $y_1 > \delta$, $0 < \delta_1 < \delta$; then $\psi = y_1^N \psi_N$ for arbitrary N and

$$\int_0^{\infty} F_{t\to\tau} W_w \psi\, dy_1 =$$

$$\int_{-\infty}^{\infty} \mathcal{F}''_{w}(y',\eta_1,\tau)\left(i\frac{\partial}{\partial\eta_1}\right)^N \hat{\psi}_N(\eta_1)d\eta_1 =$$

$$\int_{-\infty}^{\infty} \hat{\psi}_N(\eta_1)\left(-i\frac{\partial}{\partial\eta_1}\right)^N \mathcal{F}''_{w}(y',\eta_1,\tau)d\eta_1 .$$

Note that sing supp $\hat{\psi}_N=\{0\}$ provided $C\neq 0$ and $\hat{\psi}_N$ and all its derivatives decrease more quickly than every negative power of η_1 as $\eta_1\to\infty$; (4.82) implies that the last integral converges to a function belonging to $C^{\infty}(Y\times \mathbb{R}_{\tau})$; one can decompose this integral into the sum of integrals along zones $\{|\eta_1|<C_4^{-1}|\tau|\}$, $\{1/2\, C_4^{-1}|\tau|< |\eta_1|<2C_4|\tau|\}$, $\{C_4|\tau|<|\eta_1|\}$, apply (4.82) and the positive homogeneity of degree $(d-3-w-N)$ of the function $\left(\frac{\partial}{\partial\eta_1}\right)^N \mathcal{F}''^{\,hom}_{w}$ and obtain that this integral and all its derivatives with respect to (y',τ) decrease more quickly than τ^{d-N} as $\tau\to\infty$; recall that N is arbitrary.

Hence modulo functions which together with all their derivatives decrease more quickly than every negative power of τ as $\tau\to\infty$

$$F_{t\to\tau}\int_0^{\infty} W_w\,\omega_w\,dy_1 \sim 0$$

provided $\omega_w=0$ for $y_1<\delta$,

$$F_{t\to\tau}\int_0^{\infty} W_w\,\omega_w\,dy_1 \sim -\int_{-\infty}^{\infty} \mathcal{F}''^{\,hom}_{w}(y',\eta_1,\tau)(\eta_1+i0)^{-1}d\eta_1$$

provided $\omega_w=1$ for $y_1<\delta$.

The last integral converges to a function which is positively homogeneous of degree $d-w-3$ with respect to τ ; (4.82) implies that this function is infinitely smooth for $\tau\neq 0$ and hence equals to $c^{\pm}_{w}(y')|\tau|^{d-w-3}$ for $\tau\in\mathbb{R}^{\pm}$ where $c^{\pm}_{w}\in C^{\infty}(Y)$.

Thus we derived the formal asymptotics with respect to smoothness

$$\tilde{\Gamma}q_y\chi_0 u_1 \sim \sum_{w=0}^{\infty}\left(c^{+}_{w}(y')\Phi^{+}_{d-w-3}(t)+c^{-}_{w}(y')\Phi^{-}_{d-w-3}(t)\right) \tag{4.89}$$

where $\Phi_{\kappa}^{\pm}(t) = F^{-1}_{\tau \to t}\, \tau^{\kappa}_{\pm}$ and the formal asymptotics

(4.90) $$F_{t\to\tau}\tilde{\Gamma} q_y \chi_0 u_1 \sim \sum_{w=0}^{\infty} c_w^{\pm}(y')|\tau|^{d-w-3} \quad \text{as} \quad \tau \to \pm\infty$$

where $c_w^{\pm} \in C^{\infty}(Y)$.

(4.79) - (4.81) imply the formal asymptotics with respect to smoothness

(4.91) $$\tau\Gamma q_y \chi_0 u_1 \sim \sum_{w=0}^{\infty} (d_w^{+}(y')\Phi^{+}_{d-w-2}(t) + d_w^{-}(y')\Phi^{-}_{d-w-2}(t))$$

and the formal asymptotics

(4.92) $$F_{t\to\tau}\tau\Gamma q_y \chi_0 u_1 \sim \sum_{w=0}^{\infty} d_w^{\pm}(y')\,|\tau|^{d-w-2} \quad \text{as} \quad \tau \to \pm\infty$$

where $d_w^{\pm} \in C^{\infty}(Y, hom(E))$.

Note that $\Phi_{\kappa}^{\pm}(t)$ satisfy (4.41).

To justify asymptotics (4.87) - (4.92) one can repeat the arguments which were used to justify asymptotics (4.40), (4.42); one needs to replace theorem 3.1 by theorems 3.4 - 3.6 and the commutation relation (4.43) by the following commutation relations:

(4.93) $$(x_j - y_j)\overset{\pm}{G} v \equiv \overset{\pm}{G}(x_j - y_j)v + \overset{\pm}{G}[P, x_j]\overset{\pm}{G} v + \overset{\pm\prime}{G}[B, x_j]\overset{\pm}{G} v \quad ,$$

(4.93)' $$(x_j - y_j)\overset{\pm\prime}{G} w \equiv \overset{\pm\prime}{G}(x_j - y_j)w + \overset{\pm}{G}[P, x_j]\overset{\pm\prime}{G} w + \overset{\pm\prime}{G}[B, x_j]\overset{\pm}{G} w,$$
$$j = 2, \dots, d$$

(4.93)'' $$x_1 \overset{\pm}{G} v \equiv \overset{\pm}{G} x_1 v + \overset{\pm}{G}[P, x_1]\overset{\pm}{G} v + \overset{\pm\prime}{G}[B, x_1]\overset{\pm}{G} v \quad ,$$

(4.93)''' $$x_1 \overset{\pm\prime}{G} w \equiv + \overset{\pm}{G}[P, x_1]\overset{\pm\prime}{G} w + \overset{\pm\prime}{G}[B, x_1]\overset{\pm\prime}{G} w \quad (\mathrm{mod}\ C^{\infty})$$

provided $v \equiv 0$, $w \equiv 0 \pmod{C^\infty}$ for $\pm t < 0$ and the similar commutation relations for $\bar{G}^{\pm}$, $\bar{G}^{\pm\prime}$; to derive these commutation relations one needs to repeat the arguments which were used to derive the commutation relations (4.43) and replace (4.20) by (4.58)''. Finally the cutting operator $q \in OPS^{0}(X \times \mathbb{R})$ must be replaced by the cutting operator $q' \in OPS^{0\prime}(X \times \mathbb{R})$. Certainly, this operator does not equalize the smoothnesses with respect to x_1, y_1 to the (total) smoothness with respect to (x', y', t); but we have the large smoothness with respect to y_1 right from the start and the smoothness with respect to x_1 can be proved by (2.31).

Thus the following statements have been proved:

THEOREM 4.6. Let $(y^*, \eta^{*\prime}, \tau^*) \in (T^*Y\setminus 0) \times (\mathbb{R}\setminus 0)$; assume that either at this point the conditions of theorem 3.4 are fulfilled or $\Pi \in OGS^{0}(X, E)$. Then

$$\Gamma q_y \chi_0 u_1 - \sum_{w=0}^{N-1} W_w \in C^\infty(Y, \mathcal{H}^{2-d+N-\varepsilon;\infty}_{t\,;\,y_1}([0,\delta) \times [-t_3, t_3], hom(E))) \cap$$

$$C^\infty(X \times [-t_3, t] \setminus Y \times 0) \quad \forall N \in \mathbb{Z}^+,\ \varepsilon > 0$$

provided $0 < t_3$ is small enough, *cone supp* q is contained in the small enough conic neighbourhood of $(y^*, \eta^{*\prime}, \tau^*)$, $\chi \in C_0^\infty(\mathbb{R})$.

These inclusions remain true if one applies the operator $t^j y_1^k D_1^l D_t^n$ to their left parts provided $j+k \geq l+n$.

THEOREM 4.7. Let $(y^*, \eta^{*\prime}, \tau^*) \in (T^*Y \setminus 0) \times (\mathbb{R} \setminus 0)$. Then

$$\tilde{\Gamma} q_y \chi_0 u_1 - \sum_{w=0}^{N-1} (c_w^+(y')\Phi^+_{d-w-3}(t) + c_w^-(y')\Phi^-_{d-w-3}(t) \in$$

$$C^\infty(Y, H^{-d+N+5/2-\varepsilon}[-t_3, t_3]),$$

$$\tau\Gamma q_y \chi_0 u_1 - \sum_{w=0}^{N-1} (d_w^+(y')\Phi^+_{d-w-2}(t) + d_w^-(y')\Phi^-_{d-w-2}(t)) \in$$

$$C^\infty(Y, H^{-d+N+3/2-\varepsilon}[-t_3, t_3], hom(E)) \qquad \forall N \in \mathbb{Z}^+, \quad \varepsilon > 0$$

provided $0 < t_3$ is small enough, cone supp q is contained in the small enough conic neighbourhood of $(y^*, \eta^{*\prime}, \tau^*)$, $\chi_0 \in C_0^\infty(\mathbb{R})$.

These inclusions remain true if one applies the operator with an arbitrary $n \in \mathbb{Z}^+$ $t^n D_t^n$ to their left parts.

COROLLARY 4.8. If the conditions of theorem 4.6 (4.7) are fulfilled, cone supp $\chi_0 \subset [-2t_4, 2t_4]$, $\chi_0 = 1$ on $[-t_4, t_4]$, $0 < t_4$ is small enough then asymptotics (4.88) ((4.90), (4.92) respectively) hold.

4.6. In this and in the following sections we shall derive asymptotics in the normal rays zone $\{C|\xi'| \leq |\tau|, \ C|\eta'| \leq |\tau|\}$.
We shall use notations if section 3.5.

First we derive asymptotics for $f_{\mu\nu} = \tau_x \tau_y D_{x_1}^\mu D_{y_1}^\nu u_0$ in this zone; $\mu, \nu = 0,1$. We use "asymptotics" for u_0 derived in section 4.2:

$$u_0 \equiv \sum_{\omega=0}^{M-1} u_{(\omega)} \quad (mod \sum_{j=0}^{M_1} t^j H^{M-m-j}_{x,y,t})$$

where

$$u_{(\omega)} = F^{-1}_{\xi,\tau \to x,t} \mathcal{F}_\omega(y,\xi,\tau) e^{-i\langle y,\xi\rangle}(1-\zeta(\tau)),$$

$$\mathcal{F}_\omega(y,\xi,\tau) = F_\omega(y,\xi,\tau-i0) - F_\omega(y,\xi,\tau+i0),$$

$\zeta \in C_0^\infty(\mathbb{R})$, $\zeta = 0$ for $|\tau| > \frac{1}{2}$, $\zeta = 1$ for $|\tau| < \frac{1}{4}$ and the smoothness with respect to x, y, t is total.

Recall that $F_\omega(y,\xi,\tau)$ is positively homogeneous of degree $(-\omega-2)$ with respect to (ξ,τ) and is the product of factors of the following two types:

(i) $F(y',\xi,\tau)$ is homogeneous with respect to (ξ,τ) meromorphic with respect (ξ_1,τ) ; its poles coincide with the zeros of $g(y,\xi,\tau)$.

(ii) $F(y',\xi)$ is the pseudo-differential symbol with the transmission property, positively homogeneous with respect to ξ.

Hence

$$f_{\mu\nu} \equiv \sum_{\omega=0}^{M-1} f_{(\omega)\mu\nu} \ (mod \sum_{j=0}^{M_1} t^j H^{M-m-j;\infty}_{t;x',y'})$$

where

$$f_{(\omega)\mu\nu} =$$

$$F^{-1}_{\xi',\tau\to x',t}\, \mathcal{F}'_{\omega\mu\nu}(y',\xi',\tau)e^{-i\langle y',\xi'\rangle}(1-\zeta(\tau)) \ ,$$

$$\mathcal{F}'_{\omega\mu\nu}(y',\xi',\tau) = \int_{-\infty}^{\infty} \xi_1^{\mu+\nu}(-1)^{\nu}\mathcal{F}_{\omega}(y',\xi_1,\xi',\tau)\,d\xi_1 \ .$$

Then $\mathcal{F}'_{\omega\mu\nu}$ is positively homogeneous of degree $(-\omega-1+\mu+\nu)$ with respect to (ξ',τ). Note that the replacement of $\tau\in\mathbb{R}\setminus 0$ by $\tau(1\mp i0)$ in factors of the type (i) is equivalent to the replacement of ξ_1 by $\xi_1(1\pm i0)$ provided $|\xi'|/|\tau|$ is small; hence $\mathcal{F}'_{\omega\mu\nu}\in C^{\infty}$ for $|\xi'|<C^{-1}|\tau|$.

Hence we have the formal asymptotics

(4.94) $$q'_y\chi f_{\mu\nu}\sim\sum_{\omega=0}^{\infty} f_{(\omega)\mu\nu}$$

where

(4.95) $$f_{(\omega)\mu\nu}=\int_{\mathbb{R}^{d-1}_{\xi'}\times\mathbb{R}_{\tau}} \exp i(\langle x'-y',\xi'\rangle+t\tau)\,\mathcal{F}'''_{\omega\mu\nu}(y',\xi',\tau)(1-\zeta(\tau))\,d\xi'd\tau,$$

$$q'_y\in OPS^{0}(Y\times\mathbb{R}),\ \mathrm{cone\ supp}\, q\subset\{|\eta'|<C^{-1}|\tau|\},\ \chi\in C_0^{\infty}(\mathbb{R}),\ \chi=1$$

for $\chi\subset[-2t_5,\ 2t_5]$, $t_5>0$ is small enough, $\mathcal{F}'''_{\omega\mu\nu}(y',\xi',\tau)$ are pseudo-differential symbols, positively ghomogeneous of degree $(-\omega-1+\mu+\nu)$ with respect to (ξ',τ), cone supp $\mathcal{F}'''_{\omega\mu\nu}\subset\{|\xi'|<C^{-1}|\tau|\}$.

One can justify asymptotics (4.94) by means of inclusions (3.54).

REMARK. Asymptotics (4.94) - (4.95) show that for small t $q'_y\chi f_{\mu\nu}$ are Schwartz kernels of classical pseudo-differential operators on $Y\times\mathbb{R}$ which commute with D_t. Such functions will appear in future and occasionally the corresponding operators will have a simple "physical" sense.

Now we can apply lemma 3.14 and dual lemma, lemma 3.12 and (3.49) and derive asymptotics for $\tau_x\tau_y V$, where V equals to $U^{I,II}$, $U^{II,I}$, $U^{',I}$, $U^{',II}$, $U^{I,''}$, $U^{II,''}$, $U^{','''}$ (recal that other components of U belong to C^{∞} by (3.49) and lemma 3.12):

(i) If B is a differential operator then $\beta^{\cdot}_{\cdot}$ are classical pseudo-differential operators and

(4.96) $$\tau_x\tau_y q'_y\chi V\equiv V_0 =$$

$$\int_{\mathbb{R}^{d-1}_{\xi'} \times \mathbb{R}_\tau} \exp i(\langle x'-y', \xi' \rangle + t\tau) h(y', \xi', \tau)\, d\xi' d\tau$$

where h is a classical pseudo-differential symbol of degree (-1);
(ii) In the general case

$$r_x r_y q'_y \chi V = V_0 + V_1 \tag{4.97}$$

where V_0 is expressed by oscillatory integral (4.96) with $h = \bar{h} + \hat{h}$, $\bar{h}$ is a classical pseudo-differential symbol of degree (-1); $\hat{h}$ satisfies the estimates

$$| D^\alpha_{y'} D^\beta_{\xi'} D^\eta_\tau \hat{h}(y', \xi', \tau)| \leq C_{\alpha\beta n} (1 + |\tau| + |\xi'|)^{-n-2} (1 + |\xi'|)^{1-|\beta|}$$

$$\forall \alpha, \beta, n ;$$

$\hat{h}$ can be decomposed into the asymptotic series $\hat{h} \sim \sum_{w=0}^{\infty} \hat{h}_w$ where $\hat{h}_w$ are positively homogeneous of degree $(-w-1)$ and satisfy the estimates

$$| D^\alpha_{y'} D^\beta_{\xi'} D^n_\tau \hat{h}_w | \leq C_{\alpha\beta n w} (|\tau| + |\xi'|)^{-n-2} |\xi'|^{1-|\beta|-w}$$

$$\forall \alpha, \beta, n, w ,$$

in the sense that for $|\xi'| \geq 1$

$$| D^\alpha_{y'} D^\beta_{\xi'} D^n_\tau (\hat{h} - \sum_{w=0}^{N-1} \hat{h}_w)| \leq C_{\alpha\beta n N} (|\tau| + |\xi'|)^{-n-2} |\xi'|^{1-|\beta|-N}$$

$$\forall \alpha, \beta, n, N ;$$

V_1 is infinitely smooth with respect to x' and y' and satisfies the estimates

$$| D^\alpha_{x'} D^\beta_{y'} F_{t \to \tau} V_1 | \leq C_{\alpha\beta} (1 + |\tau|)^{-2} \quad \forall \alpha, \beta . \tag{4.98}$$

Then lemmas 3.7, 3.12 imply the following statement:

THEOREM 4.9. Let $q'_y \in OPS^{0'}(X \times \mathbb{R})$, $\operatorname{cone\,supp} q'_y$ be contained in the small enough conic neighbourhood of $Y \times 0 \times \tau^*$, $\chi \in C_0^\infty(\mathbb{R})$, $\operatorname{supp} \chi \subset (-2t_5, 2t_5), \chi = 1$ for $|t| < t_5$, $0 < t_5$ be small enough. Then

(i) If B is a differential operator then the following complete asymptotics hold:

$$\tilde{\Gamma} q'_y \chi u_1 \sim \sum_{\omega=0}^{\infty} (c^+_\omega(y') \Phi^+_{d-\omega-3}(t) + c^-_\omega(y') \Phi^-_{d-\omega-3}(t)), \tag{4.99}$$

$$F_{t\to\tau} \tilde{\Gamma} q'_y \chi u_1 \sim \sum_{\omega=0}^{\infty} c^\pm_\omega(y') |\tau|^{d-\omega-3} \quad \text{as} \quad \tau \to \pm\infty; \tag{4.100}$$

(ii) If B is a pseudo-differential operator then the following incomplete asymptotics holds:

$$\tilde{\Gamma} q'_y \chi u_1 \equiv \sum_{\omega=0}^{d-1} (c^+_\omega(y') \Phi^+_{d-\omega-3}(t) + c^-_\omega(y') \Phi^-_{d-\omega-3}(t)) \tag{4.99'}$$

$$(\operatorname{mod} C^\infty(Y, H^{5/2-\varepsilon}(\mathbb{R})) \quad \forall \varepsilon > 0)$$

and

$$F_{t\to\tau} \tilde{\Gamma} q'_y \chi u_1 = \sum_{\omega=0}^{d-1} c^\pm_\omega(y') |\tau|^{d-\omega-3} + O(|\tau|^{-3} \ln|\tau|) \quad \text{as} \quad \tau \to \pm\infty; \tag{4.100'}$$

here $c^\pm_\omega \in C^\infty(Y)$ and (4.99), (4.99)' remain true if one applies an operator $t^n D^n_t D^\alpha_{y'}$ to their left parts and, finally, (4.100), (4.100)' remain true if one applies an operator $\tau^n D^n_\tau D^\alpha_{y'}$ to their left parts.

We have prooved the following statement too:

THEOREM 4.10. Let the conditions of theorem 4.9 be fulfilled. Then

(i) If B is a differential operator then the following complete asymptotics hold:

$$\text{(4.101)} \qquad \tau\Gamma q'_y \chi u_1 \sim \sum_{\omega=0}^{\infty} (d_\omega^+(y')\Phi^+_{d-\omega-2}(t) + d_\omega^-(y')\Phi^-_{d-\omega-2}(t))$$

and

$$\text{(4.102)} \qquad F_{t\to\tau}\, \tau\Gamma q'_y \chi u_1 \sim \sum_{\omega=0}^{\infty} d_\omega^{\pm}(y')|\tau|^{d-\omega-2} \quad \text{as} \quad \tau \to \pm\infty ;$$

(ii) If B is a pseudo-differential operator then the following incomplete asymptotics hold:

$$\text{(4.101)}' \qquad \tau\Gamma q'_y \chi u_1 \equiv \sum_{\omega=0}^{d-1} (d_\omega^+(y')\Phi^+_{d-\omega-2}(t) + d_\omega^-(y')\Phi^-_{d-\omega-2}(t))$$

$$(\mathrm{mod}\ C^\infty(Y, H^{3/2-\varepsilon}(\mathbb{R}, \mathrm{hom}(E))) \quad \forall \varepsilon > 0)$$

and

$$\text{(4.102)}' \qquad F_{t\to\tau}\, \tau\Gamma q'_y \chi u_1 = \sum_{\omega=0}^{d-1} d_\omega^{\pm}(y')|\tau|^{d-\omega-2} + O(|\tau|^{-2} \ln|\tau|)$$

$$\text{as} \quad \tau \to \pm\infty ;$$

here $d_\omega^{\pm} \in C^\infty(Y, \mathrm{hom}(E))$ and (4.101), (4.101)' remain true if one applies an operator $t^n D_t^n D_{y'}^\alpha$ to their left parts and, finally, (4.102), (4.102)' remain true if one applies an operator $\tau^n D_\tau^n D_{y'}^\alpha$ to their left parts.

Let now V be an arbitrary component of U different from $U^{I,II}$ and $U^{II,I}$; then ΓV is expressed by the formula

$$\Gamma V \equiv \Gamma' h \tau_x \tau_y V (\mathrm{mod}\ C^\infty)$$

where $h \in OP^0([0,\delta) \times Y \times Y \times \mathbb{R}, \cdot)$, $(\Gamma' v)(y_1, y', t) = v(y_1, y', y', t)$ (see (3.53)).

Hence asymptotics (4.96) - (4.98) imply the following statement:

THEOREM 4.11. Let q'_y, χ satisfy conditions of theorem 4.8, $\mathcal{L} \in OPS^{0'}(X \times X \times \mathbb{R}, \cdot)$, V be an arbitrary component of U different from $U^{I,II}$ and $U^{II,I}$. Then

(i) If B is a differential operator then the following complete asymptotics hold;

$$\Gamma \mathcal{L} q'_y \chi V \sim \sum_{\omega=0}^{\infty} W_{\omega} \tag{4.103}$$

and

$$F_{t\to\tau} \Gamma \mathcal{L} q'_y \chi V \sim \sum_{\omega=0}^{\infty} |\tau|^{d-\omega-2} Q^{\pm}_{\omega}(y', |\tau| y_1) \tag{4.104}$$

as $\tau \to \pm\infty$;

(ii) If B is a pseudo-differential operator then the following incomplete asymptotics hold:

$$\Gamma \mathcal{L} q'_y \chi V = \sum_{\omega=0}^{d-1} W_{\omega} + \widetilde{W} \tag{4.103'}$$

and

$$F_{t\to\tau} \Gamma \mathcal{L} q'_y \chi V = \sum_{\omega=0}^{d-1} |\tau|^{d-\omega-2} Q^{\pm}_{\omega}(y', |\tau| y_1) + F_{t\to\tau} \widetilde{W} \tag{4.104'}$$

$\tau \to \pm\infty$;

here W_{ω}, Q^{+}_{ω} satisfy (4.83), (4.85), (4.86),

$$y_1^j t^{\kappa} \widetilde{W} \in C^{\infty}(Y, \mathcal{H}^{j+\kappa+2-\varepsilon;\infty}_{t;y_1}(\overline{\mathbb{R}}^+ \times \mathbb{R}, \cdot)) \quad \forall \varepsilon > 0,\ j, k \in \mathbb{Z}^+,$$

$$\| y_1^j D^{\ell}_{y_1} F_{t\to\tau} \widetilde{W}_{\omega} \|_{L_2(\overline{\mathbb{R}}^+)} \leqslant C(1+|\tau|)^{-5/2+\ell-j} \ln(2+|\tau|) ,$$

$$| y_1^j D^{\ell}_{y_1} F_{t\to\tau} \widetilde{W} | \leqslant C(1+|\tau|)^{-2+\ell-j} \ln(2+|\tau|) \quad \forall j,\ \ell \in \mathbb{Z}^+.$$

4.7. Consider now $U^{I,II}$ ($U^{II,I}$ can be considered in the same way). We use the asymptotics (4.96) - (4.98) and systems

(4.105) $$(K^{I} D_{x_1} + L^{I}_{x})\, U^{I,II} \equiv 0 \quad ,$$

(4.105)† $$U^{I,II}(-K^{II\dagger} D_{y_1} + L^{II\dagger}_{y}) \equiv 0$$

modulo functions which are infinitely smooth in the zone $\Omega=\{|t|<t_5,\ |\xi'|+|\eta'|\leq C^{-1}|\tau|,\ x_1+y_1<2\delta\}$.

Thus $U^{I,II}$ is the solution of the double Cauchy problem with the initial data $\tau_x \tau_y U^{I,II}$.

We assume that $\xi_1 = \lambda_j(x, \xi', \tau)$ — all real roots of the polynomial $g(x, \xi_1, \xi', \tau)$ — have constant multiplicities for $|\xi'|\leq C^{-1}|\tau|$.

Then without a loss of generality one can assume that $U^{I,II}=(U_{jk})$ with $j=1,\dots,s$, $k=s+1,\dots,2s$, $\lambda_j>0$, $\lambda_k<0$, $U_{jk} \in \mathcal{D}'(X\times X\times \mathbb{R},\ \mathrm{Hom}(E_j, E_k))$ satisfy equations

(4.106) $$(D_{x_1} + L_{jx})U_{jk} \equiv 0 \quad ,$$

(4.106)† $$U_{jk}(-D_{y_1} + L^{\dagger}_{ky}) \equiv 0$$

$(\mathrm{mod}\ C^{\infty})$ where L_j, L_k are scalar in principal operators:

$$L_{jx} \equiv \lambda_j(x, D'_x, D_t)(\mathrm{mod}\ OPS^{0'}(X\times\mathbb{R}, E_j)),$$

$$L_{ky} \equiv \lambda_k(y, D'_y, D_t)(\mathrm{mod}\ OPS^{0'}(X\times\mathbb{R}, E_k)).$$

Consider the eikonal equations corresponding to (4.106), (4.106)†:

(4.107) $$\varphi_{x_1} = -\lambda_j(x, \varphi_{x'}, \tau)\ ,$$

(4.107)† $$\varphi_{y_1} = \lambda_k(y, -\varphi_{y'}, \tau)$$

with the initial data

(4.108) $$\varphi\big|_{x_1=y_1=0} = \langle x'-y', \xi'\rangle .$$

Let $\varphi = \varphi_{jk}(x, y, \xi', \tau)$ be a solution of double Cauchy problem (4.107), (4.107)†, (4.108); it exists without a question.

Using the standard methods one can formally solve the double Cauchy problem with the initial data expressed by oscillatory integral (4.96):

$$(4.109)\quad \mathcal{U}_{j\kappa} = \int_{\mathbb{R}^{d-1}_{\xi'} \times \mathbb{R}_{\tau}} \exp i(t\tau + \varphi_{j\kappa}(x,y,\xi',\tau)) e_{j\kappa}(x,y,\xi',\tau)\, d\xi'\, d\tau$$

where $e_{j\kappa}$ are formal series: $e_{j\kappa} \sim \sum_{\omega=0}^{\infty} e_{\omega j\kappa}$, $e_{\omega j\kappa}$ are solutions of the double transport equations with appropriate initial datas, positively homogeneous of degree $(-\omega-1)$ with respect to (ξ',τ).

Then $e_{\omega j\kappa}$ are classical double pseudo-differential symbols provided B is a differential operator; in the general case one can easily show that $e_{\omega j\kappa} = \bar{e}_{\omega j\kappa} + \hat{e}_{\omega j\kappa}$ where $\bar{e}_{\omega j\kappa}$ are classical double pseudo-differential symbols and $\hat{e}_{\omega j\kappa}$ satisfy estimates

$$|D_x^{\alpha} D_y^{\beta} D_{\xi'}^{\gamma} D_{\tau}^{n} \hat{e}_{\omega j\kappa}| \leq C_{\alpha\beta\gamma n\omega} (|\tau| + |\xi'|)^{-2-n} |\xi'|^{1-|\beta|-\omega}$$

$$\forall \alpha, \beta, \gamma, n, \omega .$$

Summarizing $e_{\omega j\kappa}$ asymptotically we obtain classical double pseudo-differential symbols $e_{j\kappa}$ provided B is a differential operator; in the general case we obtain $e_{j\kappa} = \bar{e}_{j\kappa} + \hat{e}_{j\kappa}$, $\bar{e}_{j\kappa}$ are classical double pseudo-differential symbols and $\hat{e}_{j\kappa}$ satisfy estimates

$$| D_x^{\alpha} D_y^{\beta} D_{\xi'}^{\gamma} D_{\tau}^{n} \hat{e}_{j\kappa} | \leq$$

$$C_{\alpha\beta\gamma} (1 + |\tau| + |\xi'|)^{-2-n} (1 + |\xi'|)^{1-|\gamma|} \qquad \forall \alpha, \beta, \gamma, n$$

and for $|\xi'| \geq 1$

$$|D_x^{\alpha} D_y^{\beta} D_{\xi}^{\gamma} D_{\gamma}^{n} (\hat{e}_{j\kappa} - \sum_{\omega=0}^{N-1} \hat{e}_{\omega j\kappa})| \leq C_{\alpha\beta\gamma N} (|\tau| + |\xi'|)^{-2-n} |\xi'|^{1-|\gamma|-N}$$

$$\forall \alpha, \beta, \gamma, N .$$

Recall that $WF_f(U_{j\kappa}) \subset \{ |t| \leqslant C_0(x_1+y_1)\}$; we cut U outside of the neighbourhood of $\{ t = \xi' = \eta' = 0\}$. On the other hand $WF_f(\mathcal{U}_{j\kappa}) \subset \{ t = -\frac{\partial}{\partial\tau} \varphi_{j\kappa}(x,y,\xi',\tau)\} \subset \{ |t| \leqslant C_0(x_1+y_1)\}$.

Then $U_{j\kappa} \equiv \mathcal{U}_{j\kappa} \ (mod\ C^\infty)$ provided B is a differential operator.

Let B be a pseudo-differential operator. Let us introduce the new variable $t' = t_{j\kappa} = t + \varphi_{j\kappa}(x,y,0,\tau)\tau^{-1}$; we consider zones $\{\pm\tau \geqslant C(|\xi'|+|\eta'|\}$ separately. Then $\mathcal{U}_{j\kappa}$ satisfies (4.106), (4.106)† modulo functions f such that

$$(4.110) \quad | D_x^\alpha D_y^\beta F_{t'\to\tau} f | \leqslant C_{\alpha\beta}(1+|\tau|)^{-2} \ln(2+|\tau|) \qquad \forall \alpha,\beta ;$$

$\mathcal{U}_{j\kappa}$ satisfies the initial data at $\{x_1 = y_1 = 0\}$ modulo functions of the same nature (see (4.98)). Note that in the coordinates (x, y, t') equations (4.106), (4.106)† are

$$(D_{x_1} + M_{jx})U_{j\kappa} \equiv 0 ,$$
$$U_{j\kappa}(-D_{y_1} + M^\dagger_{\kappa y}) \equiv 0$$

where

$$M_{jx} \equiv \mu_j(x,y,D'_x,D_{t'}) (mod\ OPS^{0'}(X \times \mathbb{R}, E_j)),$$

$$M_{\kappa y} \equiv \mu_\kappa(x,y,D'_y,D_{t'}) (mod\ OPS^{0'}(X \times \mathbb{R}, E_\kappa)),$$

in the first (second) operator we consider y (x respectively) as a parameter and $\mu_j(x,y,0,\tau) = \mu_\kappa(x,y,0,\tau) = 0$; i.e.

$$(4.111) \qquad |\mu_j(x,y,\xi',\tau)| + |\mu_\kappa(x,y,\xi',\tau)| \leqslant C_6 \, |\xi'| .$$

Then one can easily show by the energy estimates methos that $\mathcal{U}_{j\kappa} \equiv U_{j\kappa}$ modulo functions which satisfy (4.110); this proof is based on (4.111).

Hence

$$(4.112) \qquad \Gamma q'_y \chi \, U_{j\kappa} \equiv W_{j\kappa}$$

modulo functions v such that $|F_{t\to\tau} v| \leqslant C(1+|\tau|)^{-2} \ln(2+|\tau|)$

where

$$(4.113)\qquad W_{jк} = \int_{\mathbb{R}^{d-1}_{\xi'}\times\mathbb{R}_\tau} \exp i(t\tau + \varphi'_{jк}(x,\xi',\tau)) e'_{jк}(x,\xi',\tau)\,d\xi'\,d\tau ,$$

$\varphi'_{jк} = \Gamma\varphi_{jк}$, $e'_{jк} = \Gamma e_{jк}$. Then

$$(4.114)\qquad F_{t\to\tau} W_{jк} = \int_{\mathbb{R}^{d-1}_{\xi'}} e^{i\varphi'_{jк}(x,\xi',\tau)} e'_{jк}(x,\xi,\tau)\,d\xi'\,d\tau .$$

Recall that $e'_{jк} \sim \sum_{w=0} e'_{wjк}$ where

$$|D^n_\tau D^\alpha_x e'_{0jк}| \leqslant C_{\alpha n} |\tau|^{-1-n} \qquad \forall \alpha, n,$$

$$|D^n_\tau D^\alpha_x D^\beta_\xi e'_{wjк}| \leqslant C_{\alpha\beta n w} |\tau|^{-2-n} |\xi'|^{1-w-|\beta|}$$

$\forall \alpha, \beta', n, w$ such that $|\beta'| + w \geqslant 1$; $\operatorname{supp} e'_{wjк} \subset \{\pm\tau \geqslant C\,|\xi'|\}$ $\forall w$.

Let $W_{(w)jк}$ be defined by (4.113), (4.114) with $e'_{jк}$ replaced by $e'_{wjк}(1-\zeta(\xi'))$, $\zeta \in C^\infty_0(\mathbb{R}^{d-1})$, $\zeta = 1$ in the neighbourhood of 0; then for $w \geqslant d-1$ $F_{t\to\tau} W^{\circ}_{(w)jк} = O(|\tau|^{-1})$ and for $w \leqslant d-2$

$$(4.115)\qquad F_{t\to\tau} W_{(w)jк} = \int_{\mathbb{R}^{d-1}_{\xi'}} e^{i\varphi'_{jк}(x,\xi',\tau)} e'_{w}(x,\xi',\tau)\,d\xi' + O(|\tau|^{-1}) =$$

$$|\tau|^{d-w-2} \int_{\mathbb{R}^{d-1}_{\eta'}} e^{i\varphi'_{jк}(x,\eta',\pm 1)|\tau|} e_{wjк}(x,\eta',\pm 1)\,d\eta' + O(|\tau|^{-1}) \text{ as } \tau \to \pm\infty;$$

hence (4.112) implies that

$$(4.116)\qquad F_{t\to\tau} \Gamma q_y \chi U_{jк} = \sum_{w=0}^{d-2} |\tau|^{d-w-2} S^{\pm}_{wjк}(x, |\tau|) + O(|\tau|^{-1}) \text{ as } \tau \to \pm\infty,$$

where

$$S^{\pm}_{\omega j\kappa}(x,\sigma)=\int_{\mathbb{R}^{d-1}} e^{i\sigma\varphi^{\pm}_{j\kappa}(x,\eta')} e^{\pm}_{\omega j\kappa}(x,\eta')d\eta',$$

(4.117)

$$\varphi^{\pm}_{j\kappa}(x,\eta')=\varphi'_{j\kappa}(x,\eta',\pm 1)\ ,$$

$$e^{\pm}_{\omega j\kappa}(x,\eta')=e'_{\omega j\kappa}(x,\eta',\pm 1)\ ;$$

remind that

(4.118)

$$|D^{\alpha}_{x}\, e^{\pm}_{oj\kappa}| \leqslant C_{\alpha} \qquad \forall\alpha\,,$$

$$|D^{\alpha}_{x} D^{\beta}_{\eta'} e^{\pm}_{\omega j\kappa}| \leqslant C_{\alpha\beta\omega}\, |\eta'|^{1-\omega-|\beta|}$$

$\forall\alpha,\beta,\omega$ such that $|\omega|+|\beta|\geqslant 1$, $e^{\pm}_{\omega j\kappa}\subset\{|\eta'|\leqslant C^{-1}\}$.

For further applications we need estimate

(4.119)

$$F_{t\to\tau}\,\Gamma q_y \chi U_{j\kappa} = O(|\tau|^{d-2})$$

which follows from (4.116) and asymptotics for

$$\int_0^{\sigma} \tau\, F_{t\to\tau}\,\Gamma q_y \chi\, U_{jk}\, d\tau \quad (\mathrm{mod}\ O(\sigma^{d-1}))$$

which will be derived.

We need to consider (4.117) for $\omega=0$. Note that $\varphi^{\pm}_{j\kappa}\big|_{x_1=0}=0$,

$$\frac{\partial}{\partial x_1}\varphi^{\pm}_{j\kappa}\Big|_{x_1=0} = -\lambda_j(x',\eta',\pm 1)+\lambda_\kappa(x',\eta',\pm 1)$$

because of (4.107), $(4.107)^{+}$, (4.108).

Then $\varphi^{\pm}_{j\kappa}(x,\eta') = x_1\psi^{\pm}_{j\kappa}(x,\eta)$. Assume that condition (H.4) is fulfilled. Then

$$\frac{\partial\psi^{\pm}_{j\kappa}}{\partial\eta_i}\Big|_{x_1=\eta'=0} = 0 \qquad i=2,\dots,d,$$

$$\left(\frac{\partial^2 \psi^{\pm}_{jk}}{\partial \eta_i \partial \eta_l} \Bigg|_{x_1=\eta'=0} \right) \quad i,l=2,\dots,d$$ are nondegenerate $(d-1)\times(d-1)$-matrices. Hence phase function $\psi^{\pm}_{jk}(x,\eta')$ has precisely one stationary point $\eta'=\eta'_{\pm jk}(x)$ in the neighbourhood of $\eta'=0$ for every x with $x_1<\delta$; this point is nondegenerate and infinitely smoothly depends on x and, finally, equals 0 for $x_1=0$.

Consider integral (4.117) with $\omega=0$:

(4.120) $$S^{\pm}_{0jk}(x,\sigma)=\int e^{i\sigma x_1 \psi^{\pm}_{jk}(x,\eta')} e^{\pm}_{0jk}(x,\eta')\,d\eta' ;$$

it contains the "slow" variables x and the "rapid" variable x_1 (with factor S). But the dependence on the "slow" variables is not as good as we wish and (0.9) does not hold; to improve this dependence we need to replace $\psi^{\pm}_{jk}(x,\eta')$ by $\bar{\psi}^{\pm}_{jk}(x',\eta')=\psi^{\pm}_{jk}(0,x',\eta')$.

We can apply the estimates of the stationary phase method to integral (4.120) and obtain the following estimate:

(4.121) $$|S^{\pm}_{0jk}(x,\sigma)|\leqslant C(1+\sigma x_1)^{-\frac{d-1}{2}}$$

Therefore $S^{\pm}_{0jk}(x,\sigma)=O(\sigma^{-1})$ outside of a boundary layer $x_1\leqslant \sigma^{-(d-3)/(d-1)}$ provided $d\geqslant 4$; we need just the same estimate because we cannot use a better estimate. Since the difference between $S^{\pm}_{0jk}(x,\sigma)$ and oscillatory integral

(4.122) $$\bar{S}^{\pm}_{jk}(x,\sigma)=\int e^{i\sigma x_1 \bar{\psi}^{\pm}_{jk}(x',\eta')} \bar{e}^{\pm}_{0}(x',\eta')\,d\eta',$$

$\bar{e}^{\pm}_{0jk}(x',\eta') = e^{\pm}_{0jk}(0,x',\eta')$, is the sum of similar oscillatory integrals with additional factor either $x_1^2\sigma$ or x_1 then $S^{\pm}_{0jk}(x,\sigma)-\bar{S}^{\pm}_{jk}(x,\sigma)=O(\sigma^{-1})$ provided $d\geqslant 5$. Thus we have the appropriate asymptotics for $F_{t\to\tau}\,\Gamma q_y \chi\, U_{jk}$ provided $d\geqslant 5$. But we need the asymptotics for

$$R^{\pm}_{jk}(x,\sigma)=\int_0^{\sigma}\tau\, F_{t\to\tau}\,\Gamma q_y \chi\, U_{jk}\,d\tau \;=$$

$$\int_0^{\sigma} \tau^{d-1} S^{\pm}_{oj\kappa}(x,\tau)\,d\tau + O(\sigma^{d-1}).$$

It is easy to prove that

$$(4.123)\qquad R^{\pm}_{j\kappa}(x,\tau) = \int_{\mathbb{R}^{d-1}} \left(\frac{\partial}{\partial x_1}\right)^{d-1} \{x_1^{-1}(e^{i\sigma x_1 \psi^{\pm}_{j\kappa}(x,\eta')} - 1)\} f^{\pm}_{j\kappa}(x,\eta')\,d\eta'$$

where $f^{\pm}_{j\kappa}$ are symbols of the same type as $e^{\pm}_{oj\kappa}$. Hence

$$(4.124)\qquad |R^{\pm}_{j\kappa}(x,\sigma)| \leqslant C\sigma^{d}(1 + x_1\sigma)^{-\frac{d+1}{2}}$$

and $R^{\pm}_{j\kappa}(x,\sigma) = O(\sigma^{d-1})$ outside of a boundary layer $x_1 \leqslant \sigma^{-(d-1)/(d+1)}$. Since the difference between $R^{\pm}_{j\kappa}(x,\sigma)$ and

$$(4.125)\qquad \bar{R}^{\pm}_{j\kappa}(x,\sigma) = \int_{\mathbb{R}^{d-1}} \left(\frac{\partial}{\partial x_1}\right)^{d-1} \{x_1^{-1}(e^{i\sigma x_1 \bar{\psi}^{\pm}_{j\kappa}(x',\eta')} - 1)\} f^{\pm}_{j\kappa}(0,x',\eta')\,d\eta'$$

is the sum of similar oscillatory integrals with additional factor either $x_1^2\sigma$ or x_1 then $R^{\pm}_{j\kappa}(x,\sigma) - \bar{R}^{\pm}_{j\kappa}(x,\sigma) = O(\sigma^{d-1})$ provided $d \geqslant 3$, $R^{\pm}_{j\kappa}(x,\sigma) - \bar{R}^{\pm}_{j\kappa}(x,\sigma) = O(\sigma^{3/2})$ provided $d = 2$.

Therefore the following statement has been proved:

THEOREM 4.12. Let condition (H.4) be fulfilled, $q_y = q(y, D'_y, D_t) \in OPS^{o'}(X \times \mathbb{R})$, *cone supp* q be contained in the small enough conic neighbourhood of $Y \times 0 \times \tau^*$, $\chi \in C_0^{\infty}(\mathbb{R})$, $\chi = 1$ for $|t| < t_6$, *supp* $\chi \subset (-2t_6, 2t_6)$, $0 < t_6$ be small enough. Then

$$F_{t\to\tau} \Gamma q_y \chi u_1 = O(\tau^{d-2})$$

and for $d \geqslant 3$

$$(4.126)\qquad \int_0^{\sigma} \tau\, F_{t\to\tau} \Gamma q_y \chi u_1\,d\tau = \sigma^{d} Q(x', x_1\sigma) + O(\sigma^{d-1})$$

where Q satisfies (0.9), Q depends only on $a(0,x',\cdot)$, $b(x',\cdot)$, $q(0,x',\cdot)$, $\Pi(0,x',\cdot)$, $\Pi'_0(x',\cdot,\cdot)$; the similar asymptotics with the remainder estimate $O(\sigma^{3/2})$ holds for $d=2$.

One can improve the previous arguments provided condition (H.2) is fulfilled and x_1 is the distance between x and Y. Let (x_1,x') be a coordinate system in which $\mathcal{M}(x,\xi)=\xi_1^2+\mathcal{M}'(x,\xi')$. Then $\psi^{\pm}=\mp 1+O(|\eta'|^2)$ as $\eta'\to 0$ and $S^{\pm}_{0jk}(x,\sigma)-\bar{S}^{\pm}_{jk}(x,\sigma)$ as well as $R^{\pm}_{jk}(x,\sigma)-\bar{R}^{\pm}_{jk}(x,\sigma)$ are the sums of similar oscillatory integrals containing additional factor either $x_1^2\sigma|\eta'|^2$ or x_1 and hence $S^{\pm}_{0jk}(x,\sigma)-\bar{S}^{\pm}_{jk}(x,\sigma)=O(\sigma^{-1})$ for $d\geqslant 3$,

$$R^{\pm}_{jk}(x,\sigma)-\bar{R}^{\pm}_{jk}(x,\sigma)=O(\sigma) \qquad \text{for } d=2.$$

Therefore the following statement has been proved:

THEOREM 4.13. Assume that condition (H.2) is fulfilled, x_1 is the distance between x and Y. Then asymptotics (4.126) holds for $d=2$ with the remainder estimate $O(\sigma)$.

4.8. Let us summarize our results for manifolds with the boundary.

THEOREM 4.14. Let $\chi\in C_0^{\infty}(\mathbb{R})$, $\chi=1$ for $|t|<t_0$, $\operatorname{supp}\chi\subset[-2t_0,2t_0]$, $0<t_0$ be small enough. Then the following complete asymptotics hold provided either B is a differential operator or $\Pi\Pi^{\pm}$ is a singular Green operator:

$$\chi\sigma_1(t)=\chi\int_X u_1(x,x,t)\,dx\sim\sum_{w=0}^{\infty}c'_w\,\Phi_{d-w-3}(t) \tag{4.127}$$

and

$$F_{t\to\tau}(\chi\sigma_1)\sim\sum_{w=0}^{\infty}c'_w|\tau|^{d-w-3}\operatorname{sign}\tau \quad \text{as} \quad \tau\to\pm\infty \tag{4.128}$$

(Recall that $u_j(x,y,t)$ and $\sigma_j(t)$ are odd with respect to t, $j=0,1$); in the general case the following incomplete asymptotics hold:

$$\chi\sigma_1(t)\equiv\sum_{w=0}^{d-1}c'_w\Phi_{d-w-3}(t) \quad (\operatorname{mod} H^{5/2-\varepsilon}(\mathbb{R}) \quad \forall\varepsilon>0) \tag{4.127'}$$

and

$$F_{t\to\tau}(\chi\sigma_1)=\sum_{w=0}^{d-1}c'_w|\tau|^{d-w-3}\operatorname{sign}\tau+O(|\tau|^{-3}\ln|\tau|) \quad \text{as} \quad \tau\to\pm\infty \tag{4.128'}$$

Here C_0' depends only on $a(0,x',\cdot)$, $b(x',\cdot)$, $\Pi(0,x',\cdot)$, $\Pi_0'(x',\cdot,\cdot)$, $x'\in Y$.

THEOREM 4.15. Let χ be the same as in theorem 4.14. Then the following complete asymptotics hold provided either B is a differential operator or $\Pi\Pi^+$ is a singular Green operator:

$$\chi r \Gamma u_1 \sim \sum_{w=0}^{\infty} d_w'(x') \Phi_{d-w-2}(t) \tag{4.129}$$

and

$$F_{t\to\tau} \chi r \Gamma u_1 \sim \sum_{w=0}^{\infty} d_w'(x') |\tau|^{d-w-2} \operatorname{sign} \tau \quad \text{as} \quad \tau \to \pm\infty; \tag{4.130}$$

in the general case the following incomplete asymptotics hold:

$$\chi r \Gamma u_1 \equiv \sum_{w=0}^{d-1} d_w'(x') \Phi_{d-w-2}(t) \tag{4.129'}$$

$$(\operatorname{mod} C^\infty(Y, H^{5/2-\varepsilon}(\mathbb{R}, \operatorname{hom}(E))) \quad \forall \varepsilon > 0)$$

and

$$F_{t\to\tau} \chi r \Gamma u_1 = \sum_{w=0}^{d-1} d_w'(x') |\tau|^{d-w-2} \operatorname{sign}\tau + O(|\tau|^{-2} \ln|\tau|)$$

$$\text{as} \quad \tau \to \pm\infty \tag{4.130'}$$

where $d_w' \in C^\infty(Y, \operatorname{hom}(E))$, $d_0'(x')$ depends only on $a(0,x',\cdot), b(x',\cdot)$, $\Pi(0,x',\cdot)$, $\Pi_0'(x',\cdot,\cdot)$.

Moreover (4.130), (4.130)' remain true if one applies an operator $D_{x'}^{\alpha}$ to their left parts.

THEOREM 4.16. Let $\Pi\Pi^+$ be a singular Green operator, $\chi \in C_0^\infty(\mathbb{R})$, $\chi = 1$ for $|t| < t_0$, $\operatorname{supp}\chi \subset [-2t_0, 2t_0]$, t_0 be small enough. Then $\Gamma\chi u_1$ is the Schwartz kernel of an operator $h \in OP^{d-2}(\overline{\mathbb{R}}^+ \times \mathbb{R})$ commuting with D_t and depending infinitely smoothly on $x' \in Y$. Moreover, the following asymptotics holds:

$$F_{t\to\tau} \Gamma \chi u_1 \sim \sum_{w=0}^{\infty} |\tau|^{d-w-2} S_w(x', x_1|\tau|) \operatorname{sign}\tau \quad \text{as} \quad \tau \to \pm\infty \tag{4.131}$$

where $D^{\alpha}_{x'} D^{n}_{s} S_{\omega}(x', s) = O(s^{-\infty})$ as $s \to +\infty$ uniformly with respect to $x' \in Y$.

Here S_0 depends only on $a(0,x',\cdot)$, $b(x',\cdot)$, $\Pi(0,x',\cdot)$, $\Pi_0'(x',\cdot,\cdot)$.

THEOREM 4.17. Let condition (H.2) be fulfilled, $\Pi\Pi^{+}$ be not a singular Green operator, $\chi \in C_0^{\infty}(\mathbb{R})$, $\chi = 1$ for $|t| < t_0$, $supp\, \chi \subset [-2t_0, 2t_0]$, to be small enough. Then the following asymptotics hold:

(4.132)
$$F_{t\to\tau} \Gamma \chi u_1 = O(\tau^{d-2})$$

as $\tau \to \pm\infty$,

(4.133)
$$\int_0^k \tau F_{t\to\tau} \Gamma \chi u_1 \, d\tau = k^d Q(x', x_1 k) + O(k^{d-1}) \text{ as } k \to +\infty$$

where Q satisfies (0.9) and depend only on $a(0,x',\cdot)$, $b(x',\cdot)$, $\Pi(0,x',\cdot)$, $\Pi_0'(x',\cdot,\cdot)$.

THEOREM 4.18. Let conditions (H.3), (H.4) be fulfilled, $\Pi\Pi^{+}$ be not a singular Green operator, $\chi \in C_0^{\infty}(\mathbb{R})$ be the same as above. Then estimate (4.132) holds; asymptotics (4.133) holds provided $d \geqslant 3$ and asymptotics (4.133) with the remainder estimate $O(k^{3/2})$ holds provided $d = 2$.

§ 5. Proofs of the main theorems

5.1. In §4 we derived asymptotics for $F_{t\to\tau} \chi v$ as $\tau \to \pm\infty$ where $\chi \in C_0^{\infty}(\mathbb{R})$, $\chi = 1$ in the neighbourhood of $t = 0$ and either

$$v = \sigma(t) = \int_X tr\, u(x,x,t)\, dx$$

or $v = \Gamma u = u(x,x,t)$ or, finally, $v = \tau\Gamma u = u(x,x,t)\ (x \in Y)$.

In all these cases

$$v = \int_0^{\infty} \tau^{-1} \sin \tau t \, d\nu(\tau)$$

where either $\nu(\tau) = N(\tau)$ or $\nu(\tau) = e(x,x,\tau)$ or, finally, $\nu(\tau) = e(x,x,\tau), (x \in Y)$; here $\nu(\tau) = O(\tau^{M})$ as $\tau \to +\infty$ (see [2], e.g.) Hence we have the asymptotics for

$$2i\tau F_{t\to\tau}\chi v = 2F_{t\to\tau}\frac{\partial}{\partial t}\chi v = \int_0^\infty \hat{\chi}(\tau-\tau')d\nu(\tau') + O(\tau^{-\infty}) \quad \text{as } \tau\to+\infty$$

because $\hat{\chi}\in \mathcal{S}(\mathbb{R})$ and $\operatorname{supp}\frac{\partial\chi}{\partial t}\cap \operatorname{sing\,supp} v = \emptyset$ by choice of χ.

We wish to derive the asymptotics for $\nu(\tau)$ as $\tau\to+\infty$ from the asymptotics for

$$\int_0^\infty \hat{\chi}(\tau-\tau')\,d\nu(\tau') \quad \text{as} \quad \tau\to+\infty .$$

To prove theorems 0.1, 0.3, 0.4, 0.9 we need only to use the Tauberian theorem of Hörmander (see below); but to prove other theorems we need to use a more general and more precise Tauberian theorem; one can prove this theorem by a one-to-one repetition of the proof of the Tauberian theorem of Hörmander (see [90], p.158-162):

THEOREM 5.1. Let $\nu(\tau)$ be a monotone nondecreasing function on $\bar{\mathbb{R}}^+$, $\nu(0)=0$, $\nu(\tau)=O(\tau^M)$ as $\tau\to+\infty$ and

$$\mu(\tau) = \int_0^\infty \hat{\beta}(\tau-\tau')\,d\nu(\tau') \tag{5.1}$$

where $\beta(\tau)$ is a Hörmander function, i.e. $\beta\in C_0^\infty(\mathbb{R})$, $\beta(0)=1$, β is even and $\hat{\beta}(\tau)>0$ for every $\tau\in\mathbb{R}$. Then

$$|\nu(\tau) - \int_0^\infty \mu(\tau')d\tau'| \leq C_0 \sup_{\tau'\in\mathbb{R}} (1+|\tau-\tau'|)^{-n}\mu(\tau') + C \tag{5.2}$$

$$\forall\tau\in\mathbb{R}^+$$

where C_0 depends only on the choice of β and $n\in\mathbb{Z}^+$ is arbitrary.

REMARKS (i) This theorem remains true if $\nu(\tau)$ is a Hermitian matrix; then $\mu(\tau')$ must be replaced by its norm on the right side of (5.2).

(ii) Let

$$\mu_1(\tau) = \int_0^\infty \hat{\chi}(\tau-\tau')\,d\nu(\tau') \tag{5.1'}$$

where $\chi\in C_0^\infty(\mathbb{R})$, $\chi(0)=1$. Then there exist a Hörmander function β and functions $\chi_1, \chi_2 \in C_0^\infty(\mathbb{R})$ such that $\beta = \chi\chi_1 = \chi(1+t\chi_2)$. Hence if μ is expressed by (5.1) then

$$|\mu(\tau)| \leq C_1 \sup_{\tau' \in \mathbb{R}^+} (1+|\tau-\tau'|)^{-n} |\mu_1(\tau')| + C,$$

$$\left|\int_0^\tau (\mu(\tau')-\mu_1(\tau'))\,d\tau'\right| \leq C_1 \sup_{\tau' \in \mathbb{R}^+} (1+|\tau-\tau'|^{-n} |\mu_1(\tau')| + C$$

and therefore

$$(5.2)' \quad |\nu(\tau) - \int_0^\tau \mu_1(\tau')\,d\tau'| \leq C_2 \sup_{\tau' \in \mathbb{R}^+} (1+|\tau-\tau'|)^{-n} |\mu_1(\tau')| + C$$

where C_1, C_2 depend only on $\beta, \chi, \chi_1, \chi_2, n$ (i.e. only on χ and n); $n \in \mathbb{Z}^+$ is arbitrary.

COROLLARY (The Tauberian theorem of Hörmander). Let $\mu(\tau) = d_0 d\tau^{d-1} + O(\tau^{d-2})$ as $\tau \to +\infty$, $d \geq 1$; then $\nu(\tau) = d_0\tau + O(\tau^{d-1})$ as $\tau \to +\infty$. Here $O(\tau^0) \overset{def}{\equiv} O(\ln\tau)$. Moreover, if $d=1$,

$$\mu(\tau) = d_0 + O(\tau^{-1-\delta}) \quad \text{as} \quad \tau \to +\infty, \ \delta > 0,$$

then $\nu(\tau) = d_0\tau + O(1)$.

Combining theorem 5.1 and the remarks on it with results of [4] we prove theorems 0.1, 0.3, 0.6, 0.7, 0.9: corollary 4.3 implies theorem 0.1; corollary 4.3, the remark on it and theorem 4.14 imply theorem 0.3; theorem 4.14 implies theorem 0.4; corollary 4.3, the remark on it and theorems 4.17, 4.18 imply theorems 0.6, 0.7.

It remains for us to prove that $\varkappa_1^\pm > 0$ in theorem 0.4. Note that $\varkappa_1^\pm = \int_Y \bar{\varkappa}_1^+(y')\,dy'$ where $\bar{\varkappa}_1^\pm(y')$ is a "corresponding" coefficient for operators with the symbols positively homogeneous and not depending on x in the half-space $\overline{\mathbb{R}}^+ \times \mathbb{R}^{d-1}$:

$$\int_0^\infty \bar{e}(x,x,\tau)\,dx_1 = \bar{\varkappa}_1^\pm \tau^{d-1} \quad \text{for} \quad \tau > 0 ;$$

$\bar{e}(x,z,\tau)$ is the Schwartz kernel of a selfadjoint projector $\bar{e}(\tau)\bar{\Pi}$, $\bar{e}(\tau)$ is the spectral selfadjoin projector in $\bar{\mathbb{R}}^+ \times \mathbb{R}^{d-1}$ for operator $\bar{A}_{\bar{B}}$, $\bar{A} = a(0,y',D_x)$, $\bar{B} = b(y',D_x)$, $\bar{\Pi} = \Pi_0(0,y',D_x,D_z)$ (certainly $\bar{e}(\tau)$, $\bar{\Pi}_1$ depend on y' too) and measures dx on X and $dx'dx_1$ on $\bar{\mathbb{R}}^+ \times \mathbb{R}^{d-1}$ coincide at the point $(0,y')$.

Then $\bar{æ}_1^{\pm}(y') \geq 0$. Assume that $æ_1^{\pm} = 0$; then $\bar{æ}_1^{\pm}(y') \equiv 0$; hence $\bar{e}(y',\tau)\bar{\Pi}(y') \equiv 0$ and therefore $(\bar{\Pi}^{\pm}\bar{\Pi})(y') \equiv 0$. Since the principal symbols of $\bar{\Pi}^{\pm}\bar{\Pi}$ and $\Pi^{\pm}\Pi$ coincide at $(0,y) \times \bar{\mathbb{R}}^+_{\xi_1,\eta_1}$ then $\Pi^{\pm}\Pi$ is a smoothing and therefore a finite-dimensional projector and hence $\pm A_B$ is semibounded from above.

REMARK. One can prove that $\varkappa_0^{\pm}(y')$ does not vanish identically in theorem 0.9: otherwise $\bar{e}(x',x',\tau;y') \equiv 0$; then $\bar{e}(x',z',\tau;y') \equiv 0$ $(x',z' \in Y)$ and therefore $\bar{u}(x',z',t;y') \equiv 0$. Then the boundary conditions imply that

$$\bar{u}_{x_1}(x',z',t;y') \equiv 0, \quad \bar{u}_{z_1}(x',z',t;y') \equiv 0$$

and $\bar{u}_{x_1 z_1}(x',z',t;y') \equiv 0$.

Since the solution of the non-characteristic Cauchy problem (on Y) for systems with constant coefficients is unique then $\bar{u}(x,z,t;y') \equiv 0$ $(x,z \in \bar{\mathbb{R}}^+ \times \mathbb{R}^{d-1})$ and hence $\bar{e}(x,z,\tau;y') \equiv 0$; then the previous arguments prove that $\pm A_B$ is semibounded from above.

PROOF OF THEOREM 0.8. Theorem 4.16 implies that

$$\int_0^{\tau} \hat{\chi}(\tau-\tau')x_1^j\, de(x,x,\tau') = (\tau x_1)^j \tau^{d-j-1} S(x',x_1\tau) + O(\tau^{d-j-2}) \quad \text{as} \quad \tau \to +\infty$$

for arbitrary $j \geq 0$. Then theorem 5.1 and the remarks on it imply that

$$|x_1^j\{e(x,x,\tau) - \int_0^{\tau} \tau'^{d-1} S(x',x_1\tau')d\tau'\}| \leq C(\tau^{d-j-1}+1) \qquad \forall j \geq 0$$

and therefore

$$e(x,x,\tau) = \int_0^{\tau} \tau'^{d-1} S(x',x_1\tau')d\tau' + O(\min(\tau^{d-1}, x_1^{-d+1})) \quad \text{as} \quad \tau \to +\infty.$$

Note that

$$\int_0^{\tau} \tau'^{\,d-1} S(x', x_1\tau')d\tau = \tau^d Q(x', x_1\tau)$$

where

$$Q(x', s) = \int_0^1 \lambda^{d-1} S(x', s\lambda)d\lambda = s^{-d}\int_0^s \sigma^{d-1} S(x', \sigma)d\sigma$$

and hence (0.14) is valid.

5.2. To derive more precise asymptotics we need to improve the Tauberian theorem of Hörmander.

Let in theorem 5.1

$$\mu_1(\tau) = \int_0^{\infty} \hat{\chi}(\tau-\tau')\, d\nu(\tau') = \varkappa_0 d\tau^{d-1} + \varkappa_1(d-1)\tau^{d-2} + o(\tau^{d-2})$$

as $\tau \to +\infty$; $d>1$; then

$$|\nu(\tau) - \varkappa_0\tau^d - \varkappa_1\tau^{d-1}| \leq C_0\varkappa_0\tau^{d-1} + o(\tau^{d-1}) \quad \text{as} \quad \tau\to+\infty$$

where C_0 depends only on $\chi \in C_0^{\infty}(\mathbb{R})$ and d.

Then by the change of variable we obtain:

THEOREM 5.2. Let $\chi_1 \in C_0^{\infty}(\mathbb{R})$, $supp\,\chi_1 \subset (-1, 1)$, $\chi_T(t) = \chi_1(t/T)$ (then $\hat{\chi}_T(\tau) = T\hat{\chi}(\tau T)$). Let $\nu(\tau)$ be a monotone nondecreasing function on $\overline{\mathbb{R}}^+$, $\nu(\tau) = O(\tau^M)$ as $\tau\to+\infty$, $\nu(0)=0$ and

$$\int_0^{\infty} \hat{\chi}_T(\tau-\tau')d\nu(\tau') = \varkappa_0 d\tau^{d-1} + \varkappa_1(d-1)\tau^{d-2} + o(\tau^{d-2})$$

as $\tau\to+\infty$, $d>1$.

Then

$$|\nu(\tau) - \varkappa_0\tau^d - \varkappa_1\tau^{d-1}| \leq \frac{C_0\varkappa_0}{T}\tau^{d-1} + o(\tau^{d-1}) \quad \text{as} \quad \tau\to+\infty$$

where C_0 depends only on χ_1 and d.

PROOF OF THEOREMS 0.2, 0.5. We shall prove a more complicated theorem 0.5; its proof repeats, with certain complications, the proof of the main theorem [56]. We shall make some remarks concerning the proof of theorem 0.2.

In corollary 4.3 and theorem 4.14 we derived the following asymptotics

$$F_{t\to\tau}\chi\sigma = \varkappa_0 d\tau^{d-1} + \varkappa_1(d-1)\tau^{d-2} + o(\tau^{d-2})$$

with $\chi = \chi_{T_0}, \chi_1$ is fixed.

Let us fix an arbitrary $T > T_0$ and consider the sets:

Σ_T is a closed nowhere dense subset of measure zero in S^*X such that through each point of $S^*X \setminus \Sigma_T$ a geodesic billiard of length T can be passed in both directions in such a way that it contains a finite number of sections and all the geodesics included in it are transversal to the boundary.

Z_0 is the set of points $\rho \in B^*Y$ *) such that at either $(\rho, 1)$ or $(\rho, -1)$ the boundary operator B does not satisfy the Lopatinskii condition with respect to the operator $P = -D_t^2 + \mathcal{A}$. Note that $g(x', \xi_1, \xi', \tau)$ has two different real roots of constant multiplicity provided $(x', \xi'/\tau) \in B^*Y$; these roots coincide provided $(x', \xi'/\tau) \in S^*Y$ all other roots are non-real. Hence the Lopatinskii determinant $\mathcal{D}(x', \xi', \tau)$ is analytic with respect to τ provided $(x', \xi'/\tau) \in B^*Y$; it is easy to show that $\mathcal{D}(x', 0, \pm 1) \neq 0$ and we obtain by the positive homogeneity of $\mathcal{D}(x', \xi', \tau)$ with respect to (ξ', τ) that $\mathcal{D}(x', \xi', \tau) \neq 0$ provided $|\xi'| \leq C^{-1}|\tau|$. Therefore $\{\mathcal{D}(x', \xi', \tau) = 0\}$ is a respectively closed nowhere dense subset of measure zero in $\{(x', \xi'/\tau) \in B^*Y\}$ and by the positive homogeneity we obtain that $Z_0 = \{\mathcal{D}(x', \xi', 1) = 0\} \cup \{\mathcal{D}(x', \xi', -1) = 0\}$ is a respectively closed nowhere dense subset of measure zero in B^*Y.

Let
$$Z_T = \bigcup_{t \in [-T,T]} \Phi(t)(j^{-1} Z_0 \cap S^*X|_Y);$$
then $Z_T \cup \Sigma_T$ is a closed nowhere dense set of measure zero in S^*X.

Let Π_T be the set of points of $S^*X \setminus \Sigma_T$ which are periodic with respect to the flow Φ with a period not exceeding T. Then $\Pi_T \cup \Sigma_T$ is a closed set and because of the conditions of theorem 0.5. $\Pi_T \cup \Sigma_T$ is a nowhere dense set of measure zero.

Thus $\Lambda_T = \Pi_T \cup \Sigma_T \cup Z_T$ is a closed nowhere dense set of measure zero in S^*X.

/In the proof of theorem 0.2 Λ_T is a respectively closed now-

---

*) B^*Y is the bundle of cotangent open unit balls over Y.

here dense conic set of measure zero in $T^*X\setminus 0$ such that through each point $\bar\rho \in T^*X\setminus(\Lambda_T \cup 0)$ for every j there passes a bicharacteristic of σ_j $\rho=\rho_j(t)$, $t\in[-T,T]$, lying in $T^*X\setminus(\Sigma_j\cup 0)$ such that $\rho_j(t)\neq\bar\rho$ for every $t\in[-T,T]\setminus 0$/.

Let $0<\delta$, $\zeta, \zeta_1 \in C^\infty(X)$, $\zeta=1$ for $x_1<\delta, \zeta=0$ for $x_1\geq 2\delta$ and $\zeta^2+\zeta_1^2=1$. Let $q_j=q_j(y,D_y)\in OPS^0(X,E)$ be operators such that modulo infinitely smoothing operators

$$|\zeta|^2+q_1^*q_1+q_2^*q_2\sim I, \tag{5.3}$$

$q_2\sim 0$ in the neighbourhood of Λ_T and $\text{mes cone supp}\, q_1<\varepsilon$ where cone supports are considered as subsets in S^*X. Such operators exist for arbitrary $\varepsilon>0$. Really, since Λ_T is a closed set of measure zero then there exist $\zeta_2\in C^\infty(X)$ and the symbols q_1^0, q_2^0 positively homogeneous of degree 0 such that

$$|\zeta_2|^2+|q_1^0|^2+|q_2^0|^2=1,$$

$q_2^0=0$ in the neighbourhood of Λ_T and $\text{mes cone supp}\, q_1^0<\varepsilon, \zeta_2=0$ for $x_1>\delta/2$, $\zeta_2=1$ for $x_1<\delta/4$. Let $q_1', q_2'\in OPS^0(X,E)$ be operators with the principal symbols q_1^0, q_2^0 respectively such that $\text{cone supp}\, q_j'=\text{cone supp}\, q_j^0$ $(j=1,2)$; then

$$q_1'^*q_1'+q_2'^*q_2'+|\zeta_2|^2=I+R,$$

$R\in OPS^{-1}(X,E)$. Then there exist operators $R', R''\in OPS^{-1}(X,E)$ such that $(I+R')^*(I+R')\equiv I+R$, $(I+R')(I+R''')\equiv$ $(I+R'')(I+R')\equiv I \pmod{OPS^{-\infty}(X,E)}$; then $q_j=q_j'(I+R'')\zeta_1$ are appropriate operators.

/In the proof of theorem 0.2 we use operators $q_1, q_2\in OPS^0(X,E)$ such that $q_1^*q_1+q_2^*q_2\sim I$, $q_2\sim 0$ in the neighbourhood of Λ_T and $\text{mes cone supp}\, q_1<\varepsilon$/.

We set

$$\nu^0(k)=\sum_{k_n<k}\|\zeta\varphi_n\|^2,$$

$$\nu^j(k)=\sum_{k_n<k}\|q_j\varphi_n\|^2 \qquad (j=1,2)$$

where φ_n are orthonormal eigenfunctions of A_B which correspond to positive eigenvalues k_n^2.

Then $\nu^\ell(\tau)$ are monotone nondecreasing functions, $\nu^\ell(\tau)=$

$O(\tau^d)$ as $\tau \to +\infty$, $\nu^{\ell}(0)=0$ $(\ell=0,1,2)$ and by (5.3)

(5.4) $$\overset{o}{\nu}(\tau)+\overset{1}{\nu}(\tau)+\overset{2}{\nu}(\tau)=N(\tau)+O(1) \quad \text{as} \quad \tau\to+\infty$$

We set

$$\overset{\ell}{\sigma}(t)=\int_0^\infty \tau^{-1}\sin\tau t \, d\overset{\ell}{\nu}(\tau);$$

then

$$\overset{o}{\sigma}(t)=\int_X tr \; |\zeta|^2 u(x,x,t)\, dx$$

and

$$\overset{j}{\sigma}(t)=\int_X tr \; \Gamma((q_j^* \, q_j)(y,D_y)u(x,y,t))dy .$$

By theorem 4.2 and the remark on it we have the asymptotics for $F_{t\to\tau}\,\chi_{T_0}\,\sigma^j$ $(j=1,2)$ as $\tau\to+\infty$; by theorems 4.2, 4.14 we have the similar asymptotics for $F_{t\to\tau}\,\chi_{T_0}\,\sigma^o$; hence

(5.5) $$\int_0^\infty \hat{\chi}_{T_0}(\tau-\tau')\,d\overset{\ell}{\nu}(\tau')\sim 2i\tau\, F_{t\to\tau}\,\chi_{T_0}\,\overset{\ell}{\sigma}\sim$$
$$\overset{\ell}{\varkappa}_0\, d\tau^{d-1}+\overset{\ell}{\varkappa}_1(d-1)\tau^{d-2}+o(\tau^{d-2}) \quad \text{as} \quad \tau\to+\infty$$

where

$$\overset{o}{\varkappa}_0=\frac{1}{d}(2\pi)^{-d}\int_{S^*X}|\zeta|^2\, tr\, \Pi\Pi^+(x,\xi)\,dx\,d\xi ,$$

$$\overset{j}{\varkappa}_0=\frac{1}{d}(2\pi)^{-d}\int_{S^*X}|\overset{o}{q}_j(x,\xi)|^2\, tr\, \Pi^+\Pi(x,\xi)\,dx\,d\xi$$

$(j=1,2)$, $\overset{o}{q}_j$ are the principal symbols of q_j.

Let us compute by the standart method $WF(\overset{2}{\sigma}(t))$ (see [24, 39, 42]). Note that through the points of $cone\; supp\, q_2$ there pass billiards which are transversal to the boundary for $t\in[-T,T]$

and the Lopatinskii condition is fulfilled at the points of reflection of these billiards at the boundary; then at the points of reflection one can express the components of solution outgoing from the boundary through the components of solution incoming to the boundary and the boundary conditions by means of pseudo-differential operators on $Y \times \mathbb{R}$; therefore the singularities of solution propagate along such billiards [91] and hence

$$WF(q_2^*(y,D_y)q_2(y,D_y)u(x,y,t)) \cap \{|t| \leqslant T\} \subset$$

$$\{(x,\xi,y,\eta,t,\tau) : (y,\eta) \in \mathit{cone\ supp}\, q_2,\ |\tau| = |\xi| = |\eta|\, (x, \xi/\tau) = \Phi_{-t}(y, -\eta/\tau),\ t \in [-T,T]\}$$

and

$$WF(\sigma^2(t)) \cap \{|t| \leqslant T\} \subset \{(t,\tau) : \exists (x,\xi) \in \mathit{cone\ supp}\, q_2, |\tau| = |\xi|,$$

$$\Phi_{-t}(x,\xi/\tau) = (x,\xi/\tau), \quad t \in [-T,T]\} = 0 \times (\mathbb{R} \setminus 0)$$

because $\mathit{cone\ supp}\, q_2$ contains no point which is periodic with respect to the flow Φ with a period not exceeding T.

Thus $\sigma^2(t)$ has no singularity different from the great singularity (at $t=0$) on the interval $(-T,T)$ and hence (5.5) with $\ell = 2$ remains true if χ_{T_0} is replaced by χ_T.

Now we can apply theorem 5.2 and obtain the following estimates:

$$|\nu^0(\tau) - \overset{0}{\varkappa}_0 \tau^d - \overset{0}{\varkappa}_1 \tau^{d-1}| \leqslant \frac{C\delta}{T_0} \tau^{d-1} + o(\tau^{d-1}),$$

$$|\nu^1(\tau) - \overset{1}{\varkappa}_1 \tau^d - \overset{1}{\varkappa}_1 \tau^{d-1}| \leqslant \frac{C\varepsilon}{T_0} \tau^{d-1} + o(\tau^{d-1}),$$

$$|\nu^2(\tau) - \overset{2}{\varkappa}_2 \tau^d - \overset{2}{\varkappa}_2 \tau^{d-1}| \leqslant \frac{C}{T} \tau^{d-1} + o(\tau^{d-1}) \quad \text{as} \quad \tau \to +\infty$$

where C does not depend on T, δ, ε.

Then by (5.4)

$$|N(\tau) - \varkappa_0 \tau^d - \varkappa_1 \tau^{d-1}| \leqslant \tag{5.6}$$

$$C\left(\frac{\delta+\varepsilon}{T_0}+\frac{1}{T}\right)\tau^{d-1}+o(\tau^{d-1}) \quad \text{as} \quad \tau\to+\infty$$

where $\varkappa_0=\varkappa_0^0+\varkappa_0^1+\varkappa_0^2$, $\varkappa_1=\varkappa_1^0+\varkappa_1^1+\varkappa_1^2$; then $d\varkappa_0$, $(d-1)\varkappa_1$ are coefficients in the asymptotics for

$$\int_0^\infty \hat{\chi}_{T_0}(\tau-\tau')\,dN(\tau')\sim 2i\,F_{t\to\tau}\chi_{T_0}\sigma;$$

$$\varkappa_0=\frac{1}{d}(2\pi)^{-d}\int_{S^*Y}\operatorname{tr}\Pi\Pi^+\,dx\,d\xi,$$

$\varkappa_1=2ic_1/(d-1)$ where c_1 is the coefficient at τ^{d-3} in the asymptotic expansion for $F_{t\to\tau}\chi_{T_0}\sigma$. Since T,δ,ε are arbitrary then (5.6) implies that

$$N(\tau)-\varkappa_0\tau^d-\varkappa_1\tau^{d-1}=o(\tau^{d-1}) \quad \text{as} \quad \tau\to+\infty$$

and theorem 0.5 has been proved.

/Theorem 0.2 is proved by similar arguments, namely, the singularities of solution propagate along the bicharacteristics of σ_j provided $\tau=\sigma_j(x,\xi)$ and $(x,\xi)\in T^*X\setminus(\Sigma_j\cup 0)$, $\varkappa_1$ is given by formula (0.6)/.

Appendices

Appendix A

We shall prove the following statement:

PROPOSITION A.1. Let $P = D_1 - L$ where $L \in OPS^{1'}(X, E)$ (we do not consider t as a separate variable now), ℓ be the principal symbol of L, $\rho^* \in T^*Y \setminus 0$ and $\ell(\rho^*)$ have precisely one eigenvalue λ and this eigenvalue is real.

Then for every $\varepsilon > 0$ there exist an operator $\omega_\varepsilon \in OPS^{0'}(X, E)$, $\mathit{cone\ supp}\ \omega_\varepsilon \ni \rho^*$, and constant C_ε such that the following estimate holds:

$$\|v\|_{Y,-1/2} \leq \varepsilon \|v\| + C_\varepsilon \{ \|v\|_{-1/2} + \sum_{j=0}^{2D-2} \| D_1^j P v \|_{-j-1} \} + \|\omega_\varepsilon v\| \tag{A.1}$$

$$\forall v \in C^\infty(X, E)$$

where $D = \dim E$, $\|\cdot\|_{Y,\sigma}$, $\|\cdot\|_\sigma$, $(\cdot,\cdot)_{Y,\sigma}$, $(\cdot,\cdot)_\sigma$ are norms and inner products in $H^\sigma(Y)$, $\mathcal{H}^{\sigma,0}(X)$ respectively.

One can assume without a loss of generality that $\lambda = 0$ (otherwise we attain this by the change of variables) and $\mathit{supp}\ v \subset \{0 \leq x_1 < \delta\}$

We shall use

LEMMA A.2. Let ℓ be $D \times D$ -matrix such that $\ell^D = 0$. Then for every $\mu > 0$ there exist γ_j $(j = 0, \dots, 2D-2)$ such that

$$\gamma_j^* = -\gamma_j \ , \tag{A.2}$$

$$-\mathrm{Re} \langle \sum_{j=0}^{D-2} \gamma_j \ell^{j+1} w, w \rangle \geq |\ell w|^2 - \mu |w|^2, \tag{A.3}$$

$$-\mathrm{Re} \sum_{j=0}^{2D-2} \sum_{\kappa=0}^{j} \mathrm{Re} \langle i \gamma_j \ell^{j-\kappa} w, \ell^\kappa w \rangle \leq \mu |w|^2 \qquad \forall w \in \mathbb{C}^D. \tag{A.4}$$

PROOF OF PROPOSITION A.1. We use the method of [50]. Let $\hat{\gamma}_j \in OPS^{-j-1'}(X, E)$, $\hat{\gamma}_j^* = -\hat{\gamma}_j$ and the principal symbols of $\hat{\gamma}_j$ at ρ^* be equal to γ_j ; let

$$Q = \sum_{j=0}^{2D-2} \sum_{\kappa=0}^{j} \hat{\gamma}_j D^{\kappa} D_1^{j-\kappa} ;$$

then

$$QP \sim \sum_{j=0}^{2D-2} \hat{\gamma}_j (D_1^{j+1} - L^{j+1})$$

modulo operators of order (-1) with respect to x ; hence

$$Re(QPv, v) \sim -\sum_{j=0}^{2D-2} Re(\hat{\gamma}_j L^{j+1} v, v) +$$

$$\sum_{j=0}^{2D-2} Re(\hat{\gamma}_j D_1^{j+1} v, v) \sim$$

$$-\sum_{j=0}^{2D-2} Re(\hat{\gamma}_1 L^{j+1} v, v) +$$

$$\frac{1}{2} \sum_{j=0}^{2D-2} \sum_{\kappa=0}^{j} Re(i\hat{\gamma}_j D_1^{j-\kappa} v, D_1^{\kappa} v)_Y$$

modulo terms not exceeding

$$C\{ \sum_{j=0}^{2D-2} \| D_1^j Pv \|^2_{-\frac{3}{2}-j} + \|v\|^2_{-\frac{1}{2}} \};$$

in the last equality we used condition $\hat{\gamma}_j^* = -\hat{\gamma}_j$.

We replace D_1^{κ}, $D_1^{j-\kappa}$ by $L^{\kappa}, L^{j-\kappa}$ in inner products on Y ; they change on terms not exceeding

$$C\{ \sum_{j=1}^{2D-2} \| D_1^{j-1} Pv \|_{Y,-\frac{1}{2}-j} +$$

$$\|v\|_{Y,-\frac{1}{2}} \sum_{j=1}^{2D-2} \| D_1^{j-1} Pv \|_{Y,-\frac{1}{2}-j} \} \leq$$

$$\mu \|v\|^2 + C_{\mu} \{ \sum_{j=0}^{2D-2} \| D_1^j Pv \|^2_{-j-1} + \|v\|^2_{-\frac{1}{2}} \}$$

by the theorems of imbedding and the estimate

$$\|D_1 v\|_{-1} \leq \|Pv\|_{-1} + C\|v\|.$$

Hence the following inequality holds

(A.5)
$$\sum_{j=0}^{2D-2} Re(-\hat{\gamma}_j L^{j+1} v, v) +$$

$$\frac{1}{2}\sum_{j=0}^{2D-2}\sum_{\kappa=0}^{j} Re(i\hat{\gamma}_j L^{j-\kappa} v, L^{\kappa} v)_Y \leq \|\mu v\|^2 + C_\mu\{\sum_{j=0}^{2D-2}\|D_1^j Pv\|^2_{-1-j} + \|v\|^2_{-\frac{1}{2}}\}$$

for every $\mu>0$.

(A.3), (A.4) and the sharp Gårding inequality imply that the following inequalities hold:

(A.6)
$$c_0\|Lv\|^2_{-1} \leq 2\mu\|v\|^2 +$$

$$Re\sum_{j=0}^{2D-2}(-\hat{\gamma}_j L^{j+1} v, v) + \|\omega_\mu v\|^2 + C_\mu\|v\|^2_{-\frac{1}{2}} \quad ,$$

(A.7)
$$\sum_{j=0}^{2D-2}\sum_{\kappa=0}^{j} Re(-i\hat{\gamma}_j L^{j-\kappa} v, L^{\kappa} v)_Y \leq 2\mu\|v\|^2_{Y,-1/2} + \|\omega_\mu v\|^2_{Y,-1/2} + C_\mu\|v\|^2_{Y,-1}$$

for an appropriate $\omega_\mu \in OPS^{0'}(X,E)$, *cone supp* $\omega_\mu \ni \rho^\kappa$; here and in what follows C_κ are positive and do not depend on μ.

(A.5) - (A.7) and the theorems of imbedding imply the estimate

$$c_1\|Lv\|^2_{-1} \leq \mu\|v\|^2 + C_\mu\{\sum_{j=0}^{2D-2}\|D_1^j Pv\|^2_{-1-j} + \|\omega_\mu v\|^2 + \|v\|^2_{-\frac{1}{2}}\}$$

for an arbitrary $\mu>0$.

On the other hand

$$\|v\|^2_{Y,-1/2} \leq 2Re(-iD_1 v, v)_{-1/2} \leq c_2\{\|Pv\|_{-1} + \|Lv\|_{-1}\}\|v\|.$$

These inequalities imply estimate (A.1) with $\varepsilon = c_3\sqrt{\mu}$.

PROOF OF LEMMA A.2. By means of induction on $q=1,\dots,D$ we shall prove that for every q there exist γ_j $(j=0,\dots,2q-2)$ such that $\gamma_j = -\gamma_j^*$ and

$$(A.8)_q \qquad \operatorname{Re}\langle -\sum_{j=0}^{2q-2} \gamma_j \ell^{j+1} w, w\rangle \geq c_q|\ell w|^2 - \mu|w|^2 - C_{q,\mu}|\ell^{2q} w|^2,$$

$$(A.9)_q \qquad \sum_{j=0}^{2q-2}\sum_{k=0}^{j} \operatorname{Re}\langle -i\gamma_j \ell^{j-k} w, \ell^{k} w\rangle \leq \mu|w|^2 + C_{q,\mu}|\ell^{q} w|^2 \qquad \forall w\in\mathbb{C}^D.$$

For $q=D$ we obtain the statement of lemma. Note that $(A.8)_1$ with $\mu=0$ and $(A.9)_1$ with arbitrary $\mu>0$ hold provided $\gamma_0 = \ell-\ell^*$.

Let $q=1,\dots,D-1$. Assume that $(A.8)_q$, $(A.9)_q$ hold for appropriate $c_q>0$ and γ_j $(j=0,\dots,2q-2)$. Then $(A.8)_{q+1}$ $(A.9)_{q+1}$ with μ replaced by 2μ are fulfilled for $c_{q+1}=c_q/2$ and γ_j $(j=0,\dots,2q)$ provided the following inequalities hold:

$$(A.10)_{q+1} \qquad \operatorname{Re}\langle(-\gamma_{2q-1}\ell^{2q} - \gamma_{2q}\gamma^{2q+1})w, w\rangle \geq$$
$$-c_q/2\,|\ell w|^2 - \mu|w|^2 + C_{q,\mu}|\ell^{2q} w|^2 - C'_{q+1,\mu}|\ell^{2q+2} w|^2,$$

$$(A.11)_{q+1} \qquad \operatorname{Re}\langle -i\gamma_{2q}\ell^{q} w, \ell^{q} w\rangle \leq \mu|w|^2 - C_{q,\mu}|\ell^{q} w|^2 + C'_{q+1,\mu}|\ell^{q+1} w|^2$$
$$\forall w\in\mathbb{C}^D.$$

Really, all the other terms in the left part of $(A.9)_{q+1}$ which are new comparatively to the left part of $(A.9)_q$ do not exceed $\delta|w|^2 + C''_{q+1,\mu,\delta}|\ell^{q+1} w|^2$ for every $\delta>0$. Recall that $C_{q,\mu}$ are fixed and $C'_{q+1,\mu}$ are not chosen yet.

It remains for us to prove that for every $\delta>0$ and $q<D$ there exist matrices $\alpha=-\alpha^*$ and $\beta=-\beta^*$ such that

$$-\operatorname{Re}\langle\alpha\ell^{2q} w, w\rangle - \operatorname{Re}\langle\beta\ell^{2q+1} w, w\rangle \geq -\delta|w|^2 + |\ell^{2q} w|^2 - C_\delta|\ell^{2q+2} w|^2$$

and

$$\mathrm{Re}\langle i\beta \ell^{q} w, \ell^{q} w\rangle \geq -\delta |w|^{2} + |\ell^{q} w|^{2} - C_{\delta} |\ell^{q+1} w|^{2} \qquad \forall w \in \mathbb{C}^{D}.$$

And finally note that $\alpha = \ell^{2q} - \ell^{*2q}$ and

$$\beta = \frac{4}{\delta}(\ell^{2q+1} - \ell^{*\,2q+1}) - i(\mathrm{Id} - \ell^{*}\ell)$$

are appropriate matrices.

REMARK. Proposition A.1 shows that the condition (D') in paper [50] by the author is connected with the non-real roots ξ_1 of the polynomial $g(x, \xi_1, \xi')$.

Appendix B

LEMMA B. Let $e'(\tau)$ be a projector to the invariant subspace of $T(\tau) = -K^{-1} L(\tau)$ corresponding to the eigenvalues lying in the complex half-plane $\{\mathrm{Im}\, z > \varepsilon\}$, $\varepsilon > 0$; we assume that there is no eigenvalue on the straight line $\{\mathrm{Im}\, z = \varepsilon\}$. Then

$$\text{(B.1)} \qquad -i e'^{*} K \frac{\partial e'}{\partial \tau} = \int_0^\infty e'^{*} e^{-itT^{*}} \frac{\partial L}{\partial \tau} e^{itT} e' dt .$$

Recall that K, L are Hermitian matrices.

PROOF. Obviously, integral on the right side of (B.1) converges. Let γ be a closed contour with a counter-clockwize orientation lying in the half-plane $\{\mathrm{Im}\, z > \varepsilon\}$; all the eigenvalues of $T(\tau)$ lying in this half-plane are contained inside γ . Let $\bar{\gamma}$ be a conjugate contour with a clockwise orientation, $\bar{\gamma}$ lies in the half-plane $\{\mathrm{Im}\, \lambda < -\varepsilon\}$. Then

$$e' e^{itT} = \frac{1}{2\pi i} \int_{\gamma} e^{itz} (z - T)^{-1} dz ,$$

$$e^{-itT^{*}} e'^{*} = -\frac{1}{2\pi i} \int_{\bar{\gamma}} e^{-itz'} (z' - T^{*})^{-1} dz'$$

and the right part of (B.1) equals

$$\frac{1}{4\pi^2}\int_0^\infty\int_\gamma\int_{\bar\gamma} e'^* e^{-itz'}(z'-T^*)^{-1}\frac{\partial L}{\partial\tau}(z-T)^{-1}e^{itz}e'\,dz\,dz'\,dt =$$

$$\frac{1}{4\pi^2 i}\int_\gamma\int_{\bar\gamma} e'^*(z'-T^*)^{-1}(z'-z)^{-1}\frac{\partial L}{\partial\tau}(z-T)^{-1}dz\,dz' =$$

$$\frac{1}{2\pi}\int_\gamma e'^*(z-T^*)^{-1}\frac{\partial L}{\partial\tau}(z-T)^{-1}dz$$

because the integrand has only one (simple) pole $z'=z$ outside $\bar\gamma$ and residue at $z'=\infty$ equals 0.

On the other hand, differentiating the equality

$$e' = \frac{1}{2\pi i}\int_\gamma (z-T)^{-1}dz$$

with respect to τ we obtain that

$$\frac{\partial e'}{\partial\tau} = \frac{1}{2\pi i}\int_\gamma (z-T)^{-1}\frac{\partial T}{\partial\tau}(z-T)^{-1}dz =$$

$$-\frac{1}{2\pi i}\int_\gamma (z-T)^{-1}K^{-1}\frac{\partial L}{\partial\tau}(z-T)^{-1}dz =$$

$$-\frac{1}{2\pi i}\int_\gamma K^{-1}(z-T^*)^{-1}\frac{\partial L}{\partial\tau}(z-T)^{-1}dz$$

because K, L are Hermitian; hence the left and the right parts of (B.1) coincide.

Appendix C

In paper [49] by the author the following statement was proved:

LEMMA C. Let $A(x)$ be a Hermitian matrix, X, Y vector fields; assume that at the point x^* the following condition is fulfilled:

(C.1) $$\langle XA(x^*)v, v\rangle \geq 0 \qquad \forall v \in \operatorname{Ker} A(x^*)\setminus 0.$$

Then

(i) $g = \det A$ has zero of order $r = \dim \operatorname{Ker} A(x^*)$ at x^* and

$$\theta \equiv X^{r} g(x^*) \neq 0 ;$$

(ii) condition (C.1) with X replaced by Y is fulfilled if and only if the following condition is fulfilled:

$$\theta^{-1} (X^{k} Y^{r-k} g)(x^*) > 0 \quad \forall k = 0, \ldots, r-1 .$$

Appendix D. Calculation of the leading coefficients for closed manifolds

We wish to calculate coefficients $c_0(y), c_1(y)$ in asymptotics (4.40), (4.42); they are given by formulas (4.37), (4.35), (4.34), (4.34)'. First of all we transform (4.37). We introduce the spherical coordinate system $(r, \sigma) \in \mathbb{R}^{+} \times S^{d-1}$ in $\mathbb{R}^{d}_{\xi} \setminus 0$. Since the replacement of $\tau = 1$ by $\tau = 1 \mp i0$ in $F^{hom}_{w}(y, \sigma, r, \tau)$ is equivalent to the replacement of r by $r(1 \pm i0)$ then

$$c_{w}(y) = -\int_{S^{d-1}} \int_{\gamma} F^{hom}_{w}(y, \sigma, r, 1) r^{d-1} dr \, d\sigma$$

where γ is a closed contour with a counter-clockwise orientation in complex half-plane $\{ \operatorname{Re} r > 0 \}$; the segment of the real axis $[C^{-1}, C]$ containing all the real poles $F^{hom}_{w}(y, \sigma, r, 1)$ must lie inside γ and imagine poles $F^{hom}_{w}(y, \sigma, r, 1)$ must lie aoutside The change of the variable $r = \tau^{-1}$ and the positive homogeneity of degree $(-w-2)$ of $F^{hom}_{w}(y, \sigma, r, \tau)$ with respect to (r, τ) imply that

$$c_{w}(y) = \int_{S^{d-1}} \int_{\gamma} F^{hom}_{w}(y, \sigma, 1, \tau) \tau^{-d+1+w} d\tau . \tag{D.1}$$

In particular (4.34) implies that

$$c_0(y) = -\frac{i}{2} (2\pi)^{-d} \int_{S^{d-1}} a(y, \sigma)^{-d/2} \Pi(y, \sigma) d\sigma . \tag{D.2}$$

Recall that $\Pi^+\Pi=\Pi$ and $\pi^+\pi=\pi$.
It follows then from (4.34)' that

$$F_1^{hom}=(2\pi)^{-d-1}\Big[((-\tau^2+A^{-1})\Pi)^s+\frac{i}{2}\sum_{j=1}^{d}\frac{\partial^2}{\partial y_j\partial\xi_j}\{(-\tau^2+a)^{-1}\pi\}+$$

$$+i\sum_{j=1}^{d}(-\tau^2+a)^{-1}\Big\{\frac{\partial a}{\partial\xi_j}(-\tau^2+a)^{-1}\frac{\partial a}{\partial y_j}-\frac{\partial a}{\partial y_j}(-\tau^2+a)^{-1}\frac{\partial a}{\partial\xi_j}\Big\}(-\tau^2+a)^{-1}\pi\Big]$$

where b^s is the principal symbol of B.

Therefore

(D.3) $$\operatorname{tr} c_1(y)=-\frac{i}{2}(2\pi)^{-d}\int_{S^{d-1}}\operatorname{tr}(A^{-\frac{d-1}{2}}\Pi)^s(y,\sigma)d\sigma+\overline{æ}_1(y)$$

where $\overline{æ}_1(y)$ depends only on a,π.

In particular $\operatorname{tr} c_1=0$ provided A is a differential operator and $\Pi=\Pi^+$.

It follows from the proof of (D.1), (D.2) that the leading coefficient in the asymptotics for $\Gamma_{qq'}\chi u$ equals

(D.4) $$\tilde c_0(y)=-\frac{i}{2}(2\pi)^{-d}\int_{S^{d-1}}a(y,\sigma)^{-d/2}\pi(y,\sigma)q_0(y,\sigma)\,q_0'(y,\sigma)d\sigma$$

provided $q=q(x,D_x)$, $q'=q'(y,D_y)\in OPS^0(X,E)$ have the scalar principal symbols, $\chi\in C_0^\infty(\mathbb{R})$, $\chi(0)=1$ and $\operatorname{supp}\chi$ is small enough.

To obtain formulas for $c_0(y),\operatorname{tr} c_1(y)$ not using the spherical coordinate system for every $\sigma\in S^{d-1}$ we introduce the basis consisting of the eigenvectors of $a(y,\tau)$ and obtain that

(D.2)' $$c_0(y)=-\frac{i}{2}d(2\pi)^{-d}\int_{T_y^*X}\pi\varepsilon^+d\xi\,,$$

(D.3)' $$\operatorname{tr} c_1(y)=\frac{i}{4}d(d-1)(2\pi)^{-d}\int_{T_y^*X}\operatorname{tr}\pi\varepsilon^+a^{-1/2}a^s d\xi+\overline{æ}_1(y)$$

where $\mathcal{E}^+$ is the selfadjoint projector to invariant subspace of corresponding to eigenvalues belonging to the segment $[0,1]$.

Appendix E. Calculation of the leading coefficients for manifolds with the boundary

E.1. As we know the leading coefficients in asymptotics (4.128), (4.130) and leading terms in asymptotics (4.131), (4.133) depend only on $a(0,y',\cdot)$, $b(y',\cdot)$, $\Pi(0,y',\cdot),\Pi_0'(y',\cdot,\cdot)$; hence one can assume without a loss of generality that X is a half-space and the symbols of A,B are positively homogeneous and do not depend on x *). These formulas follow from §4:

(E.1) $$S_0(y,\tau)=\int_{\mathbb{R}^{d+1}_{\xi,\eta_1}} \{F_0(y',\xi_1,\xi_1-\eta_1,\xi',\tau-i0)-F_0(y',\xi_1,\xi_1-\eta_1,\xi',\tau+i0)\}\cdot e^{iy_1\eta_1}\,d\xi\, d\eta_1,$$

(E.2) $$c_0(y')=\int_0^\infty \operatorname{tr} S_0(y,1)\,dy_1=$$

$$i\int_{\mathbb{R}^{d+1}_{\xi,\eta_1}} \operatorname{tr}\{F_0(y',\xi_1,\eta_1,\xi',1-i0)-F(y',\xi_1,\eta_1,\xi',1+i0)\}(\xi_1-\eta_1+i0)^{-1}\,d\xi\, d\eta_1,$$

(E.3) $$d_0(y')=S_0(0,y',1)=$$

$$\int_{\mathbb{R}^{d+1}_{\xi,\eta_1}}\{F_0(y',\xi_1,\eta_1,\xi',1-i0)-F_0(y',\xi_1,\eta_1,\xi',1+i0)\}\,d\xi\, d\eta_1,$$

(E.4) $$Q_0(y,\tau)=\int_0^\tau \tau' S_0(y,\tau')\,d\tau' \quad (\tau\in\mathbb{R}^+)$$

where

(E.5) $$F_0(y',\xi_1,\eta_1,\xi',\tau)=$$

*) Before the integration on Y in calculation c_0.

$$(2\pi)^{-d-1}(-\tau^2+a(y',\xi_1,\xi'))^{-1}b^*(y',\xi_1,\xi')R^{-1}(y',\xi',\tau)b(y',\eta_1,\xi')(-\tau^2+a(y',\eta_1,\xi'))^{-1}$$

$$\Pi(y',\eta_1,\xi')+(2\pi)^{-d-1}\{Id-(-\tau+a(y',\xi_1,\xi'))^{-1}b^*(y',\xi_1,\xi')\cdot$$

$$R^{-1}(y',\xi',\tau)\}(-\tau^2+a(y',\xi_1,\xi'))^{-1}\Pi_0'(y',\xi_1,\xi');$$

$$R(y',\xi',\tau)=\int_{-\infty}^{\infty}(-\tau^2+a(y',\xi_1,\xi'))^{-1}d\xi_1$$

provided $b(y',\xi_1,\xi')=Id$ and

$$R(y',\xi',\tau)=\int_{-\infty}^{\infty}\{b(y',\xi_1,\xi')(-\tau^2+a(y',\xi_1,\xi'))^{-1}b^*(y',\xi_1,\xi')-a_0^{-1}(y')\}\,d\xi_1$$

provided $b(y',\xi_1,\xi')=i\xi_1+b_1(y',\xi')$, $a_0(y')$ in the coefficient at ξ_1^2 in $a(y',\xi_1,\xi')$.

Author knows of no better formulas in the general case.

E.2. Let now $A=-\Delta$, Δ be the Laplace - Beltrami operator on Riemannian manifold X (perhaps Δ acts in fibering E over X). We introduce a coordinate systems in the neighbourhood of $(0,y')$ uch that

$$a(x,\xi)=\xi_1^2+a'(x,\xi'),\quad a'(0,y',\xi')=|\xi'|^2;$$

then $x_1=dist(x,Y)$. Let $\Pi=\Pi^+$. For the calculation of leading coefficients and terms we assume again that X is a half-space, symbols of Δ, B are positively homogeneous and do not depend on x and, finally, $E=X\times\mathbb{C}^D$. We shall make all calculations from the beginning without the use of formulas (E.2) - (E.5).

Note that A_B is selfadjoint provided $B=\tau$ and A_B is selfadjoint if and only if $\beta(\xi')$ is Hermitian provided $B=i\tau D_1+\beta(D')\tau$. We assume that $\beta(\xi')$ is Hermitian.

Then Π^- is a singular Green operator:
$\Pi^-=0$ provided $B=\tau$ and

$$\Pi^-=F^{-1}_{\xi'\to x'}\Pi'(\xi')F_{x'\to\xi'}$$

provided $B=i\tau D_1+\beta(D')\tau$ where $\Pi'(\xi')$ is a selfadjoint projector in $L_2(\mathbb{R}^+, E)$; $\Pi'(\xi')$ is one-dimensional integral operator with the Schwartz kernel

$$\Pi'(\xi'; x_1, y_1) = 2\beta(\xi')\varepsilon(\xi')exp(-(x_1+y_1)\beta(\xi')),$$

$\varepsilon(\xi')$ is a selfadjoint projector to the invariant subspace of $\beta(\xi')$ corresponding to eigenvalues exceeding $|\xi'|$; recall that the Šapiro - Lopatinskii condition means precisely that all the eigenvalues of $\beta(\xi')$ differ from $|\xi'|$ and that A_B is semibounded from below if and only if no eigenvalue of $\beta(\xi')$ exceeds $|\xi'|$.

Obviously

$$F_{x,t\to\xi',\tau} u_0^{\pm} = \pm(2\pi)^{-d-1}(-\tau^2+|\xi|^2)^{-1} e^{-i\langle y,\xi\rangle};$$

then

$$F_{x',t\to\xi',\tau} \tau_x D_{x_1}^{\mu} u_0^{\pm} = (2\pi)^{-d-1} vrai \int_{-\infty}^{\infty} \xi_1^{\mu}(-\tau^2+|\xi|^2)^{-1} e^{-i\langle y,\xi\rangle} d\xi_1 =$$

$$(2\pi)^{-d}\frac{i}{2}\lambda^{\mu-1}(\xi',\tau)exp(-i\lambda(\xi',\tau)-i\langle y',\xi'\rangle)$$

by the Cauchy residue theorem where $\lambda(\xi',\tau) = (\tau^2-|\xi'|^2)^{1/2}$, $\Im m\,\lambda(\xi',\tau)<0$ for $\tau\in\mathbb{C}_{\mp}\setminus\mathbb{R}$. Then $\lambda(\xi',\tau\pm i0) = -i\sqrt{|\xi'|^2-\tau^2}$ for $|\tau|\leqslant|\xi'|$, $\lambda(\xi',\tau\pm i0) = \pm\, sign\,\tau\sqrt{\tau^2-|\xi'|^2}$ for $|\tau|\geqslant|\xi'|$ $(\tau\in\mathbb{R})$; hence $\lambda(\xi',\tau)$ is holomorphic in $\mathbb{C}\setminus\{(-\infty,-|\xi'|]\cup[|\xi|,\infty)\}$.

Then

$$F_{x',t\to\xi',\tau} Bu_0^{\pm} = \pm(2\pi)^{-d}\frac{i}{2}\lambda(\xi',\tau)^{-1}exp(-iy_1\lambda(\xi',\tau)-i\langle y',\xi'\rangle)$$

provided $b = Id$ and

$$F_{x',t\to\xi',\tau} Bu_0^{\pm} = \pm(2\pi)^{-d}\frac{i}{2}(i\lambda(\xi',\tau)+\beta(\xi'))\lambda^{-1}(\xi',\tau)\cdot exp(-iy_1\lambda(\xi',\tau)-i\langle y',\xi'\rangle)$$

provided $b = i\xi_1 + \beta(\xi')$.

Hence $v^{\pm} = F_{x',t\to\xi',\tau}\, u_1^{\pm}$ is the solution of the problem

(E.6) $$(-\tau^2 + |\xi'|^2 + D_1^2)\, v^{\pm} = 0 ,$$

(E.7) $$v^{\pm}\Big|_{x_1=0} = \mp(2\pi)^{-d}\frac{i}{2}\lambda^{-1}(\xi',\tau)\exp(-iy_1\lambda(\xi',\tau) - i\langle y',\xi'\rangle)$$

provided $b = Id$ and of the problem

(E.6)' $$(-\tau^2 + |\xi'|^2 + D_1^2)v^{\pm} = \mp(2\pi)^{-d}\, 2\beta(\xi')\varepsilon(\xi')\exp(-(x_1+y_1)\beta(\xi') - i\langle y',\xi'\rangle),$$

(E.7)' $$(iD_1 + \beta(\xi'))v^{\pm}\Big|_{x_1=0} = \mp(2\pi)^{-d}\frac{i}{2}(i\lambda(\xi',\tau) + \beta(\xi'))\lambda^{-1}(\xi',\tau)\cdot$$
$$\exp(-iy_1\lambda(\xi',\tau) - i\langle y',\xi'\rangle)$$

provided $b = i\xi_1 + \beta(\xi')$ because the Schwartz kernel of $F_{x'\to\xi'}$ Π' equals

$$(2\pi)^{-d}\cdot 2\beta(\xi')\varepsilon(\xi')\exp(-(x_1+y_1)\beta(\xi') - i\langle y',\xi'\rangle).$$

A particular solution of (E.6)' is $\mp(2\pi)^{-d}\, 2\beta(\xi')\varepsilon(\xi')(-\tau^2 +$

$$|\xi'|^2 + \beta^2(\xi'))^{-1}\exp(-(x_1+y_1)\beta(\xi') - i\langle y',\xi'\rangle);$$

it satisfies the homogeneous boundary condition and rapidly decreases as $x_1 \to +\infty$; solution of (E.6) decreasing as $x_1 \to +\infty$ equals modulo constant factor $\exp(-ix_1\lambda(\xi',\tau))$. Hence the solution of problem (E.6) - (E.7) decreasing as $x_1 \to +\infty$ equals

(E.8) $$v^{\pm} = \mp(2\pi)^{-d}\frac{i}{2}\lambda^{-1}(\xi',\tau)\exp(-i(x_1+y_1)\lambda(\xi',\tau) - i\langle y',\xi'\rangle)$$

and similar solution of the problem (E.6)' - (E.7)' equals

$$\text{(E.8)'}\quad v^{\pm\prime} = \pm(2\pi)^{-d}\, 2\beta(\xi')\varepsilon(\xi')(-\tau^2+|\xi'|^2+\beta^2(\xi'))^{-1} \exp(-(x_1+y_1)\beta(\xi')-i\langle y',\xi'\rangle \mp$$

$$(2\pi)^{-d}\,\frac{i}{2}\,(-i\lambda(\xi',\tau)+\beta(\xi'))^{-1}(i\lambda(\xi',\tau)+\beta(\xi'))\,\bullet$$

$$\lambda^{-1}(\xi',\tau)\exp(-i(x_1+y_1)\lambda(\xi',\tau)-i\langle y',\xi'\rangle).$$

Obviously, $v^{\pm}$ is holomorphic in $\mathbb{C}\setminus\{(-\infty,-|\xi'|]\cup[|\xi'|,\infty)\}$ and $v^{\pm\prime}$ is meromorphic there; its poles are

$$\pm\tau_k = \pm\sqrt{|\xi'|^2-\beta_k^2(\xi')}$$

where $\beta_k(\xi')$ are eigenvalues of $\beta(\xi')$ lying between 0 and $|\xi'|$; other eigenvalues give no pole.

REMARK. The eigenvalues of $\beta(\xi')$ lying between 0 and $|\xi'|$ are connected with the nonclassical propagation of the singularities of the solutions of the corresponding hyperbolic problem.

Let $\mathcal{D}'(\mathbb{R}) \ni w = v^{+}(\tau-i0)+v^{-}(\tau+i0)$; then

$$\operatorname{supp} w \subset (-\infty,-|\xi'|]\cup[|\xi'|,\infty)\cup\bigcup_{k,\pm}\{\pm\tau_k\},$$

$$\operatorname{sing\,supp} w \subset \{-|\xi'|\}\cup\{|\xi'|\}\cup\bigcup_{k,\pm}\{\pm\tau_k\}.$$

We apply the inverse Fourier transform $F^{-1}_{\xi'\to x'}$ and set $x=y$; we introduce the spherical coordinate system $(r,\sigma)\in\mathbb{R}^{+}\times S^{d-2}$ in $\mathbb{R}^{d-1}_{\xi'}\setminus 0$; note that the replacement of $\tau\in\mathbb{R}$ by $\tau(1\mp i0)$ is equivalent to the replacement of r by $r(1\pm i0)$ and obtain after the change of variable $r=|\tau|\rho$ that

$$\text{(E.9)}\quad F_{t\to\tau}\Gamma u_1 = \frac{i}{2}|\tau|^{d-2}\operatorname{sign}\tau\,(2\pi)^{-d}\int_{S^{d-2}}\int_{\gamma}\frac{1}{\sqrt{1-\rho^2}}\exp(-2iy_1|\tau|\sqrt{1-\rho^2})\rho^{d-2}\,d\rho\, d\sigma,$$

$$\text{(E.9)'}\quad F_{t\to\tau}\Gamma u_1 = |\tau|^{d-2}\operatorname{sign}\tau\,(2\pi)^{-d}\int_{S^{d-2}}\int_{\gamma}\{2\beta(\sigma)\varepsilon(\sigma)(-1+\rho^2(1-\beta^2(\sigma)))^{-1}\,\bullet$$

$$\rho \exp(-2y_1|\tau|\beta(\sigma)) +$$

$$\frac{i}{2}(-i\sqrt{1-\rho^2}+\beta(\sigma)\rho)^{-1}(i\sqrt{1-\rho^2}+\beta(\sigma)\rho)\cdot$$

$$\frac{1}{\sqrt{1-\rho^2}}\exp(-2iy_1|\tau|\sqrt{1-\rho^2})\}\rho^{d-2}\,d\rho\, d\sigma =$$

$$\frac{i}{2}|\tau|^{d-2}\operatorname{sign}\tau \times (2\pi)^{-d}\int_{\mathbb{S}^{d-2}}\int_{\gamma}(-i\sqrt{1-\rho^2}+\beta(\sigma)\rho)^{-1}\cdot$$

$$(i\sqrt{1-\rho^2}+\beta(\sigma)\rho)\frac{1}{\sqrt{1-\rho^2}}\exp(-2iy_1|\tau|\sqrt{1-\rho^2})\rho^{d-2}\,d\rho\, d\sigma$$

respectively where γ is a closed contour with a counter-clockwise orientation lying in $\{\operatorname{Re}\rho > 0\} \cup \{0\}$; the segment $[0,1]$ and points $\rho_\kappa = (1-\beta_\kappa^2(\sigma))^{-1/2}$ with $\beta_\kappa(\sigma) \in [0,1)$ must lie inside γ; $\operatorname{Im}\sqrt{1-\rho^2} < 0$.

REMARKS. (i). The integrand has regular limits as $\rho \to \rho'(1 \pm i0)$, $\rho' \in \mathbb{R} \setminus (\{-1,1\} \cup \bigcup_\kappa \{-\rho_\kappa, \rho_\kappa\})$ these limits coincide for $|\rho'| > 1$ and differ for $|\rho'| < 1$.
(ii) Γu_1 for $b = \mathrm{Id}$ and $b = iD_1$ differ only by their signs.

After the change of variable $\rho = \sqrt{1-w^2}$ we obtain that

(E.10)
$$F_{t\to\tau}\Gamma u_1 = \frac{i}{2}|\tau|^{d-2}\operatorname{sign}\tau\,(2\pi)^{-d}\int_{\mathbb{S}^{d-2}}\int_{\gamma'}(1-w^2)^{\frac{d-3}{2}}e^{-2iy_1|\tau|w}\,dw\,d\sigma,$$

(E.10)'
$$F_{t\to\tau}\Gamma u_1 = \frac{i}{2}|\tau|^{d-2}\operatorname{sign}\tau \times (2\pi)^{-d}\cdot$$

$$(2\pi)^{-d}\int_{\mathbb{S}^{d-2}}\int_{\gamma'}(-iw+\beta(\sigma)\sqrt{1-w^2})^{-1}(iw+\beta(\sigma)\sqrt{1-w^2})(1-w^2)^{\frac{d-3}{2}}e^{-2iy_1|\tau|w}\,dw\,d\sigma$$

respectively where γ' is a contour with the starting point -1 and the endpoint $+1$ lying in the lower complex half-plane below all the points $w_\kappa = -i\beta_\kappa(\sigma)/\sqrt{1-\beta_\kappa^2(\sigma)}$ with $0 \leq \beta_\kappa(\sigma) < 1$, $\mathrm{Re}(1-w^2)^{1/2} > 0$.

Hence

$$\int_0^\kappa \tau \, F_{t\to\tau} \Gamma u_1 \, d\tau = \frac{1}{2i} \kappa^d \, Q(y', 2y_1 \kappa)$$

$(\kappa > 0)$ where

(E.11) $$Q(y',s) = (2\pi)^{-d} \omega_{d-1} (d-1) \int_{\gamma'} (1-w^2)^{\frac{d-3}{2}} (-iw)^{-d} \left(\frac{\partial}{\partial s}\right)^{d-1} \{\tfrac{1}{s}(1-e^{-isw})\} \, dw,$$

(E.11)' $$Q(y',s) = (2\pi)^{-d} \int_{S^{d-2}} \int_{\gamma'} (-iw+\beta(\sigma)\sqrt{1-w^2})^{-1} (iw+\beta(\sigma)\sqrt{1-w^2}) \cdot$$
$$(1-w^2)^{\frac{d-3}{2}} (-iw)^{-d} \left(\frac{\partial}{\partial s}\right)^{d-1} \{\tfrac{1}{s}(1-e^{-isw})\}$$

respectively;

(E.12)' $$2ic_0' = \frac{1}{4}(2\pi)^{-d} \int_{S^{d-2}} \int_{\gamma'} \mathrm{tr}(-iw+\beta(\sigma)\sqrt{1-w^2})^{-1} (iw+\beta(\sigma)\sqrt{1-w^2}) \cdot$$
$$(1-w^2)^{\frac{d-3}{2}} w^{-1} \, dw \, d\sigma$$

provided $b = i\xi_1 + \beta(D')$ and $2ic_0' = \pm \frac{1}{4}(2\pi)^{-d+1} \omega_{d-1}(d-1) D$ provided either $b = i\xi_1$ or $b = \mathrm{Id}$ (the famous coefficients of H.Weyl),

(E.13) $$2id_0 = Q(y', \sigma) = -(2\pi)^{-d} \omega_d \, \mathrm{Id},$$

(E.13)' $$2id_0 = -\frac{(2\pi)^{-d}}{d} \int_{S^{d-2}} \int_{\gamma'} (-iw+\beta(\sigma)\sqrt{1-w^2})^{-1} (iw+\beta(\sigma)\sqrt{1-w^2}) \cdot$$

$$(1-w^2)^{\frac{d-3}{2}}\,dw$$

respectively.

E.3. Let now $\mathcal{A}=\Delta$, $B=i\tau D_1+B_1\tau$, $\Pi=\Pi^+$. Then $\Pi=0$,

$$\Pi^+ = F^{-1}_{\xi'\to x'}\ \Pi'(\xi')F_{x'\to\xi'}$$

where $\Pi'(\xi')$ was introduced in the previous section, $u_0=0$ and $v^{\pm}=F_{x',t\to\xi',\tau}\ u_1^{\pm}$ is the solution of problem

(E.14) $$(-\tau^2-|\xi'|^2-D_1^2)v^{\pm}=\pm(2\pi)^{-d}2\beta(\xi')\varepsilon(\xi')exp(-(x_1+y_1)\beta(\xi')-i\langle y',\xi'\rangle).$$

(E.15) $$(iD_1+\beta(\xi'))v^{\pm}\big|_{x_1=0}=0$$

$$v^{\pm}=O(1)\ \text{as}\quad x_1\to+\infty .$$

Then

(E.16) $$F_{x',t\to\xi',\tau}u_1^{\pm}=\pm(2\pi)^{-d}(-\tau^2-|\xi'|^2+\beta^2(\xi'))^{-1}2\beta(\xi')\,\varepsilon(\xi')\cdot$$

$$exp(-(x_1+y_1)\beta(\xi')-i\langle y',\xi'\rangle)\ .$$

We apply the inverse Fourier transform $F^{-1}_{\xi'\to x'}$ and set $x=y$; we introduce the spherical coordinate system $(\tau,\sigma)\in\mathbb{R}^+\times S^{d-2}$ in $\mathbb{R}^{d-1}_{\xi'}\setminus 0$; note that the replacement of $\tau\in\mathbb{R}$ by $\tau(1\mp i0)$ is equivalent to the replacement of τ by $\tau(1\pm i0)$ and compute obtained contour integral; then

(E.17) $$F_{t\to\tau}\Gamma u_1=-i(2\pi)^{-d+1}|\tau|^{d-2}sign\,\tau\int_{S^{d-2}}\beta(\sigma)(\beta^2(\sigma)-1)^{-d/2}\cdot$$

$$exp(-2y_1|\tau|\beta(\sigma)(\beta^2(\sigma)-1)^{-1/2})\,\varepsilon(\sigma)d\sigma .$$

Hence

$$\int_0^{\kappa} \tau F_{t\to\tau} \Gamma u_1 \, d\tau = \frac{1}{2i} \kappa^d Q(y', 2y_1 \kappa)$$

$(\kappa > 0)$ where

(E.18) $$Q(y', s) = 2(2\pi)^{-d+1} \int_{S^{d-2}} \beta^{-d+1}(\sigma) \left(-\frac{\partial}{\partial s}\right)^{d-1} \left\{ \frac{1}{s} \left(1 - \exp(s\beta(\sigma) \cdot (\beta^2(\sigma) - 1)^{-1/2})\right) \right\} \varepsilon(\sigma) \, d\sigma ,$$

(E.19) $$2i c_0' = (2\pi)^{-d+1} \int_{S^{d-2}} \operatorname{tr} (\beta^2(\sigma) - 1)^{-\frac{d-1}{2}} \varepsilon(\sigma) \, d\sigma$$

and

(E.20) $$2i \, d_0'(y') = \frac{2(2\pi)^{-d+1}}{d} \int_{S^{d-2}} \beta(\sigma) (\beta^2(\sigma) - 1)^{-d/2} \varepsilon(\sigma) \, d\sigma .$$

6. The asymptotics for first-order operators

6.1. Let X be a compact d-dimensional C^∞-manifold with the boundary $Y \in C^\infty$, $d \geq 2$, dx a C^∞-density on X, E a Hermitian D-dimensional C^∞-fibering over X. Let $\mathcal{A} : C^\infty(X,E) \to C^\infty(X, E)$ be a first-order elliptic differential operator, formally selfadjoint with respect to inner product in $L_2(X, E)$. Let F be a $D/2$-dimensional C^∞-subfibering of $E|_Y$, $b: E|_Y \to E' = E|_Y/F$ a natural mapping, τ the operator of restriction to Y, $B = b\tau$ boundary operator. We assume that

(E) Operator $\{\mathcal{A}, B\} : C^\infty(X,E) \to C^\infty(X, E) \oplus C^\infty(Y, E')$ satisfies the Sapiro-Lopatinskii condition,

(S) $\mathcal{A}_B$ - the restriction of $\mathcal{A}$ to $\operatorname{Ker} B$ - is a symmetric operator in $L_2(X, E)$.

Let $A_B : L_2(X,E) \to L_2(X,E)$ be a closure of $\mathcal{A}_B$ in $L_2(X,E)$; then A_B is selfadjoints, its spectrum is discrete, with finite multiplicity and tends to $\pm\infty$. Without a loss of generality one can assume that 0 is not an eigenvalue of A_B.

REMARK. We shall not use selfadjoint projectors $\Pi^\pm$ now. But it should be noted that these projectors do not belong to Boutet de Monvel operator algebra now provided Y is non-empty because $\Pi^\pm(x,-\xi) = \Pi^\mp(x,-\xi)$ and there is no transmission property now. It is an essential distinction between differential spectral problems of even and odd orders on manifolds with the boundary.

Let H be a closed subspace in $L_2(X,E)$ such that Π - the selfadjoint projector to H - belongs to Boutet de Monvel algebra. We suppose that $\Pi A_B \subset A_B \Pi$; then $A_{B,H} : H \to H$ - restriction of A_B to H - is selfadjoint operator; its spectrum is discrete, with finite multiplicity and tends either to $\pm\infty$, or to $+\infty$ or to $-\infty$.

REMARK. If Π is not a singular Green operator then $A_{B,H}$ cannot be semibounded. Examples show that $A_{B,H}$ can be semibounded provided Π is a singular Green operator but not a smoothing operator.

Let us introduce the eigenvalue distribution functions for $A_{B,H}$: $N^\pm(k) = N^\pm_H(k)$ is the number of eigenvalues of $A_{B,H}$ lying between 0 and $\pm k$, $k > 0$.

Let $E(s)$ be spectral selfadjoint projectors of A_B, $E_H(s) =$

$\Pi E(s)$ and $e_H^{\pm}(k) = \pm(E_H(\pm k) - E_H(0))$; let $e_H^{\pm}(x,y,k)$ be Schwartz kernels of $e_H^{\pm}(k)$.

We are interested in the asymptotics of $N_H^{\pm}(k)$ and $e_H^{\pm}(x,x,k)$ as $k \to \infty$ assuming, of course, that $\pm A_{B,H}$ is not semibounded from above.

6.2. Note first that these asymptotics were derived in [57] provided X is closed and A is a pseudo-differential operator. What is to say, the following asymptotics were derived:

(6.1) $$N_H^{\pm}(k) = \varkappa_0^{\pm} k^d + O(k^{d-1}) \quad \text{as} \quad k \to \infty,$$

(6.2) $$e_H^{\pm}(x,x,k) = \alpha_0^{\pm}(x) k^d + O(k^{d-1})$$

as $k \to \infty$ uniformly with respect to $x \in X$ where

(6.3) $$\alpha_0^{\pm}(x) = \int_{T_x^*X} \varepsilon^{\pm} \, d\!\!{}^-\xi \,,$$

(6.4) $$\varkappa_0^{\pm} = \int_X \operatorname{tr} \alpha_0^{\pm}(x) dx = \int_{T^*X} \operatorname{tr} \varepsilon^{\pm} \, dx \, d\!\!{}^-\xi,$$

$\varepsilon^{\pm} = \pm(\varepsilon(\pm 1) - \varepsilon(0))\Pi$, $\varepsilon(s)$ are the spectral projectors for a; a, Π are the principal symbols of A, Π.

Under certain assumptions the following asymptotics were derived:

(6.5) $$N_H^{\pm}(k) = \varkappa_0^{\pm} k^d + \varkappa_1^{\pm} k^{d-1} + o(k^{d-1}) \quad \text{as} \quad k \to \infty$$

where

(6.6) $$\varkappa_1^{\pm} = \mp d \int_{T^*X} \operatorname{tr} \varepsilon^{\pm} a^s \, dx \, d\!\!{}^-\xi + \bar{\varkappa}_1^{\pm},$$

a^s is the subprincipal symbol of A, $\bar{\varkappa}^{\pm}$ depend only on (see Appendix F).

REMARK. If A is a differential operator and $\Pi(x,\xi) = \Pi(x,-\xi)$ then $\varkappa_0^+ = \varkappa_0^-$, $d_0^+ = d_0^-$, $\varkappa_0^+ = -\varkappa_1^-$ and assumptions of [57] providing asymptotics (6.5) for positive and negative spectrum coincide.

6.3. Let now X be a manifold with the boundary. Let $x_1 \in C^\infty$, $x_1 = 0$ and $dx_1 \neq 0$ at Y, $x_1 > 0$ in $X \setminus Y$. Consider the characteristical symbol $g(\rho, \tau) = \det(\tau + a(\rho))$.

REMARK. In contrast to the case of even-order operators we use the common characteristical symbol for positive and negative spectrum asymptotics and the conditions providing positive and negative spectrum asymptotics will coincide.

By means of a characteristical symbol g one can formulate the notion of multiplicity of the point and the concepts of positive, negative, tangential and indefinite points and, finally, formulate conditions (H.1) - (H.4) as in §0.

THEOREM 6.1. (compare with theorem 0.3). Let Π be not a singular Green operator. Then asymptotics (6.1) hold with the same coefficients $\varkappa_0^\pm > 0$.

THEOREM 6.2 (compare with theorem 0.4). Let Π be a singular Green operator. Then the following asymptotics hold:

$$N_H^\pm(k) = \varkappa_1'^{\pm} k^{d-1} + O(k^{d-2}) \text{ as } k \to \infty \tag{6.7}$$

where $\varkappa_1'^{\pm}$ depend only on $a|_Y$, b and Π_0' – the principal symbol of Π, $\varkappa_1'^{\pm} = 0$ if and only if $\pm A_{B,H}$ is semibounded from above.

THEOREM 6.3. (compare with theorem 0.5). Let Π be not a singular Green operator and condition (H.2) be fulfilled. Assume that the set of points periodic with respect to geodesic flow Φ with reflection at the boundary has measure zero. Then the following asymptotics hold:

$$N_H^\pm(k) = \varkappa_0^\pm k^d + (\varkappa_1^\pm + \varkappa_1''^{\pm}) k^{d-1} + o(k^{d-1}) \quad \text{as} \quad k \to \infty \tag{6.5$'$}$$

where $\varkappa_0^\pm$, $\varkappa_1^\pm$ are given by (6.4), (6.6) and $\varkappa_1''^{\pm}$ depend only on $a|_Y$, b, $\Pi|_Y$ and Π_0'.

6.4. In this section we give the asymptotics for $e_H^\pm(x, x, k)$ near the boundary. Identify some neighbourhood of Y with $[0,\delta) \times Y$; then point x will be identified with (x_1, x'). If condition (H.4) is fulfilled then $x_1 = \operatorname{dist}(x, Y)$ and this identification is canonical.

THEOREM 6.4 (compare with theorem 0.6). Let Π be not a singular Green operator and condition (H.2) be fulfilled. Then in the neighbourhood of Y the following asymptotics hold:

(6.8) $$e_H^{\pm}(x,x,k) = k^d(\mathscr{d}_0^{\pm}(x) + Q_0(x',x_1 k)) + O(k^{d-1})$$

as $k \to \infty$ uniformly with respect to x where $\mathscr{d}_0^{\pm}(x)$ is given by (6.3), $Q_0^{\pm} \in C^{\infty}(Y \times \bar{\mathbb{R}}^+)$ satisfy (0.10).

THEOREM 6.5 (compare with theorem 0.7). Let Π be not a singular Green operator and conditions (H.3), (H.4) be fulfilled. Then for $d \geqslant 3$ the asymptotics (6.8) hold, for $d=2$ the following asymptotics hold:

(6.8)' $$e_H^{\pm}(x,x,k) = k^2(\mathscr{d}_0^{\pm}(x) + Q^{\pm}(x',x_1 k)) + O(k^{3/2})$$

as $k \to \infty$ uniformly with respect to x where $Q_0^{\pm} \in C^{\infty}(Y \times \bar{\mathbb{R}}^+)$ satisfy (0.10).

THEOREM 6.6 (compare with theorem 0.8). Let Π be a singular Green operator. Then the following asymptotics hold:

(6.9) $$e_H^{\pm}(x,x,k) = k^d Q^{\pm}(x',x_1 k) + O(\min(k^{d-1}, x_1^{-d+1}))$$

as $k \to \infty$ uniformly with respect to x where $Q_0^{\pm} \in C^{\infty}(Y \times \bar{\mathbb{R}}^+)$ satisfy (0.14).

THEOREM 6.7 (compare with theorem 0.9). The following asymptotics hold:

(6.10) $$e_H^{\pm}(x,x,k) = \mathscr{d}_0'^{\pm}(x) k^d + O(k^{d-1})$$

as $k \to \infty$ uniformly with respect to $x \in Y$ where $\mathscr{d}'^{\pm}(x)$ depend only on $a(x,\cdot)$, $b(x,\cdot)$, $\Pi_0'(x,\cdot,\cdot)$, $\pi(x,\cdot)$.

REMARKS (i). Recall that if Π is a singular Green operator then $A_{B,H}$ can be semibounded.

(ii) All remarks to theorems 0.3 - 0.9 remain true.

(iii) In contrast to the case of even -order operators asymptotics (6.7), (6.9) cannot appear unless Π is a singular Green operator.

6.5. One can prove theorems 6.1 - 6.4 by repeating the arguments of Part I with essential simplifications and inessential modifications:

(i) §1 must be omitted.

(ii) $u(x,y,t)$ is the Schwartz kernel of operator $exp(-iA_B t)\Pi$; $u(x,y,t)$ satisfies boundary problem

$$u_t + i\mathcal{A}_x u = 0, \tag{6.11}$$

$$B_x u = 0, \tag{6.12}$$

$$u|_{t=0} = K_\Pi(x,y) \tag{6.13}$$

as well as the dual problem with respect to y,t (compare with (0.18) - (0.20), $(2.3)^\dagger$ - $(2.4)^\dagger$). It should be noted that instead of two equalities $u(x,y,t) = u(x,y,-t)$ and $u^\dagger(x,y,t) = u(y,x,t)$ we have no only one equality $u^\dagger(x,y,t) = u(y,x,-t)$. It also should be noted that problem (6.11) - (6.13) is well-posed a fortiori and problem (0.18) - (0.20) could be ill-posed.

(iii) Theorem 2.3 concerning the finite speed of propagation of singularities can be proved by repeating the arguments of §2 with certain simplifications and obvious modifications; the proof of theorem 2.4 does not change.

(iv) The proofs of the theorems about normality of the great singularity do not change at all; but we have the first-order differential system at the beginning.

(v) Subspaces E', E'' are mising in the decomposition of E in the normal rays zone and hence we consider only the hyperbolic systems in this zone; we have no need to consider pseudo-differential operators with respect to x' and hence the complete asymptotics are derived.

(vi) We need to consider only hyperbolic parametrices in §4 because of the well-posedness of the problem (6.11) - (6.13); hence all factors are holomorphic with respect to $\tau \in \mathbb{C} \setminus \mathbb{R}$ (in part I these factors were holomorphic with respect to $\tau \in \mathbb{C} \setminus (\mathbb{R} \cup i\mathbb{R})$ only unless $a(x,\xi)$ was definite). The parametrices in the half-space for operators with constant coefficients and homogeneous symbols are given by formulas

$$\bar{G}_0^\pm = F^{-1}_{\xi,\tau\to x,t} (\tau \mp i0 + a(\xi))^{-1} (1-\zeta(\xi)) F_{x,t\to\xi,\tau} ,$$

$$\bar{G}'^{\pm} = F^{-1}_{\xi',\tau\to x',t}\, \mathcal{G}'(x_1,\xi',\tau \mp i0)(1-\zeta(\xi))\, F_{x',t\to\xi',\tau}$$

where $\zeta \in C_0^{\infty}(\mathbb{R}^d)$, $\zeta' \in C_0^{\infty}(\mathbb{R}^{d-1})$, $\zeta = 1$, $\zeta' = 1$ in the neighbourhoods of 0,

$$\mathcal{G}'(x_1,\xi',\tau) = \int_{\gamma} (\tau + a(\xi_1,\xi'))^{-1} e^{ix_1\xi_1} d\xi_1 \cdot b^* S^{-1}(\xi',\tau),$$

γ is a closed contour with a counter-clockwise orientation lying in $\mathbb{C}_+$; all the roots of the polynomial $g(\xi_1, \xi', \tau)$ lying in $\mathbb{C}_+ \setminus \mathbb{R}$ must be contained inside γ,

$$S(\xi',\tau) = b\int_{\gamma} (\tau + a(\xi_1,\xi'))^{-1} b^* d\xi_1 =$$

$$\int_{-\infty}^{\infty} b(\tau + a(\xi_1,\xi'))^{-1} b^* d\xi_1 ;$$

note that $b a_0^{-1} b^* = 0$ because $\bar{A}_{\bar{B}}$ is symmetrical operator; $a(\xi_1, \xi') = a_0 \xi_1 + a_1(\xi')$; moreover

$$|S^{-1}(\xi',\tau)| \leq C(|\tau| + |\xi'|)\, |\mathrm{Im}\, \tau|^{-1},$$

$\mathcal{G}'$ satisfies the problem

$$(a(D_1,\xi') + \tau)\, \mathcal{G}' = 0,$$

$$b r \mathcal{G}' = \mathrm{Id}, \quad \mathcal{G}' = O(1) \text{ as } x_1 \to +\infty ;$$

all other formulas do not change.

7. The asymptotics for second-order spectral problems

7.1. Let X be a compact d-dimensional C^∞-manifold with the boundary Y, $d \geq 2$, dx a C^∞-density on X, E a Hermitian D-dimensional C^∞-fibering over X. Let $\mathcal{A}: C^\infty(X,E) \to C^\infty(X,E)$ be a second-order elliptic differential operator, formally selfadjoint with respect to inner product in $L_2(X,E)$. Let ν be a C^∞ vector field transversal to Y at every point, τ the operator of restriction to Y, $B=\tau$ or $B=\tau\nu+B_1\tau$ a boundary operator where $B_1: C^\infty(Y,E) \to C^\infty(Y,E)$ is a first-order classical pseudo-differential operator on Y.

We suppose that

(E) Operator $\{\mathcal{A},B\}: C^\infty(X,E) \to C^\infty(X,E) \oplus C^\infty(Y,E)$ satisfies the Šapiro - Lopatinskii condition,

(S) $\mathcal{A}_B$ - the restriction of $\mathcal{A}$ to $\operatorname{Ker} B$ - is a symmetric operator in $L_2(X,E)$.

(P) $\mathcal{A}_B$ is a positive definite in $L_2(X,E)$ operator.

Let $A_B: L_2(X,E) \to L_2(X,E)$ be a closure of $\mathcal{A}_B$ in $L_2(X,E)$; then A_B is selfadjoint positive definite operator. Let $\mathcal{H} = \mathcal{D}(A_B^{1/2})$.

Let $\mathcal{J} = \mathcal{J}(x): E_x \to E_x$ be an invertible Hermitian C^∞-matrix, $\mathcal{U} = \mathcal{J}^{-1}A_B: \mathcal{H} \to \mathcal{H}$, $\mathcal{D}(\mathcal{U}) = A_B^{-1}\mathcal{J}\mathcal{H}$. Then $\mathcal{U}$ is a selfadjoint in $\mathcal{H}$ operator, its spectrum is discrete, with finite multiplicity and tends either to $\pm\infty$, or to $+\infty$ or to $-\infty$.

PROPOSITION 7.1. (i) $\mathcal{U}$ is semibounded from below (above) if and only if $\mathcal{J}$ is positive (negative) definite; then $\mathcal{U}$ is positive (negative) definite.

(ii) $\Pi^\pm$ - selfadjoint projectors to positive and negative invariant subspaces of $\mathcal{U}$ in $\mathcal{H}$ - belong to Boutet de Monvel operator algebra. Moreover, $\pi^\pm$ - the principal symbols of pseudo-differential operators $\Pi^{0\pm}$ are projectors to positive and negative invariant subspaces of $\alpha = \mathcal{J}^{-1}a$; a, α are the principal symbols of $\mathcal{U}, \mathcal{A}, \alpha$ and $\pi^\pm$ are Hermitian with respect to inner product $[v,w] = \langle av, w\rangle$.

Let H be a closed subspace in $\mathcal{H}$ such that Π - selfadjoint projector to H - belongs to Boutet de Monvel operator algebra. We suppose that $\Pi\mathcal{U} \subset \mathcal{U}\Pi$; then $\mathcal{U}_H: H \to H$ - a restriction of $\mathcal{U}$ to H - is a selfadjoint operator; its spectrum is discrete, with finite multiplicity and tends either to $\pm\infty$, or to $+\infty$, or to $-\infty$.

Let us introduce the eigenvalue distribution functions for $\mathcal{U}_H : N^{\pm}(k) = N^{\pm}_H(k)$ is the number of eigenvalues of $\mathcal{U}_H$ lying between 0 and $\pm k^2, k>0$.

Let $E(s)$ be spectral selfadjoint projectors of $\mathcal{U}$, $E_H(s) = \Pi E(s)$ and $e^{\pm}_H(k) = \pm(E_H(\pm k^2) - E_H(0))$; let $e^{\pm}_H(x, y, k)$ be Schwartz kernels of $e^{\pm}_H(k)$.

We are interested in the asymptotics of $N^{\pm}_H(k)$ and $e^{\pm}_H(x, x, k)$ as $k \to \infty$; note that $\pm e^{\pm}_H(x,x,k)\mathcal{J}^{-1}(x)$ are monotone non-decreasing functions of $k>0$.

Let $g_{\pm}(\rho,\tau) = \det(\tau^2 \mp \alpha(\rho))$ be the characteristical symbols, α^s the subprincipal symbol of $\mathcal{U}$.

The main result of this paragraph:

THEOREM 7.2. The statements of theorems 0.1 - 0.9 remain true in the situation described; in formulas (0.3), (0.4), (0.6) a and a^s must be replaced by α and α^s respectively.

REMARKS. (i) It goes without saying, we consider the selfadjoint spectral problem

$$(7.1) \qquad (A_B - \mu \mathcal{J})u = 0 \qquad u \in \mathcal{D}(A_B).$$

(ii) As we mentioned above, the case " $\mathcal{J}$ is positive definite and A_B satisfies (E), (S)" can be reduced to the case " $\mathcal{J}=I$, $A'_B = \mathcal{J}^{-1/2} A_B \mathcal{J}^{-1/2}$ satisfies (E), (S)" which was investigated in Part I.

(iii) If X is a closed manifold then operator $A^{1/2}$ is an elliptic first-order classical pseudo-differential operator, hence $A' = A^{1/2}\mathcal{U}A^{-1/2} = A^{1/2}\mathcal{J}^{-1}A^{1/2}$ is an elliptic second-order classical pseudo-differential operator selfadjoint in $L_2(X, E)$; $\Pi' = A^{1/2}\Pi A^{-1/2}$ is a selfadjoint projector in $L_2(X,E)$ and a classical pseudo-differential operator of order 0 . Thus the case of indefinite $\mathcal{J}$ and positive definite A can be reduced to the case of $\mathcal{J} = I$ and indefinite A provided X is a closed manifold.

(iv) Let $\Pi = I$. Then either $N^{\pm}(k) = O(1)$ (provided $\mp\mathcal{J}$ is positive definite) or $N^{\pm}(k) = \varkappa^{\pm}_0 k^d + O(k^{d-1})$, $\varkappa^{\pm}_0 > 0$ (provided $\mp\mathcal{J}$ is not positive definite). Thus in contrast to Part I the asymptotics $N^{\pm}(k) = \varkappa^{\pm}_1 k^{d-1} + O(k^{d-2})$, $\varkappa^{\pm}_1 > 0$ can appear only provided $\Pi \neq I$; but in contrast to § 6 this asymptotics appears provided $\Pi\Pi^{\pm}$ is a singular Green operator; Π can be not a singular Green operator.

7.2. One can prove theorems 0.3 - 0.9 in a new situation by repeating the arguments of Part I with certain modifications:

In §1 we need to change the proofs of theorem 1.10 and proposition 1.16:

(i) THEOREM 1.10 ($\Pi^{\pm}$ belong to Boutet de Monvel operator algebra). We shall assume here that $\mathcal{A}$ is an operator of order m, m is even, boundary conditions are of the same type as in theorem 1.10 and A_B is a positive definite selfadjoint operator and, finally, $\mathfrak{I}$ is an invertible Hermitian matrix. Then the selfadjoint projectirs $\Pi^{\pm}$ are given by formulas

$$\Pi^{\pm}=\frac{1}{2}\pm\frac{m}{\pi}\int_0^{\infty}(A_B+i\mathfrak{I}z^m)^{-1}\mathfrak{I}(A_B-i\mathfrak{I}z^m)^{-1}z^{m-1}dz\cdot A_B \tag{7.2}$$

(compare with (1.4')). Without a loss of generality one can assume that in the neighbourhood of Y

$$\mathcal{A}=\mathrm{Id}\,D_1^m+a_0'(x)x_1D_1^m+a_1(x,D')D_1^{m-1}+\dots+a_m(x,D'),$$

$$\mathfrak{I}=\begin{pmatrix}\mathfrak{I}^+(x') & 0\\ 0 & -\mathfrak{I}^-(x')\end{pmatrix}+x_1\mathfrak{I}'(x)$$

where a_j are differential operators of order j, $\mathfrak{I}^{\pm}(x')$ are positive definite Hermitian matrices. Then in corresponding notations

$$\bar{\mathcal{A}}=\bar{\bar{\mathcal{A}}}=\begin{pmatrix}\bar{\mathcal{A}}^+ & 0\\ 0 & \bar{\mathcal{A}}^-\end{pmatrix},\quad \bar{\mathcal{A}}^{\pm}=D_1^m+\Lambda^{\pm}(x',D'),$$

$$\bar{\mathfrak{I}}=\begin{pmatrix}\mathfrak{I}^+(x') & 0\\ 0 & -\mathfrak{I}^-(x')\end{pmatrix},\quad \bar{\bar{\mathfrak{I}}}=\begin{pmatrix}\mathfrak{I}^+(x') & 0\\ 0 & \mathfrak{I}^-(x')\end{pmatrix}$$

and after these modifications the proof of the theorem 1.10 fits the new situation.

(ii) PROPOSITION 1.16. Inequality (1.12) must be replaced by inequality

(7.3) $$\mathrm{Re}(\mathcal{A}v,v) \geqslant c_0 \|v\|_1^2 - C\|Bv\|^2_{Y,\frac{1}{2}-\mu}$$

for every $v \in H^2(X,E)$ where $\mu = \mathrm{ord}\, B$ and $c_0 > 0$. By the same method one can prove inequality

(7.4) $$\mathrm{Re}\,(A\mathcal{U}\Pi^+ w, w) \geqslant c_1 \|\Pi^+ w\|_2^2$$

$(c_1 > 0)$ for every $w \in \mathcal{D}(\mathcal{U})$; this inequality combined with equalities $A_B = \mathcal{J}\mathcal{U}$, $\Pi^+\mathcal{U} \subset \mathcal{U}\Pi^+$ implies inequality

(7.5) $$\mathrm{Re}\,(\mathcal{J}\Pi^+ v, \Pi^+ v) \geqslant c_2 \|\Pi^+ v\|_0^2 \qquad (c_2 > 0)$$

7.3. (iii) $u(x,y,t)$ is the Schwartz kernel of operator

$$\mathcal{U}^{-\frac{1}{2}} \sin(\mathcal{U}^{\frac{1}{2}} t)\, \Pi\Pi^+ \mathcal{J}^{-1};$$

then $u(x,y,t)$ satisfies boundary problem

(7.6) $$P_x u \equiv \mathcal{J}_x u_{tt} + \mathcal{A}_x u = 0,$$

(7.7). $$B_x u = 0,$$

(7.8) $$u\big|_{t=0} = 0, \quad u_t\big|_{t=0} = K_{\Pi\Pi^+\mathcal{J}^{-1}}(x,y)$$

as well as the dual problem with respect to (y,t) (compare with (0.18) - (0.20), $(2.3)^\dagger$ - $(2.4)^\dagger$, (6.11) - (6.13)).

(iv) Theorem 2.3 about the finite speed of propagation of singularities can be proved by repeating the arguments of §2 using inequalities (7.3), (7.5) instead of (1.12).

(v) The arguments of §3 need no change; we need only to prove inequality

(7.9) $$-i\langle e'^* K \frac{\partial e'}{\partial \tau} w, w \rangle > c_0 |w|^2$$

$$\forall w \in \mathrm{Ran}\, e' \cap \mathrm{Ker}\, \mathcal{B}_0, \quad c_0 > 0$$

(see the proof of proposition 3.9). This inequality is equivalent to the following inequality:

(7.10) If $V \in \mathcal{S}(\overline{\mathbb{R}}^+, E \oplus E)$, $V \neq 0$ is the solution of problem

$$(KD_1 + L_1(x', \xi', \tau))V(x_1) = 0,$$

$$\mathcal{B}_0(x', D_1, \xi')V|_{x_1=0} = 0,$$

then

$$\int_0^\infty < \frac{\partial L_1}{\partial \tau}(x', \xi', \tau)V(x_1), V(x_1) > dx_1 > 0;$$

we used proposition B.

Let v_0 be the first component of V, $V = \begin{pmatrix} v_0 \\ v_1 \end{pmatrix}$; then (7.10) is equivalent to the following inequality:

(7.11) If $v \in \mathcal{S}(\overline{\mathbb{R}}^+, E)$, $v_0 \neq 0$ is the solution of problem

$$(a(x', D_1, \xi') - \tau^2 \mathcal{J}(x'))\, v_0(x_1) = 0,$$

$$b(x', D_1, \xi')\, v_0\, |_{x_1=0} = 0$$

then

$$\int_0^\infty < \mathcal{J}(x')v_0(x_1), \; v_0(x_1) > dx_1 > 0.$$

Since $\mathcal{J}(x')v_0 = \tau^{-2} a(x', D_1, \xi')\, v_0(x_1)$ then the left part of (7.12) equals

$$\tau^{-2} \int_0^\infty < a(x', D_1, \xi')\, v_0, v_0 > dx_1$$

and inequality (7.11) holds because of (7.3); (7.9) has been proved.
(vi) The arguments of §4 need no change excepting of the construction of parametrices for the problems in the half-space. Let $B = \tau$; then we set

$$\mathcal{G}'(x_1,\xi',\tau)=\int_\gamma(-\mathfrak{I}\tau^2+a(\xi_1,\xi'))^{-1}e^{ix_1\xi_1}d\xi_1\cdot S^{-1}(\xi',\tau) \tag{7.12}$$

where

$$S(\xi',\tau)=\int_\gamma(-\mathfrak{I}\tau^2+a(\xi_1,\xi'))^{-1}d\xi_1=\int_{-\infty}^{\infty}(-\mathfrak{I}\tau^2+a(\xi_1,\xi'))^{-1}d\xi_1. \tag{7.13}$$

Then

$$i\{\tau^2 S^*(\xi',\tau)-\bar{\tau}^2 S(\xi',\tau)\} =$$

$$\operatorname{Im}\tau^2\int_{-\infty}^{\infty}(-\mathfrak{I}\tau^2+a(\xi_1,\xi'))^{-1}\,a(\xi_1,\xi')$$

$$(-\mathfrak{I}\bar{\tau}^2+a(\xi_1,\xi'))^{-1}\,d\xi_1$$

and hence $S(\xi',\tau)$ is invertible for $\tau\in\mathbb{R}\cup i\mathbb{R}$ and

$$\|S^{-1}(\xi',\tau)\|\leqslant C\,|\operatorname{Im}\tau^2|^{-1}(|\tau|^2+|\xi'|)^3. \tag{7.14}$$

Let now $B = i\tau D_1 + B_1\tau$; then we set

$$\mathcal{G}'(x_1,\xi',\tau)=\int_\gamma(-\mathfrak{I}\tau^2+a(\xi_1,\xi'))^{-1}e^{ix_1\xi_1}d\xi_1\cdot\tilde{S}^{-1}(\xi',\tau) \tag{7.12'}$$

where

$$\widetilde{S}(\xi',\tau)=\int_\gamma (i\xi_1+b_1(\xi'))(-\mathcal{J}\tau^2+a(\xi_1,\xi'))^{-1}\,d\xi_1 . \tag{7.13)'}$$

We wish to prove that $\widetilde{S}(\xi',\tau')$ is invertible for $\tau \in \mathbb{R}\cup i\mathbb{R}$. Note that the operator $(-\tau^2\mathcal{J}+A_B)$ is invertible for $\tau \in \mathbb{R}\cup i\mathbb{R}$ and the following estimate hold:

$$\|u\|_2+|\tau|^2\|u\|_0 \leqslant C\,|\mathcal{J}m\tau^2|^{-1}|\tau|^2\,\|(-\tau^2\mathcal{J}+A_B)u\|_0$$

$\forall u \in \mathcal{D}(A_B)$ because A_B is positive definite.
Hence the following estimates hold:

$$\|u\|_2+|\tau|^2\|u\|_0 \leqslant C\,|\mathrm{Im}\,\tau^2|^{-1}|\tau|^2\cdot$$

$$\{\|(-\tau^2\mathcal{J}+\mathcal{A})u\|_0+\|Bu\|_{Y,1/2}\} \qquad \forall u\in H^2$$

and

$$\|u\|_{Y,3/2}+|\tau|\,\|u\|_{Y,1/2}\leqslant$$

$$C|\mathrm{Im}\,\tau^2|^{-1}|\tau|^2\,\|Bu\|_{Y,1/2}\quad \forall u:(-\mathcal{J}\tau^2+\mathcal{A})u=0.$$

We set

$$u=\Big(\int_\gamma(-\mathcal{J}\tau^2+a(\xi_1,\xi'))^{-1}e^{ix_1\xi_1}d\xi_1\Big)v;$$

then $(-\mathcal{J}\tau^2+\mathcal{A})u=0$ and we obtain the following inequality:

$$(|\xi'|+|\tau|)\,\|S(\xi',\tau)v\| \leqslant C\,|\operatorname{Im}\tau^2|^{-1}\,|\tau|^2\,\|\tilde{S}(\xi',\tau)v\| \quad \forall v\,;$$

combining this estimate with (7.14) we obtain the estimate

$$\|\tilde{S}^{-1}(\xi',\tau)\| \leqslant C\,|\operatorname{Im}\tau^2|^{-2}\,(|\tau|+|\xi'|)^4 .$$

Other formulas do not change.

8. The asymptotics for higher order spectral problems related to the Laplace - Beltrami operator

8.1. Let X be a compact d-dimensional Riemannian C^∞-manifold with the boundary Y, $d \geqslant 2$, E a Hermitian D-dimensional C^∞-fibering over X, Δ the Laplace - Beltrami operator on X. Let

$$\mathcal{A}\mathcal{J} : C^\infty(X,E) \to C^\infty(X,E),$$

$$\mathcal{A} = (-\Delta)^p\,\mathrm{Id} + \mathcal{A}', \quad \mathcal{J} = (-\Delta)^q\,\mathrm{Id} + \mathcal{J}'$$

where $\mathcal{A}', \mathcal{J}'$ are differential operators of orders $2p-1$, $2q-1$ respectively, $0 \leqslant q < p$ are integers, $m = 2(p-q)$; certainly neither $(-\Delta)^s\mathrm{Id}$ nor $\mathcal{A}', \mathcal{J}'$ are invariant objects. Let ν be a C^∞ vector field normal to Y at every point, τ the operator of restriction to Y, $B = \{B_1, \ldots, B_p\}$,

$$B_j = \tau\nu^{\mu_j} + \sum_{l<\mu_j} b_{jl}\,\tau\nu^l$$

where $b_{jl} : C^\infty(Y,E) \to C^\infty(Y,E)$ are classical pseudo-differential operator of order $\mu_j - l$ on Y, $0 \leqslant \mu_1 < \mu_2 < \ldots < \mu_p \leqslant 2p-1$ and $\mu_j = j-1$ for $j = 1, \ldots, q$; then $\operatorname{Ker} B \subset \operatorname{Ker} D$ where $D = \{\tau, \tau\nu, \ldots, \tau\nu^{q-1}\}$ is the Dirichlet boundary operator.

We assume that

(E) Operator $\{\mathcal{A}, B\} : C^\infty(X,E) \to C^\infty(X,E) \oplus$

$C^{\infty}(Y, E^p)$ satisfies the Šapiro - Lopatinskii condition,

(S.1) $\mathcal{J}_D$ – the restriction of $\mathcal{J}$ to $Ker\, D$ – is an operator symmetric in $L(X,E)$.

(S.2) $\mathcal{A}_B$ – the restriction of $\mathcal{A}$ to $Ker\, B$ – is an operator symmetric in $L_2(X,E)$ *).

Note that $\{\mathcal{J}, D\}$ satisfies the Šapiro - Lopatinskii condition a fortiori and that condition (S.1) is fulfilled provided $\mathcal{J}$ is formally selfadjoint.

Let A_B, $J_D : L_2(X,E) \to L_2(X,E)$ be a closures of $\mathcal{A}_B$, $\mathcal{J}_D$ in $L_2(X,E)$; then A_B and J_D are selfadjoint operators in $L_2(X,E)$; J_D is semibounded from below and $D(A_B) \subset D(J_D)$.

Finally we assume that either

(P.1) J_D is positive definite or

(P.2) A_B is positive definite and $Ker\, J_D = 0$.

Let us consider the spectral problem

$$(8.1) \qquad (A_B - \lambda J_D)u = 0, \quad u \in \mathcal{D}(A_B),$$

its spectrum is real, discrete, with finite multiplicity and tends either to $\pm\infty$, or to $+\infty$; its eigenfunctions are orthogonal with respect to sesquilinear forms $(A_B u, v)$ and $(J_D u, v)$.

Assume first that condition (P.1) is fulfilled. Then $(J_D u, v)$ is the inner product and $(J_D u, u)^{1/2}$ is the norm in the Hilbert space $\mathcal{H}_1 = D(J_D^{1/2}) = H_0^q(X,E) = H^q(X,E) \cap Ker\, D$ and operator $\mathcal{U} = J_D^{-1} A_B$ with the domain $\mathcal{D}(\mathcal{U}) = H^{2p-q}(X,E) \cap Ker\, B$ is selfadjoint in this space (remind that $\mu_p \leq 2p-q-1$ and hence Bu is meaningful for $u \in H^{2p-q}(X,E)$; obviously, $J_D^{-1} v$ is meaningful for $v \in H^{-q}(X,E)$; certainly, A_B and J_D are replaced here by their natural extensions). Hence the eigenfunctions of problem (8.1) are complete in $\mathcal{H}_1$ **).

---

*) Then the set $\{\mu_1, \dots, \mu_p\}$ contains a number $k = 0, \dots, 2p-1$ if and only if it does not contain $(2p-1-k)$; hence $\mu_{2p-1} \leq 2p-q-1$.

**) This can also be proved by variational methods.

It is easy to prove that the spectrum of (8.1) is semibounded from below if and only if A_B is semibounded from below; the necessary and sufficient condition of the semiboundedness of A_B can be easily obtained by the methods of §1 (it is the Šapiro - Lopatinskii condition for operator $(-\Delta)^p + z^{2p}$ with a parameter $z \in \mathbb{R}$).

Assume now that condition (P.2) is fulfilled. Then $(A_B u, v)$ is the inner product and $(A_B u, u)^{1/2}$ is the norm in the Hilbert space $\mathcal{H}_2 = \mathcal{D}(A_B^{1/2}) = \{ u \in H^p(X,E),\ B_j u = 0 \quad \forall j$ with $\mu_j \leq p-1\}$ and operator $\mathcal{U} = \mathcal{I}_D^{-1} A_B$ with the domain $\mathcal{D}(\mathcal{U}) = A_B^{-1} \mathcal{I}_D \mathcal{H}_2$ is selfadjoint in this space (as was expained above, all operators are replaced by their natural extensions). Hence the eigenfunctions of the problem (8.1) are complete in $\mathcal{H}_2$ *). In this case the spectrum of problem (8.1) is semibounded from below.

In any case the eigenfunctions of (8.1) are complete in $\overset{q}{H}_0(X,E)$.

One can assume without a loss of generality that 0 is not an eigenvalue of the problem (8.1). One can introduce $\Pi^{\pm}$ – the projectors to positive and negative invariant subspaces of $\mathcal{U}$, $\Pi^+ + \Pi^- = I$; these projectors are selfadjoint with respect to both sesquilinear forms $(\mathcal{I}_D u, v)$, $(A_B u, v)$.

PROPOSITION. Π^- is a singular Green operator.

Let us introduce the eigenvalue distribution functions for problem (8.1) : $N^{\pm}(k)$ is the number of eigenvalues lying between 0 and $\pm k^m$, $k>0$.

Let $E(s)$ be a spectral selfadjoint projector of A_B, $e^{\pm}(k) = \pm(E(\pm k^m) - E(0))$; let $e^{\pm}(x,y,k)$ be the Schwartz kernels of $e^{\pm}(k)$; here $m = 2p$, $q = 0$.

8.2. We are interested both in the asymptotics of $N^{\pm}(k)$ as $k \to \infty$ and in the asymptotics of $e^{\pm}(x,x,k)$ as $k \to \infty$, provided $q = 0$; otherwise we do not know whether $e^{\pm}(x,x,k)$ are monotone functions of k or not.

For $q = 0$, $p = 1$ these asymptotics were derived in [56, 58], see also §0. It appears that these asymptotics remain true in the general case too.

THEOREM 8.1. The following asymptotics holds

$$N^{+}(k) = \varkappa_0^{+} k^{d} + O(k^{d-1}) \text{ as } k \to \infty \tag{8.2}$$

---

*) This can also be proved by variational methods.

where $\varkappa_0^+ = (2\pi)^{-d}\omega_d D \operatorname{vol} X$.

THEOREM 8.2. If $q=0$ then the following asymptotics holds:

$$e^+(x,x,k) = k^d(\alpha_0^+ + Q^+(x',x_1k)) + O(k^{d-1}) \quad \text{as} \quad k\to\infty \tag{8.3}$$

uniformly with respect to x where $\alpha_0^+ = (2\pi)^{-d}\omega_d \, Id$, $C^\infty(Y\times\bar{\mathbb{R}}^+) \ni Q^+$ satisfies (0.9). Recall that $x_1 = dist(x,Y)$.

THEOREM 8.3. If the spectrum of problem (8.1) is not semibounded from below then the following asymptotics holds:

$$N^-(k) = \varkappa_1^- k^{d-1} + O(k^{d-2}) \quad \text{as} \quad k\to\infty, \tag{8.4}$$

where $\varkappa_1^- > 0$.

THEOREM 8.4. If the spectrum of problem (8.1) is not semibounded from below and $q=0$ then the following asymptotics holds:

$$e^-(x,x,k) = k^{d-1} Q^-(x',x_1k) + O(\min(k^{d-1}, x_1^{-d+1})) \tag{8.5}$$

as $k\to\infty$ uniformly with respect to x where $C^\infty(Y\times\bar{\mathbb{R}}^+) \ni Q^-$ satisfies (0.14).

THEOREM 8.5. If the set of points periodic with respect to geodesic flow with reflection at the boundary has measure zero then the following asymptotics holds:

$$N^+(k) = \varkappa_0^+ k^d + \varkappa_1^+ k^{d-1} + o(k^{d-1}) \quad \text{as} \quad k\to\infty. \tag{8.6}$$

8.3. To prove theorems 8.1 - 8.5 we need to essentially modificate the arguments of Part I.

Note first that projectors $\Pi^\pm$ are given by formulas

$$\Pi^\pm = \frac{1}{2} \pm \mathcal{U}\int_0^\infty (\mathcal{U}^2+\lambda^2)^{-1}d\lambda = \pm\frac{m}{\pi}\mathcal{U}\int_0^\infty (A_B + iz^m J_D)^{-1} J_D \cdot \tag{8.7}$$

$$(A_B - iz^m J_D)^{-1} z^{m-1} dz \cdot J_D .$$

Similarly for $s \in (-2,0)$, $n \in \mathbb{Z}^+$ operators $(\pm \mathcal{U})^s \Pi^{\pm}$ are given by formulas

$$(\pm \mathcal{U})^s \Pi^{\pm} = c_{s,n} \int_0^{\infty} (\mathcal{U}^2 + \lambda^2)^{-n-1} \lambda^{s+1} d\lambda \cdot \Pi^{\pm} =$$

(8.8)

$$c_{s,n} m \int_0^{\infty} \{ (A_B + i z^m \mathcal{J}_D)^{-1} \mathcal{J}_D \cdot$$

$$(A_B - i z^m \mathcal{J}_D)^{-1} \mathcal{J}_D \}^{n+1} z^{sm+2m-1} dz \cdot \Pi^{\pm}$$

where

$$c_{s,n}^{-1} = \int_0^{\infty} (1 + \lambda^2)^{-n-1} \lambda^{s+1} d\lambda .$$

Let us introduce the classes of generalized singular Green operators: $G \in \widetilde{GOGS}^{\mu,h}(X,E)$, $\mu \in \mathbb{R}$, $h \in \mathbb{Z}^+$ if the Schwartz kernel of G is expressed modulo Schwartz kernel of a neglible operator by oscillatory integral

$$K_G(x,y) \sim$$

(8.9)

$$(2\pi)^{-d+1} \int e^{i\langle x'-y',\xi'\rangle} k_G(x', x_1, y_1, \xi') d\xi'$$

where $k_G(x', x_1, y_1, \xi')$ satisfies the inequalities

$$|D_x^{\alpha} D_{y_1}^{\alpha_0} D_{\xi'}^{\beta} k_G | x_1^i y_1^j \leq C_{\alpha\alpha_0\beta ij\ell} (1+|\xi'|)^{\mu+1+\alpha_0+\alpha_1-|\beta|-i-j} \cdot$$

$$(1+(x_1+y_1)(1+|\xi'|))^{-\ell}$$

$\forall \alpha, \alpha_0, \beta, i, j, \ell$ such that $\alpha_0+\alpha_0 \leq i+j+h$, $\alpha_0 \leq i+h$, $\alpha_1 \leq j+h$,

k_G can be decomposed into asymptotical series $k_G \sim \sum_{h=0}^{\infty} k_n$,
where

$$k_n(x', \lambda^{-1} x_1, \lambda^{-1} y_1, \lambda \xi') = \lambda^{\mu-n+1} k_n(x', x_1, y_1, \xi') \qquad \forall \lambda > 0,$$

$$|D_x^{\alpha} D_{y_1}^{\alpha_0} D_{\xi'}^{\beta} k_n| x_1^i y_1^j \leq C_{\alpha\alpha_0\beta ij\ell n} |\xi'|^{\mu+1+\alpha_0+\alpha_1-|\beta|-i-j-n} (1+(x_1+y_1)|\xi'|)^{-\ell}$$

$\forall \alpha, \alpha_0, \beta, i, j, \ell, n$ such that $\alpha_0+\alpha_1 \leq i+j+h$, $\alpha_0 \leq i+h$, $\alpha_1 \leq j+h$,

in the sense that for $|\xi'| \geq 1$

$$|D_x^{\alpha} D_{y_1}^{\alpha_0} D_{\xi'}^{\beta} (k_G - \sum_{n=0}^{N-1} k_n)| x_1^i y_1^j \leq$$

$$C_{\alpha\alpha_0\beta ij\ell N} |\xi'|^{\mu+1+\alpha_0+\alpha_1-|\beta|-i-j-N} (1+(x_1+y_1)|\xi'|)^{-\ell}$$

$\forall \alpha, \alpha_0, \beta, i, j, \ell, N$ such that $\alpha_0+\alpha_1 \leq i+j+h$, $\alpha_0 \leq i+h$, $\alpha_1 \leq j+h$;

neglible operators are operators G such that the following estimates hold: $\| x_1^i D_1^p G x_1^j D_1^q u \|_{s,0} \leq C_{stijpq} \| u \|_{t,0}$ $\forall u \in C_0^\infty(\mathring{X})$
$\forall i, j, p, q$ such that $p+q \leq i+j+h$, $p \leq i+h$, $q \leq j+h$, $s, t \in \mathbb{R}$.

REMARK. Classes $\widetilde{GOGS}^{\mu,h}(X, E)$ are more extensive than classes $GOGS^{\mu,h}(X, E)$ introduced in §1; there α_0, α_1, i, j were connected only by the inequality $\alpha_0+\alpha_1 \leq i+j+h$ and therefore p, q, i, j were connected only by the inequality $p+q \leq i+j+h$. We introduce new classes to consider the case $q \geq 1$.

PROPOSITION 8.6. (i) $\Pi^- \in OGS^0(X, E)$.
(ii) $(-\mathfrak{U})^s \Pi^- \in OGS^{sm}(X, E)$ $\forall s \in \mathbb{R}$.
(iii) Let $Q_\pm \in OPS^{\pm q}(X, E) + OGS^{\pm q}(X, E)$;

then for $j \in \mathbb{Z}^+ \setminus 0$

$$Q_+ \mathcal{U}^{-2j/m} \Pi^+ Q_- \in OPS^{-2j}(X,E) + \widetilde{GOGS}^{-2j,\,2j-1}(X,E).$$

(iv) Let $q=0$; then for $j \in \mathbb{Z}^+ \setminus 0$

$$\mathcal{U}^{-2j/m} \Pi^+ \in OPS^{-2j}(X,E) + GOGS^{-2j,\,2j-1}(X,E).$$

(v) For $j \in \frac{m}{2}\mathbb{Z}^+$

$$\mathcal{U}^{-2j/m} \Pi^+ \in OPS^{-2j}(X,E) + OGS^{-2j}(X,E).$$

REMARKS. (i) For every $s \in \mathbb{R}$ (and even $s \in \mathbb{C}$) $\mathcal{U}^s \Pi^+$ is very likely to belong to Rempel - Schulze operator algebra [84].
(ii) For $s \overline{\in} \frac{2}{m}\mathbb{Z}^+$ the pseudo-differential component of $\mathcal{U}^s \Pi^+$ has no transmission property.
(iii) For $j \in \mathbb{Z}^+$, $s = -2j/m \overline{\in} \mathbb{Z}^+$ the singular Green component of $\mathcal{U}^s \Pi^+$ does not necessarily belong to $OGS^{-2j}(X,E)$. G.Grubb drew our attention to the following fact: let $\mathcal{A} = \Delta^2$, $\mathcal{J} = I$, $D = \{\tau, \tau D_1\}$, $s = \frac{1}{2}$; then $A_D^{1/2} + \Delta \overline{\in} OGS^2(X)$.

PROOF OF PROPOSITION 8.6. Consider first the simplest case: $q=0$, i.e. $\mathcal{J} = I$, $\mathcal{U} = A_B$. Then (i) follows from theorem 1.10 and implies (v). Repeating the first part of the proof of theorem 1.10 but using formula (8.8) instead of (8.7) and noting that $(-\mathcal{U})^s \Pi^- = \Pi^-(-\mathcal{U})^s \Pi^-$ we easily prove (ii) and (iv).

Now let $q \geq 1$. Then the new difficulty appears: operators $\mathcal{A} \pm i z^m \mathcal{J}$, $\{\mathcal{A} \pm i z^m \mathcal{J}, B\}$ are not elliptic with a parameter z. Ellipticity of $\mathcal{A} \pm i z^m \mathcal{J}$ is violated at $\xi = 0$ and the parametrices of $\mathcal{A} \pm z^m \mathcal{J}$ are pseudo-differential operators with the symbols $c^{\pm}(x,\xi,z)$ which satisfy the estimates

$$|D_x^{\alpha} D_{\xi}^{\beta} c^{\pm}(x,\xi,z)| \leq C_{\alpha\beta}(1+|\xi|)^{-2q-|\beta|}(1+|\xi|+z)^{-m} \qquad \forall \alpha,\beta$$

and can be decomposed into asymptotical series $c^{\pm}(x,\xi,z) \sim \sum_{n=0}^{\infty} c_n^{\pm}(x,\xi,z)$ where $c_n^{\pm}$ are homogeneous of degree $(-2p-n)$ with respect to (ξ, z) and satisfy the estimates

$$|D_x^{\alpha} D_\xi^{\beta} c_n^{\pm}(x,\xi,z)| \leq C_{\alpha\beta n} |\xi|^{-2q-|\beta|-n} (|\xi|+z)^{-m} \quad \forall \alpha,\beta,n ,$$

in the sense that for $|\xi| \geq 1$

$$|D_x^{\alpha} D_\xi^{\beta} (c^{\pm} - \sum_{n=0}^{N-1} c_n^{\pm})| \leq C_{\alpha\beta N} |\xi|^{-2q-|\beta|-N} (|\xi|+z)^{-m} \quad \forall \alpha,\beta,N.$$

Now consider operators $\{\mathcal{A} \pm i z^m \mathcal{J},\ \mathcal{B}\}$. They satisfy the following estimate

$$(8.10) \quad \sum_{j=0}^{m} z^j \|u\|_{2p-j} \leq C\{\|(\mathcal{A} \pm i z^m \mathcal{J})u\|_0 +$$

$$\sum_{j=1}^{p} \|(\Lambda^2 + z^2)^{(2p-\mu_j-\frac{1}{2})/2} B_j u\|_{Y,0}\}$$

where $OPS^1(Y,E) \ni \Lambda$ is elliptic, selfadjoint and invertible; (8.10) can be proved by the same arguments as (1.5). Note that in the left part of (8.10) we sum from $j=0$ to $j=m$ instead of $j=2p$; hence (8.10) is weaker than ellipticity with a parameter; but it implies that the Šapiro - Lopatinskii condition with a parameter z is fulfilled for $\xi' \neq 0$ and therefore we need to examine only zone $1+|\xi'| \leq C^{-1} z$ where C is arbitrary large.

We introduce the special coordinate system in the neighbourhood of Y; then the metric equals $g(x,\xi) = \xi_1^2 + g'(x,\xi')$; let $\rho = \sqrt{g'(x,\xi')}$. Note that the principal symbols of $\mathcal{A} \pm i z^m \mathcal{J}$ are $g^q(x,\xi)(g^{m/2}(x,\xi) \pm i z^m \mathrm{Id})$; recall that $\mu_1 < \mu_2 < \ldots < \mu_p$ and $\mu_j = j-1$ for $j=1,\ldots,q$; then it is easy to prove that the singular Green components of $(A_B \pm i z^m J_D)^{-1}$ are given by oscillatory integrals (8.9) where $k_G(x', x_1, y_1, \xi', z)$ can be decomposed into sum

(8.11) $$k_G(x', x_1, y_1, \xi', z) = k'(x', x_1, y_1, \xi', z) +$$

$$\sum_{s=0}^{q-1} x_1^s k_s''(x', x_1, y_1, \xi', z) e^{-x_1 \rho} +$$

$$\sum_{t=0}^{q-1} y_1^t k_t'''(x', x_1, y_1, \xi', z) e^{-y_1 \rho} +$$

$$\sum_{s,t=0}^{q-1} x_1^s y_1^t k_{st}^{IV}(x', x_1, y_1, \xi', z) e^{-(x_1+y_1)\rho}$$

and the following estimates hold:

$$|D_x^{\alpha} D_{y_1}^{\alpha_0} D_{\xi'}^{\beta} k'| x_1^i y_1^j \leqslant$$

$$C_{\alpha\alpha_0\beta ij\ell} (1+\rho)^{-|\beta|} z^{-2\rho+1+\alpha_0+\alpha_1-i-j} (1+(x_1+y_1)z)^{-\ell}$$

$$\forall \alpha, \alpha_0, \beta, i, j, \ell ,$$

$$|D_x^{\alpha} D_{y_1}^{\alpha_0} D_{\xi'}^{\beta} k_s''| y_1^j \leqslant$$

$$C_{\alpha\alpha_0\beta j\ell} (1+\rho)^{-|\beta|} z^{-2\rho+1+\alpha_0-j+s} (1+y_1 z)^{-\ell} \quad \forall \alpha, \alpha_0, \beta, j, \ell, s ,$$

$$|D_x^{\alpha} D_{y_1}^{\alpha_0} D_{\xi'}^{\beta} k_t'''| x_1^i \leqslant$$

$$C_{\alpha\alpha_0\beta i\ell} (1+\rho)^{-|\beta|-2q+1+t} z^{-m+\alpha_1-i} (1+x_1 z)^{-\ell} \quad \forall \alpha, \alpha_0, \beta, i, \ell, t ,$$

$$|D_x^{\alpha} D_{y_1}^{\alpha_0} D_{\xi'}^{\beta} k_{s,t}^{IV}| \leqslant$$

$$C_{\alpha\alpha_0\beta}(1+\rho)^{-|\beta|+s+t-2q+1} z^{-m}$$

$$\forall \alpha, \alpha_0, \beta, s, t$$

and $k', k''_s, k'''_t, k^{IV}_{st}$ can be expanded into asymptotical series with positively homogeneous terms *).

In reality the third estimate can be improved: since operators $(A_B \pm iz^m J_D)^{-1}$ are mutually adjoint then variables x and y are equal in rights and hence

$$|D^{\alpha}_{x} D^{\alpha_0}_{y_1} D^{\beta}_{\xi'} k'''_t| x_1^i \leqslant$$

$$C_{\alpha\alpha_0\beta i l}(1+\rho)^{-|\beta|} z^{-2\rho+1+\alpha_1-i+t} (1+x_1 z)^{-l}$$

$$\forall \alpha, \alpha_0, \beta, i, l, t.$$

Consider now the operators $D^{\mu}(A_B \pm iz^m J_D)^{-1} D^{\nu}$ with $|\mu| \leqslant q$, $|\nu| \leqslant q$. Since the Schwartz kernels of operators $(A_B \pm iz^m J_D)^{-1}$ vanish for $y \in Y$ as well as all their derivatives up to the order $(q-1)$ then

$$K_{D^{\mu}(A_B \pm iz^m J_D)^{-1} D^{\nu}}(x,y) =$$

$$D^{\mu}_{x}(-D_y)^{\nu} K_{(A_B \pm iz^m J_D)^{-1}}(x,y);$$

hence $D^{\mu}(A_B \pm iz^m J_D)^{-1} D^{\nu}$ are sums of pseudo-differential operators and singular Green operators; the symbols of pseudo-differential operators satisfy the estimates

*) I.e. $k'_n(x', \lambda^{-1}x_1, \lambda^{-1}y_1, \lambda\xi', \lambda z) = \lambda^{-2\rho-n+1} k'_n(x', x_1, y_1, \xi', z)$ $\forall \lambda > 0$ etc.

$$|D_x^\alpha D_\xi^\beta c(x,\xi,z)| \leq C_{\alpha\beta}(1+|\xi|)^{-|\beta|}(|\xi|+z)^{-m} \qquad \forall \alpha,\beta$$

and can be expanded into asymptotical series with positively homogeneous terms and the Schwartz kernels of singular Green operators are given by formulas (8.9), (8.11) with k', k''_s, k'''_t, k^{IV}_{st} satisfying the estimates

$$|D_x^\alpha D_{y_1}^{\alpha_0} D_{\xi'}^\beta k'|\, x_1^i y_1^j \leq$$

$$C_{\alpha\alpha_0\beta ij\ell}(1+\rho)^{-|\beta|} z^{-m+1+\alpha_0+\alpha_1-i-j}(1+(x_1+y_1)z)^{-\ell} \qquad \forall \alpha,\alpha_0,\beta,i,j,\ell,$$

$$|D_x^\alpha D_{y_1}^{\alpha_0} D_{\xi'}^\beta k''_s|\, y_s^j \leq$$

$$C_{\alpha\alpha_0\beta j\ell}(1+\rho)^{-|\beta|+q} z^{-m+1+\alpha_0-j+s-q}(1+y_1 z)^{-\ell} \qquad \forall \alpha,\alpha_0,\beta,j,\ell,s,$$

$$|D_x^\alpha D_{y_1}^{\alpha_0} D_{\xi'}^\beta k'''_t|\, x_1^i \leq$$

$$C_{\alpha\alpha_0\beta i\ell}(1+\rho)^{-|\beta|+q} z^{-m+1+\alpha_1-i+t-q}(1+x_1 z)^{-\ell} \qquad \forall \alpha,\alpha_0,\beta,i,\ell,t,$$

$$|D_x^\alpha D_{y_1}^{\alpha_0} D_{\xi'}^\beta k^{IV}_{st}| \leq$$

$$C_{\alpha\alpha 0\beta}(1+\rho)^{-|\beta|+s+t+1} z^{-m}$$

$$\forall \alpha,\alpha_0,\beta,s,t$$

and expanding into the asymptotical series with positively homogeneous terms. Hence for $|\mu| \leq q$ operators

$$D^\mu \{(A_B+iz^m \mathcal{J}_D)^{-1}\mathcal{J}_D(A_B-iz^m\mathcal{J}_D)^{-1}\mathcal{J}_D\}^{-n-1}$$

can be decomposed into sums $\sum_{|\nu|\leq q} R_{\mu\nu} D^{\nu}$ where $R_{\mu\nu}$ are sums of pseudo-differential operators with symbols satisfying the estimates

$$|D_x^{\alpha} D_{\xi}^{\beta} c(x,\xi,z)| \leq C_{\alpha\beta}(1-|\xi|)^{-|\beta|}(z+|\xi|)^{-m(n+1)}$$

and singular Green operators with the Schwartz kernels given by formulas (8.9), (8.11) with $k', k_s'', k_t''', k_{st}^{IV}$ satisfying the estimates

$$|D_x^{\alpha} D_{y_1}^{\alpha_0} D_{\xi'}^{\beta} k'| x_1^i y_1^j \leq$$

$$C_{\alpha\alpha_0\beta ij}(1+\rho)^{-|\beta|} z^{-2m(n+1)+1+\alpha_0+\alpha_1-i-j}(1+(x_1+y_1)z)^{-\ell}$$

$$\forall \alpha, \alpha_0, \beta, i, j, \ell,$$

$$| D_x^{\alpha} D_{y_1}^{\alpha_0} D_{\xi'}^{\beta} k_s'' | y_1^j \leq$$

$$C_{\alpha\alpha_0\beta j\ell}(1+\rho)^{-|\beta|+q} z^{-2m(n+1)+\alpha_0-j+s-q}(1+y_1 z)^{-\ell} \quad \forall \alpha, \alpha_0, \beta, j, \ell, s,$$

$$| D_x^{\alpha} D_{y_1}^{\alpha_0} D_{\xi'}^{\beta} k_t''' | x_1^i \leq$$

$$C_{\alpha\alpha_0\beta i\ell}(1+\rho)^{-|\beta|+q} z^{-2m(n+1)+1+\alpha_1-i+t-q}(1+x_1 z)^{-\ell} \quad \forall \alpha, \alpha_0, \beta, i, \ell, t,$$

$$| D_x^{\alpha} D_{y_1}^{\alpha_0} D_{\xi'}^{\beta} k_{st}^{IV} | \leq$$

$$C_{\alpha\alpha_0\beta}(1+\rho)^{-|\beta|+s+t+1} z^{-2m(n+1)} \quad \forall \alpha, \alpha_0, \beta, s, t$$

and expanding into the asymptotical series with positively homogeneous terms.

Further proof of the proposition repeats the proof of theorem 1.10: to prove i we need to repeat this proof completely, i implies (v); to prove (ii), (iii) we need to repeat only the first part of the proof; to prove (ii) we need to use the equality $\Pi^-(-\mathcal{U})^s\Pi^- = (-\mathcal{U})^s\Pi^-$.

REMARK. One can also prove (i), (ii) by arguments used in the proofs of propositions 1.13 - 1.15.

Classes $GOGS^{\mu,h}$, $\widetilde{GOGS}^{\mu,h}$ are introduced in the local coordinate systems yet. One can prove by standard methods the following statements which justify the use of these classes in fiberings over manifolds:

PROPOSITION 8.7. (i) Classes $GOGS^{\mu,h}$, $\widetilde{GOGS}^{\mu,h}$ are invariant with respect to the change of variables preserving X and Y.
(ii) If $G \in GOGS^{\mu,h}$, $\tilde{G} \in \widetilde{GOGS}^{\mu,h}$ and $f, f' \in C^\infty$ then $fGf' \in GOGS^{\mu,h}$, $f\tilde{G}f' \in \widetilde{GOGS}^{\mu,h}$.

PROPOSITION 8.8. If $\tilde{G} \in \widetilde{GOGS}^{\mu,h}(X, E)$ then the mapping

$$\tilde{G}: \mathcal{H}^{s,0}(X, E) \cup \mathcal{E}'(\mathring{X}, E) \to \mathcal{H}^{s-\mu,h}(X, E) \cap C^\infty(\mathring{X}, E)$$

is continuous for every $s \in \mathbb{R}$.

PROPOSITION 8.9. (i) If $G \in GOGS^{\mu,h}(X,E)$, $\tilde{G} \in \widetilde{GOGS}^{\mu,h}(X, E)$ then

$$G^* \in GOGS^{\mu,h}(X,E), \ \tilde{G}^* \in \widetilde{GOGS}^{\mu,h}(X,E).$$

(ii) If $\tilde{G}_j \in \widetilde{GOGS}^{\mu_j,h}(X,E)$ then $\tilde{G}_1 \tilde{G}_2 \in \widetilde{GOGS}^{\mu_1+\mu_2,h}(X,E)$

(this statement is not valid for classes $GOGS^{\mu,h}$).
(iii) If $G_j \in OGS^{\mu_j}(X,E)$, $\tilde{G} \in \widetilde{GOGS}^{\mu,0}(X,E)$ then $G_1 \tilde{G} G_2 \in OGS^{\mu_1+\mu_2+\mu}(X,E)$.
(iv) $\bigcap_h \widetilde{GOGS}^{\mu,h}(X,E) = OGS^{\mu}(X,E)$.

PROPOSITION 8.10. If $G \in GOGS^{\mu,h}(X,E)$, $\tilde{G} \in \widetilde{GOGS}^{\mu,h}(X, E)$ then

(i) $\quad K_G \in C^\infty(X \times X \setminus Y \times Y)$, $\ K_{\tilde{G}} \in C^\infty((X \setminus Y) \times (X \setminus Y))$,

(ii) $\quad {}^{T}WF_f^{\prime,h}(K_{\tilde{G}}) \subset j_x^{-1} j_y^{-1} \operatorname{diag}(T^*Y \setminus 0)^2$.

PROPOSITION 8.11. If $G \in GOGS^{\mu,h}(X,E)$, $\tilde{G} \in \widetilde{GOGS}^{\mu,h}(X, E)$ then

(i) $\quad (x'-y')^{\alpha'} x_1^{\alpha_1} y_1^{\alpha_0} K_G \in C(\mathbb{R}^{d-1}_{y'}, \mathcal{H}^{-\mu+|\alpha|-\frac{d-1}{2}-\varepsilon,\, h+\alpha_0+\alpha_1}$

$$(\mathbb{R}^{d-1}_{x'} \times \overline{\mathbb{R}}^+_{x_1} \times \overline{\mathbb{R}}^+_{y_1}, \operatorname{Hom}(E))) \qquad \forall \alpha \in \mathbb{Z}^{+(d+1)} \qquad \forall \varepsilon > 0;$$

(ii) $$(x'-y')^{\alpha'} x_1^{\alpha_1} y_1^{\alpha_0} D_{x_1}^{\beta_1} D_{y_1}^{\beta_0} K_{\tilde{G}} \in$$

$$C(\mathbb{R}^{d-1}_{y'}, \mathcal{H}^{-\mu+|\alpha|-|\beta|-\frac{d-1}{2}-\varepsilon,0}(\mathbb{R}^{d-1}_{x'} \times \overline{\mathbb{R}}^{+}_{x_1} \times \overline{\mathbb{R}}^{+}_{y_1}, \mathrm{Hom}(E)))$$

$\forall \alpha \in \mathbb{Z}^{+(d+1)}$, $\beta \in \mathbb{Z}^{+2}$ such that $\beta_1 \leqslant \alpha_1 + h$, $\beta_0 \leqslant \alpha_0 + h$, $\beta_0 + \beta_1 \leqslant \alpha_0 + \alpha_1 + h$ $\forall \varepsilon > 0$; here $|\alpha| = \alpha_0 + \alpha_1 + \cdots + \alpha_d$, $|\beta| = \beta_0 + \beta_1$; the same inclusions with permutated x_1 and y_1 hold.

One can easily prove the inequalities for A_B and $\mathfrak{I}_D$ similar to inequality (1.12) but we do not need them.

8.4. In this section we discuss the proofs of the finiteness of the speed of propagation of singularities and the normality of the great singularity outside of the normal rays zone.

Let $u_{\pm}(x,y,t)$ be the Schwartz kernel of the operator

$$(\pm \mathcal{U})^{\frac{1}{m}-n} \sin((\pm \mathcal{U})^{\frac{1}{m}} t) \Pi^{\pm} \mathfrak{I}_D^{-1}$$

where $n \in \mathbb{Z}^+$ is arbitrary; then $u_{\pm}(x,y,t)$ satisfies the problem

$$(\mathcal{J}_x D_t^m \mp \mathcal{A}_x) u_{\pm} = 0 , \tag{8.12}$$

$$B_x u_{\pm} = 0, \tag{8.13}$$

$$D_t^{2j} u_{\pm}\big|_{t=0} = 0, \tag{8.14}$$

$$D^{2j+1} u_{\pm}\big|_{t=0} = -i K_{(\pm\mathcal{U})^{\frac{2j}{m}} \Pi^{\pm} \mathfrak{I}_D^{-1}} (x,y)$$

$$j = 0, \ldots, \frac{m}{2} - 1$$

as well as the dual problem.

Moreover, $u_{\pm}(x,y,t)$ satisfies another problem

$$(D_t^2 \mp \mathcal{U}) u_{\pm} = 0 , \tag{8.15}$$

(8.16) $$u_\pm \big|_{t=0} = 0 ,$$

$$D_t u_\pm \big|_{t=0} = -iK_{(\pm\mathfrak{U})^{-n}\Pi^\pm \mathfrak{I}_D^{-1}}(x,y)$$

as well as the dual problem.

Let $u_{0\pm}$ be a solution of problem (8.15) - (8.16) on a more extensive manifold. Then $u_{0-} \equiv 0 \ (\mathrm{mod}\ C^\infty(X\times X\times \mathbb{R}))$ and $u_1 = u_+ - u_{0+}$ satisfies (8.12) with non-homogeneous boundary conditions

(8.13)' $$B_x u_1 = -B_x u_{0+}$$

and initial conditions

(8.17) $$D_t^{2j} u_1 \big|_{t=0} = 0 ,$$

$$D_t^{2j+1} u_1 \big|_{t=0} = -iK_{(j)}(x,y) \qquad j = 0, \ldots, \frac{m}{2} - 1$$

where $K_{(j)}(x,y)$ are the Schwartz kernels of the singular Green components of the operators $\mathfrak{U}^{\frac{2j}{m}-n}\Pi^\pm \mathfrak{I}_D^{-1}$.

All we need is to reduce problem (8.12) - (8.13) to an appropriate problem for first-order symmetric t-hyperbolic (in the sense of §3) system; then one can easily repeat the arguments of § 2,3. Certainly, $K_{(j)}$ are not infinitely smooth with respect to x_1, y_1; but in reality we need only the finite smoothness *) of $K_{(j)}$ which is provided by an appropriate choice of n.

First we reduce problem (8.12) - (8.13) to the problem for the second-order system. We use the special coordinate system again; then $g(x,\xi) = \xi_1^2 + \rho^2(x,\xi')$.

We set $U = (u_1, \ldots, u_p)^T$, $u_j = (-\Delta)^{j-1} \Lambda^{-j+1} u$ where $OPS^{2'}(X,E) \ni \Lambda$ is elliptic, selfadjoint, invertible and its principal symbol equals $\rho^2(x,\xi')$. Then problem (8.12) - (8.13) is reduced to problem

(8.18)$_\pm$ $$(M\Delta + M' D_1 + N)\,\overline{U} = 0$$

(8.19) $$\mathcal{B}\overline{U} = 0$$

---

*) Which depends only on D.

where $M, M' \in OPS^{0'}(X\times\mathbb{R}, E^{p}),\ N = MS,\ S \in OPS^{2'}(X\times\mathbb{R}, E^{p})$ and the principal symbols of M, N, S have the following components:

$$M_{0(ij)} = \delta_{i+j,\,p+1} \mp \tau^{m}\rho^{-m}\delta_{i+j,q+1}\ ,$$

$$N_{2(ij)} = \rho^{2}\delta_{i+j,p+2} \pm \tau^{m}\rho^{2-m}\left(\delta_{i1}\delta_{j,q+1} + \delta_{i,q+1}\delta_{j1} - \delta_{i+j,q+2}\right),$$

$$S_{2(ij)} = \rho^{2}\delta_{i,j-1} \pm \tau^{m}\rho^{2-m}\delta_{ip}\delta_{j,q+1}$$

$(i, j = 1, \ldots, p)$ respectively.

Both systems $(8.18)_{\pm}$ are symmetric in principal and they are t-hyperbolic in the sense of §3:

$(8.20)_{+}$
$$\langle \tau M_0 \frac{\partial S_2}{\partial \tau}\, v, v\rangle > 0$$
$$\forall v \in \mathrm{Ker}\,(\xi_1^2 + \rho^2 - S_2)\setminus 0,\quad \xi_1 \in \mathbb{R}$$

for the first system,

$(8.20)_{-}$
$$\mathrm{Ker}\,(\xi_1^2 + \rho^2 - S_2) = 0\ ,\quad \forall \xi_1 \in \mathbb{R}$$

for the second system.

Since the boundary condition $Bu = 0$ provided for the symmetricity of $\mathcal{A}$ and $\mathcal{J}$ then the boundary condition $\bar{\mathcal{B}}\bar{U} = 0$ provides for the symmetricity of $(M\Delta + N)$ in principal (i.e. modulo lower order terms).

We set $U = (\bar{U}\ \ \Lambda^{-1/2}D_1\bar{U})^{T}$; then problem $(8.18)_{\pm}$ - (8-19) is reduced to problem

$(8.21)_{\pm}$
$$(KD_1 + L)U = 0\ ,$$

(8.22)
$$\mathcal{B}U = 0$$

where $K \in OPS^{0'}(X \times \mathbb{R}, E^{2p}),\ L = -KT,\ T \in OPS^{1'}(X \times \mathbb{R}, E^{2p})$

have the principal symbols

$$K_0 = \begin{pmatrix} 0 & M_0 \\ M_0 & 0 \end{pmatrix},$$

$$L_1 = \begin{pmatrix} \rho^{-1}(\rho^2 M_0 - N_2) & 0 \\ 0 & -\rho M_0 \end{pmatrix},$$

$$T_1 = \begin{pmatrix} 0 & Id \\ \rho^{-1}(\rho^2 - S_2) & 0 \end{pmatrix}$$

respectively.

Both systems $(8.21)_\pm$ are symmetric in principal; (8.20) imply that $(8.21)_\pm$ are t-hyperbolic:

$$(8.23)_+ \qquad -\tau \langle K_0 \frac{\partial T_1}{\partial \tau} V, V \rangle > 0$$

$$\forall V \in Ker(\xi_1 - T_1) \setminus 0, \quad \xi \in \mathbb{R}$$

for the first system and

$$(8.23)_- \qquad Ker(\xi_1 - T_1) = 0 \qquad \forall \xi_1 \in \mathbb{R}$$

for the second system.

It remains to prove that

$$(8.24) \qquad \mp Re\ i\tau \langle e'^{*} K_0 \frac{\partial e'}{\partial \tau} W, W \rangle > 0$$

$$\forall W \in Ker\, \mathcal{B}_0 \cap Ran\, e' \setminus 0.$$

It should be noted that now K_0 depends on τ. Repeating the proof of proposition B one can easily show that

$$-ie'^{*} K_0 \frac{\partial e'}{\partial \tau} = -\int_0^\infty e'^{*} e^{isT_1^{*}} K_0 \frac{\partial T_1}{\partial \tau} e^{-isT_1} e' ds.$$

Therefore we need to prove that if $U \in \mathcal{S}(\overline{\mathbb{R}}^+, E^{2p}) \setminus 0$ is the solution of problem

(8.25) $$(K_0 D_1 + L_1) U(x_1) = 0 ,$$

(8.26) $$\mathcal{B}_0 U = 0 \qquad *)$$

then

$$\mp \tau \operatorname{Re} \int_0^\infty \langle K_0 \frac{\partial T_1}{\partial \tau} U(x_1), U(x_1) \rangle dx_1 > 0 .$$

But if U is the solution of problem (8.25) - (8.26) then $U = (\overline{U} \ \rho^{-1} D_1 \overline{U})^T$ where $\overline{U}$ is the solution of problem

(8.27) $$(-M_0(D_1^2 + \rho^2) + N_2) \overline{U} = 0 ,$$

(8.28) $$\overline{\mathcal{B}}_0 \overline{U} = 0 .$$

Note that

$$-K_0 \frac{\partial T_1}{\partial \tau} = \begin{pmatrix} \rho^{-1} M_0 \frac{\partial S_2}{\partial \tau} & 0 \\ 0 & 0 \end{pmatrix} .$$

Therefore we need to prove that if $U \in \mathcal{S}(\overline{\mathbb{R}}^+, E^p) \setminus 0$ is the solution of problem (8.27) - (8.28) then

(8.29) $$\pm \tau \operatorname{Re} \int_0^\infty \langle M_0 \frac{\partial S_2}{\partial \tau} \overline{U}(x_1), \overline{U}(x_1) \rangle dx_1 > 0 .$$

But $\overline{U} \in \mathcal{S}(\overline{\mathbb{R}}^+, E^p)$ is the solution of problem (8.27) - (8.28) if and only if $\overline{U}_j = (D_1^2 + \rho^2)^{j-1} \rho^{2-2j} u$ where $u \in \mathcal{S}(\overline{\mathbb{R}}^+, E)$ is the solution of problem

---

*) $\mathcal{B}_0$, $\overline{\mathcal{B}}_0$, $B_{(0)}$ denote the principal parts of boundary operator with D', D_t replaced by ξ', τ.

$$(8.30) \qquad [(D_1^2+\rho^2)^p \mp \tau^m (D_1^2+\rho^2)^q]u = 0,$$

$$(8.31) \qquad B_{(0)} u = 0 .$$

Since

$$(\tau M_0 \frac{\partial S_2}{\partial \tau})_{ij} = \pm m\tau^m \rho^{2-m} \delta_{i1} \delta_{j,q+1}$$
$$(i, j = 1, \ldots, p)$$

then the left part of (8.29) equals

$$m\tau^m \rho^{2-m} \operatorname{Re} \int_0^\infty \langle u, (D_1^2+\rho^2)^q u \rangle dx_1 ;$$

conditions (8.31) imply that $\tau D_1^j u = 0$ for $j = 0, \ldots, q-1$ and since the operator $(-\Delta)^q$ with the Dirichlet boundary conditions is positive definite then (8.29) holds true.

REMARK. It is easy to prove that in the chosen special coordinate system there are more precise inclusions for $u = u_+$ then given by theorems 2.3(iii), 2.4:

$$WF_f^{',mn}(u_1^{\pm}) \cap \{\pm t \leq t_0\} \subset$$

$$\{dist(x', y') + |\xi' + \eta'|(|\xi'|^{-1} + |\eta'|^{-1}) \leq C(\pm t - x_1 - y_1)\} \cup$$

$$\{\xi' = \eta' = 0, \; x_1 = y_1 = \pm t\} \cup \{\xi' = \eta' = t = 0\},$$

$$WF_f(u_1) \cap \{|t| \leq t_0\} \subset$$

$$\{dist(x', y') + |\xi' + \eta'|(|\xi'|^{-1} + |\eta|^{-1}) \leq C(|t| - x_1 - y_1)\} \cup$$

$$\{ \xi' = \eta' = 0 , \quad x_1 = y_1 = |t| \} .$$

8.5. Now we need to discuss the construction of the successive approximations in the zone $\{ C^{-1} |\tau| \leqslant |\xi'| \leqslant C|\tau| \}$.
It is convenient to use the special coordinate system introduced above; then the construction of successive approximations is simpler than in §4. We need to discuss only the construction of the parametrices.

Consider first problem $(8.12)_+$ - (8.13). Then the parametrices $G^{\pm}$ are given by formulas

$$G^{\pm} = F^{-1}_{\tau\to t}\left(-(\tau\mp i0)^m \mathcal{J}_D + A_B\right)^{-1} F_{t\to\tau} = \tag{8.32}$$

$$F^{-1}_{\tau\to t}\left(\mathfrak{U} - (\tau \mp i0)^m\right)^{-1} F_{t\to\tau}\, \mathcal{J}_D^{-1} =$$

$$(G_e^{(1)} \Pi^- + G_e^{(2)} G_h \Pi^+)\, \mathcal{J}_D^{-1}$$

where

$$G_e^{(1)} = F^{-1}_{\tau\to t}\left(\mathfrak{U} - \tau^m\right)^{-1} F_{t\to\tau}$$

on $\operatorname{Ran} \Pi^-$,

$$G_e^{(2)} = F^{-1}_{\tau\to t}\left(\sum_{k=0}^{\frac{m}{2}-1} \mathfrak{U}^{\frac{2k}{m}} \tau^{m-2k} \right)^{-1} F_{t\to\tau}$$

and

$$G_h = F^{-1}_{\tau\to t}\left(\mathfrak{U} - (\tau \mp i0)^2\right)^{-1} F_{t\to\tau}$$

on $\operatorname{Ran} \Pi^+$

It is easy to show that the parametrices $G^{\pm}$ are appropriate from the point of view of §4 and hence the appropriate parametrices

$G_j^{\prime\pm}$ exist too $(j = 1,\dots,p)$; remind that we have p boundary conditions now.

Consider now the operators with positively homogeneous symbols which do not depend on x in the whole space and in the half-space; then the parametrices are given by formulas:

(8.33) $$\bar{G}_0^{\pm} = F^{-1}_{\xi,\tau\to x,t}\, a(\xi,\tau \mp i0)^{-1}(1-\zeta(\xi))^{-1} F_{x,t\to\xi,\tau}$$

where $a(\xi,\tau) = |\xi|^{2q}(|\xi|^{m} - \tau^{m})$ and

(8.34) $$\bar{G}_j^{\prime\pm} = F^{-1}_{\xi',\tau\to x',t}\, \mathcal{G}_j'(\xi',\tau \mp i0, x_1)(1-\zeta'(\xi')) F_{x',t\to\xi',\tau}$$

$$(j = 1,\dots,p),$$

(8.35) $$\bar{G}^{\pm} = F^{-1}_{\xi',\tau\to x,t}\, \mathcal{G}'(\xi',\tau \mp i0)(1-\zeta'(\xi')) F_{x',t\to\xi',\tau}$$

where $\mathcal{G}_j'(\xi',\tau,x_1)$ are solutions of the boundary value problems

(8.36) $$(D_1^2+\rho^2)^q[(D_1^2+\rho^2)^{m/2} - \tau^m]\,\mathcal{G}_j' = 0,$$

(8.37) $$r B_k(D_1,\xi')\,\mathcal{G}_j' = \delta_{jk}\, Id \qquad k = 1,\dots,p,$$

(8.38) $$\mathcal{G}_j' = o(1) \quad \text{as} \quad x_1 \to +\infty$$

and $\mathcal{G}(\xi',\tau)$ is an integral operator in $L_2(\mathbb{R}^+, E)$ with the Schwartz kernel

(8.39) $$K_{\mathcal{G}}(\xi',\tau,x_1,y_1) = \frac{1}{2\pi}\int_{-\infty}^{\infty} [e^{i(x_1-y_1)\xi_1} - \sum_{k=1}^{p} \mathcal{G}_k'(\xi',\tau,x_1)\cdot$$

$$b_\kappa(\xi_1,\xi')e^{-iy_1\xi_1}]a^{-1}(\xi_1,\xi',\tau)d\xi_1\ ;$$

remind that $\rho(x,\xi')=|\xi'|$.

Describe now the construction of $\mathcal{G}'_j$. Assume first that we have the Dirichlet boundary conditions:

(8.37)' $$\tau D_1^{\kappa-1}\mathcal{G}'_j=\delta_{j\kappa}Id \qquad k=1,\dots,p\ .$$

Then $\mathcal{G}'_j=\mathcal{G}'_{jD}$ can be constructed by means of the procedure suggested by Agmon, Douglis, Nirenberg [4]. Note that for $|arg\,\tau^2|<4\pi/m$ the polynomial $a(\xi_1,\xi',\tau)$ has $2p-2$ roots $\pm\lambda_j(\rho,\tau)$ $(j=1,\dots,p-1)$ which are non-real everywhere ($Im\,\lambda_j>0$, $\pm\lambda_1=\dots\pm\lambda_q=\pm i\rho$; other roots are simple) and two roots $\pm\lambda=\pm\sqrt{\tau^2-\rho^2}$, $Im\,\lambda\geqslant 0$, which are real for $\tau\in\mathbb{R}$, $|\tau|\geqslant\rho$ and coincide for $\tau=\pm\rho$.

We set

$$a_+(\xi_1,\rho,\tau)=\prod_{j=1}^{p-1}(\xi_1-\lambda_j(\rho,\tau))=\sum_{j=0}^{p-1}c_j(\rho,\tau)\xi_1^{p-j-1}\quad ;$$

then

$$\mathcal{G}'_{jD}(\rho,\tau,x_1)=\frac{1}{2\pi i}\int_{-\infty}^{\infty}a_+^{-1}(\xi_1,\rho,\tau)\Big[\sum_{\kappa=0}^{p-j-1}c_\kappa(\rho,\tau)\xi_1^{p-j-\kappa-1}+$$

$$(\xi_1^2+\rho^2-\tau^2)^{-1}(\xi_1+\lambda(\rho,\tau))c_{p-j}(\rho,\tau)\Big]e^{ix_1\xi_1}d\xi_1\,.$$

The solution of problem (8.36) - (8.38) in a general case is given by formulas

(8.40) $$\mathcal{G}'_j(\xi',\tau,x_1)=\sum_{\kappa=1}^{p}\mathcal{G}'_{\kappa D}(\xi',\tau,x_1)S_{\kappa j}(\xi',\tau)$$

where $S(\xi',\tau)=(S_{\kappa j}(\xi',\tau))_{\kappa,j=1,\dots,p}$ is the matrix inverse to the matrix $R(\xi',\tau)=(R_{jk}(\xi',\tau))_{j,k=1,\dots,p}$, $R_{jk}(\xi',\tau)=\tau B_j(D_1,\xi')\mathcal{G}_{kD}(\xi',\tau,x_1)$.

It is easy to prove by repeating the arguments of section 7.3 (vii) that $S(\xi',\tau)$ is holomorphic for $0<|\arg\tau^2|<\frac{4\pi}{m}$ and has the degree-type singularities at $\mathbb{R}$.

The construction of parametrices for problem $(8.12)_-$ - (8.13) is similar: in (8.32), (8.33) and in the definition of $a(\xi,\tau)$ τ^m must be replaced by $-\tau^m$ and all $2p$ roots of the polynomial $a(\xi_1,\rho,\tau)$ are non-real everywhere.

8.6. Consider now the normal rays zone $\{|\xi'|\leqslant C^{-1}|\tau|\}$; at this zone we need to consider only $u=u_+$ because

$$WF_f(u_-)\cap\{|\xi'|\leqslant C^{-1}|\tau|\}=\emptyset;$$

one can prove it by means of simplified arguments which will be used below. Thus let u be the Schwartz kernel of the operator $\mathfrak{A}^{-1/m}\sin(\mathfrak{A}^{1/m}t)\Pi^+J_D^{-1}$; then u satisfies

$$(8.12)_+ \qquad (\mathcal{A}_x-\mathcal{J}_x D_t^m)u_1\equiv 0,$$

$$(8.12)_+^\dagger \qquad u_1(\mathcal{A}_y-\mathcal{J}_y D_t^m)^\dagger\equiv 0 \quad (\mathrm{mod}\, C^\infty),$$

$$(8.41) \qquad B_x u_1=-B_x u_0,$$

$$(8.41)^\dagger \qquad u_1 B_y^\dagger=-u_0 B_y^\dagger$$

where u_0 satisfies (8.15), $(8.15)^\dagger$ and

$$(8.42) \qquad u_0|_{t=0}=0,\quad u_{0t}|_{t=0}\equiv K_{J_0^{-1}}(x,y) \quad (\mathrm{mod}\, C^\infty)$$

on a more extensive manifold.

Remind that

$$(8.43) \qquad WF_f(u_1)\cap\{|t|\leqslant t_0\}\subset\{x_1+y_1\leqslant|t|\}.$$

It is easy to show that the operator $P=\mathcal{J}D_t^m-\mathcal{A}$ can be decomposed into product

$$(8.44) \qquad P\equiv\Lambda_+\Lambda_-\prod_{j=1}^{\frac{m}{2}-1}\Lambda_j^+\prod_{j=1}^{\frac{m}{2}-1}\Lambda_j^- M$$

$$(\mathrm{mod}\ OPS^{-\infty,2p-1'}(X\times\mathbb{R},E))$$

where $OPS^{\mu,n'}$ consist of operators

$$b=\sum_{\kappa=0}^{n} b_\kappa D_1^\kappa \quad \text{with} \quad b_\kappa\in OPS^{\mu-\kappa'},$$

$$\Lambda_\pm = D_1-\lambda_\pm(x,D',D_t),$$

$$\Lambda_j^\pm = D_1-\lambda_j^\pm(x,D',D_t),$$

$\lambda_\pm$, $\lambda_j^\pm\in OPS^{1'}(X\times\mathbb{R},E)$ have the principal symbols $\pm(\tau^2-\rho^2)^{1/2}$, $\pm(\varepsilon_j\tau^2-\rho^2)^{1/2}$ respectively; here ε_j are non-real roots of degree $\frac{m}{2}$ from $1\ (j=1,\ldots,\frac{m}{2}-1)$, $\pm\lambda_j^\pm>0$, $M\equiv(-\Delta)^q$ $(\mathrm{mod}\ OPS^{2q-1,2q-1'}(X\times\mathbb{R},E))$ has the complete symbol $M(x,\xi,\tau)$ such that

$$D_x^\alpha D_\xi^\beta D_\tau^\gamma M=O(|\tau|^{-\gamma}) \text{ as } \tau\to\infty$$

and ξ', ξ_1 are bounded $\forall\alpha,\beta,\gamma$.

It should be noted that the factors in decomposition (8.44) can be permutated with the change of lower order terms; if we permutate two neighbouring factors then we need to change lower order terms only in these factors. We shall not pay attention to lower order terms in general; if R is some factor in decomposition (8.44) then $\bar{R}$ denotes this factor with the same lower order terms which it has on the left position on the right.

$(8.12)_+$, $(8.12)_+^\dagger$, (8.43), (8.44) imply that modulo functions which are infinitely smooth at zone $\{|t|<t_0,|\xi'|+|\eta'|<C^{-1}|\tau|\}$

$$(8.45)\qquad u_1\equiv T_+V^{+-}T_-^\dagger+T_-V^{-+}T_+^\dagger+T_+\sum_{j=1}^{p-1}V_j^+T_j^\dagger+$$

$$T_-\sum_{j=1}^{p-1}V_j^-T_j^\dagger+\sum_{\kappa=1}^{p-1}T_\kappa V_\kappa^{'+}T_+^\dagger+\sum_{\kappa=1}^{p-1}T_\kappa V_\kappa^{'-}T_-^\dagger+$$

$$\sum_{j,\kappa=1}^{p-1}T_\kappa V_{\kappa j}T_j^\dagger$$

where $T_\pm$, T_j $(j=1,\dots,\frac{m}{2}-1)$, T_ℓ $(\ell=\frac{m}{2},\dots,p-1)$ are operators such that
$\bar{\Lambda}_\pm T_\pm = 0,\ \bar{\Lambda}_j^\pm T_j = 0,\ \bar{M} T_\ell = 0,\ \tau T_\pm = I,\ \tau T_j = I,\ \tau D_1^s T_\ell = \delta_{s,\ell-\frac{m}{2}} I$. $(s=0,\dots,q-1)$; all operators $T.$ $(T.^\dagger)$ are to the left (right) of $V..$ and act with respect to x, t (y, t - respectively), $V. = V.(x', y', t)$.

One can prove (8.45) by means of arguments used in the proof of lemma 3.12.

Since u_0 satisfies (8.15), $(8.15)^\dagger$, (8.42) at a more extensive manifold than modulo functions which are infinitely smooth at $X \cap \{|\xi'| + |\eta'| \leqslant \frac{1}{2}|\tau|,\ |t| < t_0\}$

$$u_0 \equiv T_+ W^+ T_+^\dagger + T_- W^- T_-^\dagger \tag{8.46}$$

where $W^\pm = W^\pm(x', y', t)$.

Moreover, modulo functions which are infinitely smooth in this zone

$$W^\pm \equiv (2\pi)^{-d} \int e^{i\langle x'-y', \xi'\rangle + it\tau} c^\pm(x', \xi', \tau)\, d\xi'\, d\tau \tag{8.47}$$

where $c^\pm \in S^{-2q-1}(Y \times \mathbb{R} \times \mathbb{R}^{d-1}, E)$, i.e. $W^\pm$ are the Schwartz kernels of pseudo-differential operators belonging to $OPS^{-2q-1}(Y \times \mathbb{R}, E)$ and commuting with D_t . One can prove it either by methods of sections 3.5, 4.6 or simply expressing $u_0(x, y, t)$ by the sum of oscillatory integrals.

We substitute (8.45), (8.46) into the boundary conditions (8.41), $(8.41)^\dagger$ and obtain the following equalities:

$$B_x\Big(T_- V^{-+} + \sum_{k=1}^{p-1} T_k V_k^{\prime +}\Big) \equiv -B_x T_+ W_+ , \tag{8.48$_+$}$$

$$B_x\Big(T_+ V^{+-} + \sum_{k=1}^{p-1} T_k V_k^{\prime -}\Big) \equiv -B_x T_- W^- , \tag{8.48$_-$}$$

$$B_x\Big(T_+ V_j^+ + T_- V_j^- + \sum_{k=1}^{p-1} T_j V_{jk}\Big) \equiv 0 \qquad (j=1,\dots,p-1) , \tag{8.49$_j$}$$

$$(8.48)^{\dagger}_{+} \quad \Big(V^{+-}T^{\dagger}_{-} + \sum_{j=1}^{p-1} V^{+}_{j}\, T^{\dagger}_{j}\Big)\, B^{\dagger}_{y} \equiv -W^{+} T^{\dagger}_{-}\, B^{\dagger}_{y} \,,$$

$$(8.48)^{\dagger}_{-} \quad \Big(V^{-+}T^{\dagger}_{+} + \sum_{j=1}^{p-1} V^{-}_{j}\, T^{\dagger}_{j}\Big)\, B^{\dagger}_{y} \equiv -W^{-} T^{\dagger}_{-}\, B^{\dagger}_{y} \,,$$

$$(8.49)^{\dagger}_{k} \quad \Big(V'^{+}_{k}\, T^{\dagger}_{+} + V'^{-}_{k}\, T^{\dagger}_{-} + \sum_{j=1}^{p-1} V_{kj}\, T^{\dagger}_{j}\Big)\, B^{\dagger}_{y} \equiv 0$$

$$(k = 1, \ldots, p-1).$$

It is easy to prove that for every $s \in \mathbb{Z}^{+}$ $\tau D_{1}^{s} T_{\pm}$, $\tau D_{1}^{s} T_{j} \in OPS^{s}(Y \times \mathbb{R}, E)_{\pm s}$ $(j = 1, \ldots, \frac{m}{2} - 1)$ have the principal symbols $\lambda^{s}_{\pm}(\rho, \tau)$, $\lambda^{\pm s}_{j}(\rho, \tau)$ respectively and $\tau D_{1}^{s} T_{\frac{m}{2}+k} \in \widetilde{OPS}^{s-k}(Y \times \mathbb{R}, E)$ $(k = 0, \ldots, q-1)$ have the principal symbols $c_{ks}(i\rho)^{s-k}$, $c_{ks} = const$, $c_{ks} = \delta_{ks}$ for $s \leq q-1$, where $\widetilde{OPS}^{\mu, h}(Y \times \mathbb{R}, E)$ is the class of pseudo-differential operators with symbols $b(x', \xi', \tau)$ satisfying the estimates

$$| D^{\alpha}_{x'} D^{\beta}_{\xi'} D^{\gamma}_{\tau}\, b | \leq C_{\alpha\beta\gamma} (1 + |\xi'|)^{\mu - |\beta|} (1 + |\xi'| + |\tau|)^{h - \gamma} \qquad \forall \alpha, \beta, \gamma,$$

expanding into asymptotical series with positively homogeneous terms b_{n},

$$| D^{\alpha}_{x'} D^{\beta}_{\xi'} D^{\gamma}_{\tau}\, b_{n} | \leq C_{\alpha\beta\gamma n} |\xi'|^{\mu - |\beta| - n} (|\xi'| + |\tau|)^{h - \gamma} \qquad \forall \alpha, \beta, \gamma, n,$$

in the natural sense; neglible operators are operators B such that for every $n \in \mathbb{Z}^{+}$ operators $(Ad\, t)^{n} B$ have order $-\infty$ with respect to x' and $(h-n)$ with respect to t.

Thus every equality $(8.48)_{\pm}$ as well as $(8.48)^{\dagger}_{\pm}$ is the system of p pseudo-differential equations (or subsystems) with p unknown (E-valued) functions and every equality $(8.49)_{j}$ as well as $(8.49)^{\dagger}_{k}$ is the system of p pseudo-differential equations (subsystems) with $(p+1)$ unknown (E-valued) functions. So systems $(8.48)^{\bullet}_{\cdot}$ are determinate and systems $(8.49)^{\bullet}$ are underdeterminate.

Consider systems $(8.48)_{\pm}$ in greater detail; systems $(8.48)^{\dagger}_{\pm}$ can be considered similarly; we omit the second index $(\pm)$ and obtain systems

$$(8.50)_{\pm} \qquad \mathcal{B}_{\pm} V \equiv -B_x W^{\pm}$$

where $V = (V_0, \ldots, V_{p-1})$ and $\mathcal{B}_{\pm} V = B_x (T_{\mp} V_0 + \sum_{K=1}^{p-1} T_K V_K)$;

$B_{\pm}$ is $p \times p$-matrix operator with with indices (l,k), $l = 1, \ldots, p$, $k = 0, \ldots, p-1$.

Note that (8.50) are Leray - Volevič systems with order $(\mu_1, \ldots, \mu_p; 0, \ldots, 0, -1, \ldots, -q+1)$.

LEMMA 8.12. (8.50) are Leray - Volevič elliptic systems in the neighbourhood of $\xi' = 0$.

PROOF. Consider the principal symbols of $\mathcal{B}_{\pm}$ at $\xi'=0$, $\tau>0$; it equals

$$(8.51)_{\pm} \qquad \left(\begin{array}{cccc|cccc} (\theta_{\pm}\tau)^{\mu_1} & (\theta_1\tau)^{\mu_1} & \ldots & (\theta_{\frac{m}{2}-1}\tau)^{\mu_1} & 1 & & & 0 \\ (\theta_{\pm}\tau)^{\mu_2} & (\theta_1\tau)^{\mu_2} & \ldots & (\theta_{\frac{m}{2}-1}\tau)^{\mu_2} & & 1 & & \\ \ldots & \ldots & \ldots & \ldots & 0 & & 1 & \\ & & & & & & & 1 \\ (\theta_{\pm}\tau)^{\mu_p} & (\theta_1\tau)^{\mu_p} & \ldots & (\theta_{\frac{m}{2}-1}\tau)^{\mu_p} & & & 0 & \end{array} \right)$$

(the right block has q columns)

$$(\mu_1 = 0, \ldots, \mu_q = q-1)$$

where $\theta_{\pm} = \mp 1$, θ_K are roots of degree m from 1 with positive imaginary parts $(k = 1, \ldots, \frac{m}{2} - 1)$. The determinant of matrix $(8.51)_{\pm}$ equals $C_{\pm} \tau^K$, $K = \mu_{q+1} + \cdots + \mu_p$, $C_{\pm}$ are constants; hence if $\Delta(\tau_0) = 0$ for some $\tau_0 \in \mathbb{R}^+$ then $\Delta(\tau) \equiv 0$ and the problem

$$(\bar{A} - \tau^m \bar{J}) v = 0, \quad \bar{B} v = 0$$

with $\bar{A} = D_1^{2p}$, $\bar{J} = D_1^{2q}$, $\tau^m \in \mathbb{C} \setminus \mathbb{R}$, $\bar{B} = \{\tau D_1^{\mu_l}\}_{l=1,\ldots,p}$ has non-trivial solution $v = v_1 + v_2$, $v_1 \in \mathcal{S}(\bar{\mathbb{R}}^+)$, v_2 is a polynomial of degree not exceeding $q-1$; it is impossible because $\bar{A}_{\bar{B}}$, $\bar{J}_D$ are selfadjoint operators and $\bar{J}_D$ is positive definite. The lemma has been proved.

Thus operators $\mathcal{B}_{\pm}$ have parametrices in the neighbourhood of $\xi'=0$ modulo operators which are infinitely smoothing with respect to (x',t) ; these parametrices are

$$\mathcal{R}^{\pm}:(F_1,\ldots,F_p)\to(V_0,\ldots,V_{p-1}),$$

$$V_k=\sum_{\ell=1}^{p}\mathcal{R}^{\pm}_{k\ell}F_\ell$$

where $\mathcal{R}^{\pm}_{k\ell}\in OPS^{-\ell}(Y\times\mathbb{R},E)\oplus\widetilde{OPS}^{1,-\ell-1}(Y\times\mathbb{R},E)$

for $k=0,\ldots,\frac{m}{2}-1$ and $\mathcal{R}^{\pm}_{\frac{m}{2}+s,\ell}\in OPS^{s-\ell}(Y\times\mathbb{R},E)\oplus\widetilde{OPS}^{1,s-1-\ell}(Y\times\mathbb{R},E)$ for $s=0,\ldots,q-1$.

Hence $(8.48)_{\pm}$, $(8.48)^{+}_{\pm}$, (8.47) imply that modulo functions which are infinitely smooth in $\{|t|<t_0,|\xi'|\leq C^{-1}|\tau|\}$

(8.52) $$V^{-+},V^{+-},V'^{\pm}_{k},V^{\pm}_{k}\in KOPS^{-2q-1}(Y\times\mathbb{R},E)\oplus K\widetilde{OPS}^{1,-2q-2}(Y\times\mathbb{R},E)$$

$$k=1,\ldots,\frac{m}{2}-1,$$

(8.52)' $$V'^{\pm}_{\frac{m}{2}+s},V^{\pm}_{\frac{m}{2}+s}\in KOPS^{s-2q-1}(Y\times\mathbb{R},E)\oplus K\widetilde{OPS}^{1,s-2q-2}(Y\times\mathbb{R},E)$$

$$s=0,\ldots,q-1,$$

where $K\mathcal{L}$ denotes the class of functions which are Schwartz kernels of operators belonging to the operator class $\mathcal{L}$.

Now $(8.49)_j$, $(8.49)^{+}_{k}$ are overdeterminate elliptic systems and we have

(8.52)'' $$V_{jk}\in KOPS^{-2q-1}(Y\times\mathbb{R},E)\oplus K\widetilde{OPS}^{1,-2q-2}(Y\times\mathbb{R},E),$$

(8.52)''' $$V_{k,\frac{m}{2}+s},V_{\frac{m}{2}+s,k}\in KOPS^{s-2q-1}\oplus K\widetilde{OPS}^{1,s-2q-2}(Y\times\mathbb{R},E),$$

(8.52)^{IV} $$V_{\frac{m}{2}-h,\frac{m}{2}+s}\in KOPS^{s+h-2q-1}(Y\times\mathbb{R},E)\oplus K\widetilde{OPS}^{1,s+h-2q-2}(Y\times\mathbb{R},E)$$

$$(j,k=1,\dots,\tfrac{m}{2}-1\,,\quad s,h=0,\dots,q-1)\,.$$

Thus modulo

$$K\widetilde{OPS}^{-\infty,-2q-2}(Y\times\mathbb{R},E)\subset\bigcap_{\varepsilon>0}C^{\infty}(Y\times Y,H^{2q+\frac{3}{2}-\varepsilon}(\mathbb{R},\,Hom(E)))$$

$$V^{+-},\,V^{-+},\,V_k'^{\pm}\,,\;V_j^{\pm}\,,\;V_{jk}\qquad(j,k=1,\dots,\tfrac{m}{2}-1)$$

can be expressed by oscillatory integrals (8.47) with amplitudes $c\in S^{-2q-1}\oplus\widetilde{S}^{1,-2q-2}$.

Hence

(8.53) $$T_{\pm}V^{\pm\mp}T_{\mp}^{\dagger}\equiv\int exp\,i(t\tau+\varphi_{\pm}(x,y,\xi',\tau))c^{\pm}(x,y,\xi',\tau)\,d\xi'd\tau\,,$$

(8.53)' $$T_{\pm}V_j^{\pm}T_j^{\dagger}\equiv$$

$$\int exp\,i(t\tau+\varphi_{\pm}(x,y',\xi',\tau)+\lambda_j^{+}(x',\xi',\tau)y_1)c_j^{\pm}(x,y,\xi',\tau)\,d\xi'd\tau\,,$$

(8.53)'' $$T_jV_j'^{\pm}T_{\pm}^{\dagger}\equiv$$

$$\int exp\,i(t\tau+\varphi_{\pm}(x,y',\xi',\tau)+\lambda_j^{+}(x',\xi',\tau)x_1)c_j'^{\pm}(x,y,\xi',\tau)\,d\xi'd\tau\,,$$

(8.53)''' $$T_jV_{jk}T_k^{\dagger}\equiv$$

$$\int \exp i(\langle x'-y',\xi'\rangle + t\tau + \lambda_j^+(x',\xi',\tau)x_1 +$$

$$\lambda_k^+(x',\xi',\tau)y_1)\, c_{jk}(x,y,\xi',\tau)\, d\xi' d\tau$$

$$(\mathrm{mod}\ C^\infty(Y\times Y, C^n(\overline{\mathbb{R}}^+\times\overline{\mathbb{R}}^+, H^{2q-\frac{3}{2}-n-\varepsilon}(\mathbb{R}, \mathrm{Hom}(E)))$$

$$\forall n \in \mathbb{Z}^+, \varepsilon>0)$$

where $\varphi_\pm(x,y,\xi',\tau)$ is the solution of the double Cauchy problem

(8.54) $$\varphi_{x_1} = \pm\sqrt{\tau^2-\rho^2(x,\varphi_{x'})}\ ,$$

$$\varphi_{y_1} = \pm\sqrt{\tau^2-\rho^2(y,\varphi_{y'})}\ ,$$

$$\varphi\big|_{x_1=y_1=0} = \langle x'-y',\xi'\rangle$$

and $\varphi_\pm(x,y',\xi',\tau) = \varphi_\pm(x,0,y',\xi',\tau)$, $\varphi_\pm(x',y,\xi',\tau) = \varphi_\pm(0,x',y',\xi',\tau)$ are the solutions of the ordinary Cauchy problems for eikonal equations; amplitudes $c^\pm$, $c_j^\pm$, $c_k'^\pm$, c_{jk} belong to $C^\infty(\overline{\mathbb{R}}^+\times\overline{\mathbb{R}}^+\times Y, S^{-2q-1}\oplus\widetilde{S}^{1,-2q-2})$.

Similarly modulo

$$K\widetilde{OPS}^{-\infty,h-2q-2}(Y\times\mathbb{R},E)\subset\bigcap_{\varepsilon>0} C^\infty(Y\times Y, H^{2q+\frac{3}{2}-\varepsilon-h}(\mathbb{R},\mathrm{Hom}(E)))$$

$$V^\pm_{\frac{m}{2}+h}\ ,\quad V'^\pm_{\frac{m}{2}+h}\ ,\quad V_{\frac{m}{2}+h,j}\ ,\quad V_{j,\frac{m}{2}+h}$$

$$(h=0,\ldots,q-1,\ j=1,\ldots,\tfrac{m}{2}-1)$$

can be expressed by oscillatory integrals (8.47) with amplitudes $c\in S^{h-2q-1}\oplus\widetilde{S}^{1,h-2q-2}$. Hence

(8.55) $$\sum_{h=0}^{q-1} T_{\pm} V^{\pm}_{\frac{m}{2}+h} \; T^{\dagger}_{\frac{m}{2}+h} \equiv$$

$$\sum_{h=0}^{q-1} \int exp(i(t\tau + \varphi_{\pm}(x,y',\xi',\tau)) - \rho(x',\xi)y_1)\, y_1^{h} c_h^{\pm}(x,y,\xi',\tau)\, d\xi' d\tau ,$$

(8.55)' $$\sum_{h=0}^{q-1} T_j \, V_{j,\frac{m}{2}+h} \; T^{\dagger}_{\frac{m}{2}+h} \equiv$$

$$\sum_{h=0}^{q-1} \int exp(i(\langle x'-y',\xi'\rangle + t\tau + \lambda_j^{+}(x',\xi',\tau)x_1) - \rho(x',\xi')y_1)\, y_1^{h} c_{j(h)}(x,y,\xi',\tau)\, d\xi' d\tau$$

$$(mod\ C^{\infty}(Y_x \times X_y, C^{n}(\overline{\mathbb{R}}^{+}_{x_1}, H^{q+\frac{3}{2}-\varepsilon-n}(\mathbb{R}, Hom(E)))) \forall n \in \mathbb{Z}^{+}, \varepsilon > 0)$$

and

(8.55)'' $$\sum_{h=0}^{q-1} T_{\frac{m}{2}+h} \, V'^{\pm}_{\frac{m}{2}+h} \; T^{\dagger}_{\pm} \equiv$$

$$\sum_{h=0}^{q-1} \int exp(i(t\tau + \varphi_{\pm}(x',y,\xi',\tau)) - \rho(x',\xi')x_1)\, x_1^{h} c^{\pm}_{(h)}(x,y,\xi',\tau)\, d\xi' d\tau ,$$

(8.55)'' $$\sum_{h=0}^{q-1} T_{\frac{m}{2}+h} \, V_{\frac{m}{2}+h} \; T_j^{\dagger} \equiv$$

$$\sum_{h=0}^{q-1}\int exp(i(\langle x'-y',\xi'\rangle+t\tau+\lambda_j^+(x',\xi',\tau)y_1)-\rho(x',\xi')x_1)x_1^h c_{(h)j}(x,y,\xi',\tau)\,d\xi' d\tau$$

$$(mod\, C^\infty(X_x\times Y_y, C^n(\overline{\mathbb{R}}^+_{y_1}, H^{q+\frac{3}{2}-\varepsilon-n}(\mathbb{R}, Hom(E)))) \;\forall n\in\mathbb{Z}^+,\; \varepsilon>0)$$

where amplitudes $c^{\pm}_{(h)}$, $c'^{\pm}_{(h)}$, $c_{(h)j}$, $c_{j(h)}$ belong to $C^\infty(\overline{\mathbb{R}}^+\times\overline{\mathbb{R}}^+\times Y, S^{h-2q-1}\oplus\widetilde{S}^{1,h-2q-2})$, $h=0,\dots,q-1$, $j=1,\dots,\frac{m}{2}-1$.

Finally, modulo

$$C^\infty(Y\times Y, H^{2q+\frac{3}{2}-h-s-\varepsilon}(\mathbb{R}, Hom(E)))$$

$$(\;h,\; s = 0,\; .\;.\;.,\; q-1\;)$$

$V_{\frac{m}{2}+h,\,\frac{m}{2}+s}$ can be expressed by oscillatory integrals (8.47) with amplitudes $c\in S^{h+s-2q-1}\oplus\widetilde{S}^{1,h+s-2q-2}$. Hence

$$\sum_{h,s=0}^{q-1} T_{\frac{m}{2}+h}\, V_{\frac{m}{2}+h,\,\frac{m}{2}+s}\, T^\dagger_{\frac{m}{2}+s} \equiv \tag{8.56}$$

$$\sum_{h,s=0}^{q-1}\int exp(i(\langle x'-y',\xi'\rangle+t\tau)-\rho(x',\xi')(x_1+y_1))\,x_1^h y_1^s c_{(hs)}(x,y,\xi',\tau)\,d\xi' d\tau$$

$$(mod\, C^\infty(X\times X, H^{\frac{3}{2}-\varepsilon}(\mathbb{R}, Hom(E)))\quad\forall\varepsilon>0)$$

where amplitudes $c_{(hs)}$ belong to $C^\infty(\overline{\mathbb{R}}^+\times\overline{\mathbb{R}}^+\times Y, S^{h+s-2q-1}\oplus\widetilde{S}^{h+s-2q-2})$.

8.7. Now we are able to finish our calculations. First of all we consider

$$\sigma(t)=\int_X tr\,\Gamma(u\,\mathcal{J}_y^\dagger)\,dy\,.$$

Since u and all its derivatives up to the order $q-1$ vanish for $x\in Y$ or $y\in Y$ then

$$\sigma(t) = \sum_{|\alpha|,|\beta|\leq q} \int_X tr\, a_{\alpha\beta}(y)\Gamma(D_x^{\alpha} D_y^{\beta} u)\, dy =$$

$$\sum_{|\alpha|,|\beta|\leq q} \int_X tr\, a_{\alpha\beta}(y)\Gamma(D_x^{\alpha} D_y^{\beta} u_0)\, dy +$$

$$\sum_{|\alpha|,|\beta|\leq q} \int_X tr\, a_{\alpha\beta}(y)\Gamma(D_x^{\alpha} D_y^{\beta} u_1)\, dy = \sigma^0(t) + \sigma^1(t).$$

These equalities remain true for u replaced by Qu, $Q \in OPS'(X\times\mathbb{R}, E)$, $Q \equiv 0$ for $x_1 > \delta$.

We know that

$$\Gamma D_x^{\alpha} D_y^{\beta} u_0 \sim \sum_{\omega=0}^{\infty} d_\omega(y)\, \Phi_{d-2q+|\alpha|+|\beta|-\omega-2}(t)$$

where $d_\omega \in C^{\infty}(X, hom(E))$; see section 4.2 e.g.

Hence

$$\sigma^0(t) \sim \sum_{\omega=0}^{\infty} c_\omega^0\, \Phi_{d-\omega-2}(t) .$$

We also know that

$$\sigma_Q^1(t) = \sum_{|\alpha|,|\beta|\leq q} \int_X tr\, a_{\alpha\beta}(y)\Gamma(D_x^{\alpha} D_y^{\beta} Q u_1)\, dy \sim \sum_{\omega=0}^{\infty} c'_\omega\, \Phi_{d-\omega-3}(t)$$

provided $Q \in OPS^{0'}(X\times\mathbb{R}, E)$, $\text{cone supp}\, Q \subset \{x_1 < \delta, |\xi'| \geq C^{-1}|\tau|\}$.

Finally (8.45), (8.53) - (8.53)''', (8.55) - (8.55)''', (8.56) imply that

$$\sigma_Q^1(t) \equiv \sum_{\omega=0}^{d-2} c'_\omega \Phi_{d-\omega-3}(t)$$

$$(\mathrm{mod}\ H^{\frac{3}{2}-\varepsilon}([-t_0,t]) \quad \forall\varepsilon>0)$$

provided t_0 is small enough,

$$\text{cone supp } Q \subset \{x_1<\delta,\ |\xi'| \leq 2C^{-1}|\tau|\}.$$

Thus the following statement has been proved:

THEOREM 8.13. In the neighbourhood of $t=0$ the following incomplete asymptotics holds:

$$\sigma_+(t) = \mathrm{tr}\, \mathcal{U}^{-\frac{1}{m}} \sin(\mathcal{U}^{\frac{1}{m}} t)\Pi^+ \equiv \sum_{\omega=0}^{d-2} c_\omega \Phi_{d-\omega-2}(t) \tag{8.57}$$

$$(\mathrm{mod}\ H^{\frac{3}{2}-\varepsilon} \quad \forall\varepsilon>0)$$

where $c_0 = \frac{d}{2i}(2\pi)^{-d}\omega_d\, D\, \mathrm{vol}\, X$, c_1 depends only on ρ, q, Y and the principal symbol of B.

(8.57) remains true if one applies an operator $(tD_t)^n$ to its left part.

COROLLARY 8.14. If $\chi \in C_0^\infty(\mathbb{R})$ has a small enough support and $\chi=1$ in the neighbourhood of $t=0$ then the following incomplete asymptotics holds:

$$F_{t\to\tau}(\chi\sigma_+) = \sum_{\omega=0}^{d-2} c_\omega |\tau|^{d-\omega-2}\, \mathrm{sign}\,\tau + O(|\tau|^{\varepsilon-2})$$

$$\text{as} \quad \tau \to \pm\infty \quad \forall\varepsilon>0. \tag{8.58}$$

REMARK. The complete asymptotics for $\sigma_+(t)$ and $F_{t\to\tau}(\chi\sigma_+)$ can be derived provided $q=0$ and the boundary operators are differential.

The following statement follows from section 8.5 and the fact that

$u_- \in C^\infty$ in the zone $\{ |\xi'| \leqslant C^{-1} |\tau| \}$:

THEOREM 8.15. In the neighbourhood of $t=0$ the following asymptotics holds:

$$(8.59) \quad \sigma_-(t) = tr(-\mathcal{U})^{-\frac{1}{m}} \sin(-\mathcal{U})^{\frac{1}{m}} t \cdot \Pi^- \sim \sum_{\omega=0}^{\infty} c'_\omega \Phi_{d-\omega-3}(t)$$

where C'_0 depends only on ρ, q, Y and the principal symbol of B.

COROLLARY 8.16. If $\chi \in C_0^\infty(\mathbb{R})$ has a small enough support and $\chi = 1$ in the neighbourhood of $t=0$ then the following asymptotics holds:

$$(8.60) \quad F_{t\to\tau}(\chi\sigma_-) \sim \sum_{\omega=0}^{\infty} c'_\omega |\tau|^{d\ \omega\ 3} \operatorname{sign}\tau \quad \text{as} \quad \tau \to \pm\infty .$$

8.8. Let now $q=0$. Then in the decomposition of u_1 only the terms of types (8.53) - (8.53)''' are included and there are no terms of types (8.55) - (8.55)''', (8.56). Therefore we can apply the arguments of section 4.7 and obtain the following statement:

THEOREM 8.17. Let $q=0$, $\chi \in C_0^\infty(\mathbb{R})$ has a small enough support, $\chi = 1$ in the neighbourhood of $t=0$. Then the following asymptotics hold:

$$(8.61) \quad F_{t\to\tau} \Gamma\chi u_1 = O(\tau^{d-2}) \quad \text{as} \quad \tau \to \pm\infty ,$$

$$(8.62) \quad 2i\int_0^k \tau F_{t\to\tau} \Gamma\chi u_1 d\tau = k^d Q^+(x', x_1 k) + O(k^{d-1}) \quad \text{as} \quad k \to +\infty$$

uniformly with respect to x where $Q^+ \in C^\infty(Y \times \overline{\mathbb{R}}^+)$ satisfies (0.9).

REMARK. $Q^+ \in C^\infty(Y \times \overline{\mathbb{R}}^+)$ because $\chi\Gamma u_1 \in \mathcal{H}^{s;\infty}_{t;x}$ for some $s \in \mathbb{R}$.

This statement follows from section 8.5:

THEOREM 8.18. If $\chi \in C_0^\infty(\mathbb{R})$ has a small enough support, $\chi = 1$ in the neighbourhood of $t=0$ then the following asymptotics hold:

$$(8.63) \qquad F_{t\to\tau} \Gamma \chi u_- \sim \sum_{w=0}^{\infty} |\tau|^{d-w-2} \operatorname{sign}\tau \, S_w(x', x_1|\tau|)$$

as $\tau \to \pm\infty$,

$$(8.64) \qquad 2i \int_0^k \tau F_{t\to\tau} \Gamma \chi u_- d\tau = k^d Q^-(x', x_1 k) + O(\min(k^{d-1}, x_1^{-d+1}))$$

as $k \to +\infty$

uniformly with respect to x where $S_w, Q^- \in C^\infty(Y \times \overline{\mathbb{R}}^+)$,

$$D_{x'}^\alpha D_s^l S_w(x', s) = O(s^{-\infty}) \quad \text{as } s \to +\infty$$

uniformly with respect to $x' \in Y$, Q^- satisfies (0.14).

REMARK. $Q^\pm$ depend only on ρ, q, Y and the principal symbol of B.

8.9. Theorems 8.1 - 8.5 are proved using the arguments of §5.

Appendix F

1. In my papers [55, 57] the incorrect famulas for the second term coefficient were given; namely, the following item was lost:

$$\bar{\varkappa} = -\frac{(2\pi)^{-d-1}}{d-1} \int_{T^*X} \left(\bar{F}(1-i0,y,\xi) - \bar{F}(1+i0,y,\xi) \right) dy\, d\xi ,$$

$$\bar{F}_{(\tau,y,\xi)} = 2\,\mathrm{tr} \sum_{j=1}^{d} (-\tau^2+a)^{-1} \left(\frac{\partial a}{\partial \xi_j} (-\tau^2+a)^{-1} \frac{\partial a}{\partial y_j} - \frac{\partial a}{\partial y_j} (-\tau^2+a)^{-1} \frac{\partial a}{\partial \xi_j} \right) (-\tau^2+a)^{-1} \Pi ,$$

$$a = a(y,\xi), \quad \Pi = \Pi(y,\xi)$$

If a, Π are not only Hermitian but also symmetric matrices this item equals 0.

Similar items were lost in my papers [63, 101] devoted to quasi-classical asymptotics and in [103, 106] on the asymptotics for operators in R^d .

2. The text of this manuscript was written in 1982. This appendix and References 100-11 were added subsequently.

List of notations

$\mathbb{R}$ - real axis,

$\mathbb{R}^{\pm} = \{ t \in \mathbb{R}, \ \pm t > 0 \}$

$\overline{\mathbb{R}}^{\pm} = \{ t \in \mathbb{R}, \ \pm t \geqslant 0 \}$,

$\mathbb{C}$ - complex plane,

$\mathbb{C}_{\pm} = \{ z \in \mathbb{C}, \ \pm \operatorname{Im} z \geqslant 0 \}$ - upper (lower) complex half-plane,

$\mathbb{Z}$ - the set of integers,

$\mathbb{Z}^{+}$ - the set of nonnegative integers.

T^*X - cotangent bundle,

S^*X, B^*X - unit spheres (balls) bundle.

$(\, , \,)$, $\langle \, , \, \rangle$ - inner products.

$[\, , \,]$ - commutator,

$\{ \, , \, \}$ - Poisson brackets.

Subject index

REFERENCES

1. S. Agmon, Lectures on elliptic boundary problems, Princeton, N.Y., a.o., 1965, (Van Nostrand Math.Studies).
2. S. Agmon, On kernels, eigenvalues and eigenfunctions of operators, related to elliptic problems, Communs on Pure Appl.Math., 1965, v.18, No 4, p.627-663.
3. S. Agmon, Asymptotic formulas with remainder estimates for eigenvalues of elliptic operators, Arch.Rat.Mech.Anal., 1968, v.28, No 3, p.168-183.
4. S. Agmon, A.Douglis, L.Nirenberg, Estimates near the boundary for solutions of elliptic partial differential equations satisfying general boundary conditions, I, Communs on Pure and Appl.Math., 1959, v.12, p.623-727.
5. S.Agmon, Ya.Kannai, On the asymptotic behaviour of spectral functions and resolvent kernels of elliptic operators, Israel J. of Math., 1967, v.5, No 1, p.1-30.
6. M.S.Agranovič, Boundary-value problems for systems of pseudo-differential operators of first order (in Russian), Uspechi Matem.Nauk, 1969, v.24, No 1, p.61-125; English translation: Russian Math.Surveys, 1969, v.24, No 1, p.45-126.
7. V.G.Avakumovič, Über die Eigenfunktionen auf geschlossenen Riemannischen Mannigfaltigkeiten, Math.Z., 1956, v.65, s.327-344.
8. V.M.Babič, The focussing problem and spectral function asymptotics of Laplace - Beltrami operator (in Russian), Math.problems of waves propagation theory. 10(LOMI proceedings, v.89), Nauka. Leningrad, 1979, p.14-53.
9. V.M.Babič, V.S.Buldyrev, Asymptotic methods in the short waves diffraction problems (in Russian), Nauka, Moscow, 1972.
10. V.M.Babič, B.M.Levitan, The focussing problem and spectral function asymptotics of Laplace-Beltrami operator (in Russian), Dokl.AN SSSR, 1964, v.157, No 3, p.536-538.
11. V.M.Babič, B.M.Levitan, The focussing problem and spectral function asymptotics of Laplace-Beltrami operator (in Russian), Math.problems of waves propagation theory. 9 (LOMI proceedings, v.78), Nauka, Leningrad, 1978, p.3-19.
12. P.Berard, On the wave equation on a compact Riemannian manifold without conjugate points, Math.Z., 1977, b.155, h.2, s.249-276.
13. M.S.Birman, M.Z.Solomjak, Asymptotics of the spec-

trum for differential equations (in Russian), Itogy Nauki. Ser. Matem.Analys. v.14, VINITI, Moscov, 1977, p.5-58.

14. L.Boutet de Monvel, Boundary problems for pseudo-differential operators, Acta Math., 1971, v.126, No 1-2, p.11-51.
15. J.Chazarain, Formule de Poisson pour les variétés Riemanniennes, Invent.Math., 1974, v.24, p.65-82.
16. J.Chazarain, Spectre d'un Hamiltonien quantique et mécanique classique, Communs in Part.Different Equats., 1980, v.5, No 6, p.695-644.
17. C.Clark, The asymptotics distributions of eigenvalues and eigenfunctions for elliptic boundary value problems, SIAM Rev., 1967, v.9, No 4, p.627-646.
18. Y.Colin de Verdière, Spectre du Laplacien et longueurs des géodésiques périodiques, Composito Math., 1974, v.27, p.86-106.
19. Y.Colin de Verdière, Spectre du Laplacien et longueurs des géodésiques périodiques, II, Composito Math., 1973, v.27, p.159-184.
20. Y.Colin de Verdière, Sur le spectre des opérateurs elliptiques à bicharactéristiques toutes périodiques, Comment.Math.Helv., 1979, v.54, No 3, p.508-522.
21. I.P.Cornfeld, S.V.Fomin, J.G.Sinai, Ergodic theory (in Russian), Nauka, Moscow, 1970. English translation: Springer-Verlag, 1982 (Grundlehren 245).
22. R.Courant, Über die Eigenwerte bei den Differentialgleichungen der mathematischen Physik, Math.Z., 1920, b.7, s.1-57.
23. J.J.Duistermaat, On the spectrum of positive elliptic operators and periodic bicharacteristics, Lecture Notes in Math., 1975, v.459, p.15-22.
24. J.J.Duistermaat, V.Guillemin, The spectrum of positive elliptic operators, Invent.Math., 1975, v.29, p.39-79.
25. J.J.Duistermaat, L.Hörmander, Fourier integral operators. II, Acta Math., 1972, v.128, No 3-4, p.183-269.
26. B.V.Fedosov, Asymptotics formulae for eigenvalues of the Laplace operator in the case of polygonal domains (in Russian), Dokl.AN SSSR, 1963, v.151, No 4, p.786-789.
27. B.V.Fedosov, Asymptotic formulae for eigenvalues of the Laplace operator in the case of polyhedral domains (in Russian), Dokl.AN SSSR, 1964, v.157, No 3, p.536-538.
28. V.I.Feigin, On the spectral asymptotics for boundary value problems and the asymptotics of the negative spectrum (in Russian)

Dokl.AN SSSR, 1977, v.232, No 6, p.1269-1272;
English translation: Soviet Math.Dokl., 1977, v.18, No 1.

29. V.I.F e i g i n , Estimates of the length of gaps in the spectrum of a first-order system (in Russian), Uspechi Matem.Nauk, 1979, v.34, No 3, p.215-216; English translation: Russian Math.Surveys, 1979, v.34, No 3, p.198-199.

30. V.I.F e i g i n , Asymptotic distribution of eigenvalues and Bohr-Sommerfeld formula (in Russian) Matem.Sbornik, 1979, v.110, No 1, p.68-87.

31. I.M.G e l f a n d , G.E.S h i l o v , Generalized functions, v.I: properties and operators (in Russian), Fizmatgiz, Moscow, 1958; English translation: Academic press, N.Y., 1964.

32. G.G r u b b , Le spectre négatif des problèmes aux limites auto-adjoints fortement elliptiques, C.r.Acad.sci.Paris, 1972, t 274, No 5, p.A409-412.

33. G.G r u b b , Spectral asymptotics for Douglis-Nirenberg elliptic systems and pseudo-differential boundary problems. Comm. in Part.Different.Equats., 1977, v.2, p.1071-1150.

34. G.G r u b b , On the spectral theory of pseudo-differential boundary problems, Kobenhavn Universitet, Mat.Inst., 1978, Preprint No 33.

35. G.G r u b b , Remainder estimate for eigenvalues and kernels of pseudo-differential elliptic systems, Math.Scand., 1978, v.43, No 2. p.275-307.

36. V.G u i l l e m i n , Lectures on spectral theory of elliptic operators, Duke Math.J., 1977, v.44, No 3, p.485-517.

37. V.G u i l l e m i n , Some spectral results for the Laplace operator with potential on the n-sphere, Adv. in Math., 1978, v.27, No 3, p.273-286.

38. V.G u i l l e m i n , Some classical theorems in spectral theory revised, Seminar on singularities of solutions of linear partial different.equats., Ann.Math.Studies, v.91, Princeton,Univ., 1979, p.219-259.

39. V.G u i l l e m i n , R.M e l r o s e , The Poisson summation formula for manifolds with boundary, Adv. in Math., 1979, v.32, No 3, p.204-232.

40. V.G u i l l e m i n , S.S t e r n b e r g , On the spectra of commuting pseudo-differential operators: recent work of Kac-Spencer, Weinstein and others, Part.Different.Equats, and Geometry, Proc.Park-City Conf., 1977, N.Y. - Basel, 1979, p.149-165.

41. V.G u i l l e m i n , S.S t e r n b e r g , Geometric asymptotics

Math.Surveys, v.14, Amer.Math.Soc., Prov., R.I., 1977.

42. L.H ö r m a n d e r , Fourier integral operators, I, Acta Math., 1971, v.127, No 1-2, p.79-183.

43. L.H ö r m a n d e r , The spectral function of an elliptic operator, Acta Math., 1968, v.124, No 1-2, p.193-218.

44. L.H ö r m a n d e r , On the Riesz means of spectral functions and eigenfunction expansion for elliptic differential operators, Recent Adv. in the Basic Sci., 2, Belfer Graduate School of Sci., Yeshiva Univ., 1966, p.155-202.

45. L.H ö r m a n d e r , Pseudo-differential operators, Comm. on Pure and Appl.Math., 1965, v.18, p.501-517.

46. L.H ö r m a n d e r , Pseudo-differential operators and hypoelliptic equations, Proc.Symp. on singular integrals, Chicago, 1966.

47. L.H ö r m a n d e r , Pseudo-differential operators and non-elliptic boundary problems, Ann. of Math., 1966, v.83, No 1, p.129-209.

48. L.H ö r m a n d e r , Linear partial differential operators, Springer-Verlag, 1963.

49. V.I v r i i , Wave fronts of solutions of symmetric pseudo-differential systems (in Russian), Sibirskii Matem.J., 1979, v.20, No 3, p.557-578: English translation: Siberian Math.J., 1979, v.20, No 3, p.390-405.

50. V.I v r i i , Wave fronts of solutions of boundary-value problems for symmetric hyperbolic systems.I. The basic theorem (in Russian) Sibirskii Matem.J., 1979, v.20, No 4, p.741-751; English translation: Siberian Math.J., 1979, v.20, No 3, p.516-523.

51. V.I v r i i , Wave fronts of solutions of boundary-value problems for symmetric hyperbolic systems.II. Systems with characteristics of constant multiplicity (in Russian), Sibirskii Matem.J., 1979, v.20, No 5, p.1022-1038; English translation: Siberian Math.J., 1979, v.20, No 5, p.722-734.

52. V.I v r i i , Wave fronts of solutions of boundary value- problems for symmetric hyperbolic systems.III. Systems with characteristics of variable multiplicity (in Russian); Sibirskii Matem.J., 1980, v.21, No 1, p.74-81; English translation: Siberian Math.J., 1980, v.21, No 1, p.54-60.

53. V.I v r i i , Wave fromts of solutions of boundary-value problems for a class of symmetric hyperbolic systems (in Russian), Sibirskii Matem.J., 1980, v.21, No 4, p.62-71; English translation: Siberian Math.J., 1980, v.21, No 4, p.527-534.

54. V.I v r i i , Propagation of the singularities of solutions of non-classical boundary-value problems for second-order hyperbolic

equations (in Russian), Trudy Moskovskogo Matem.Obs., 1981,v.43, p.83-91; English translation will appear in Trans.Moscow Math.Soc.

55. V.I v r i i , On the second term of the spectral asymptotics for the Laplace-Beltrami operator on manifolds with boundary and for elliptic operators acting in fiberings (in Russian), Dokl.AN SSSR, 1980, v.250, No 6, p.1300-1302; English translation: Soviet Math. Dokl., 1980, v.21, No 1, p.300-302.

56. V.I v r i i , Second term of the spectral asymptotics expansion of the Laplace-Beltrami operator for manifold with boundary (in Russian), Funkt.Analys i Ego Pril., 1980, v.14, No 2, p.25-34; English translation: Funct.Anal. Applic., 1980, v.14, p.98-106.

57. V.I v r i i , Precise spectral asymptotics for elliptic operators acting in fiberings (in Russian), Funkt.Analys i Ego Pril., 1982, v.16, No 2, p.30-38; English translation will appear in Funct.Anal. Applic.

58. V.I v r i i , Exact spectral asymptotics for the Laplace-Beltrami operator in the case of general elliptic boundary conditions (in Russian), Funkt.Analys i Ego Pril., 1981, v.15, No 1, p.74-75; English translation: Funct.Anal.Applic., 1981, v.15, No 1, p.59-60.

59. V.I v r i i , On the asymptotics for certain spectral problem related to the Laplace - Beltrami operator on manifold with the boundary (in Russian), Funkt.Analys i Ego Pril., 1983, v.17, No 1, p.72-73; English translation will appear in Funct.Anal.Applic.

60. V.I v r i i , Asymptotics of the eigenvalues for some elliptic operators acting in bundles over a manifold with a boundary (in Russian), Dokl.AN SSSR, 1981, v.258, No 5, p.1045-1046; English translation: Soviet Math.Dokl., 1981, v.23, No 3, p.623-625.

61. V.I v r i i , On asymptotics of eigenvalues for certain class of elliptic operators acting in fiberings over a manifold with a boundary (in Russian), Dokl.AN SSSR, 1982, v.263, No 3, p.283-284; English translation will appear in Soviet Math.Dokl.

62. V.I v r i i , Global and partially global operators, Propagation of singularities and spectral asymptotics (in Russian) Uspechi Matem.nauk, 1982, v.37, No 4, p.115-116.

63. V.I v r i i , On quasi-classical spectral asymptotics for the Schrodinger operator on manifold with the boundary and for h-pseudo-differential operators acting in fiberings (in Russian), Dokl.AN SSSR, 1982, v.266, No 1, p.14-18; English translation will appear in Soviet.Math.Dokl.

64. V.I v r i i , M.A.Š u b i n , On the asymptotics of the spectral-shift function (in Russian); Dokl.AN SSSR, 1982, v.263, No 2,

p.283-284; English translation will appear in Soviet Math.Dokl.

65. Ja.V.K u r y l e v , Asymptotics close to the boundary of a spectral function of an elliptic second-order differential operator (in Russian), Funkt.Analys i Ego Pril., 1980, v.14, No 3, p.85-86; English translation: Funkt.Anal. Applic., 1980, v.14, No 3, p. 236-237.

66. P.D.L a x , L.N i r e n b e r g , On stability for difference schemes; a sharp form of Gårding inequality, Comm. on Pure and Appl.Math., 1966, v.19, p.473-492.

67. V.F.L a z u t k i n , Convex billiard and eigenvalues of the Laplace operator (in Russian), Publ.Leningrad Univ., 1981.

68. V.F.L a z u t k i n , D.J.T e r m a n , On the remainder estimate in the H Weyl formula (in Russian), Funkt.Analys i Ego Pril., 1981, v.15, No 4, p.81-82; English translation will appear in Funct.Anal.Applic.

69. B.M.L e v i t a n , On the asymptotical behaviour of the spectral function of selfadjoint second-order elliptic equation, I, (in Russian),Izvestija Acad.Nauk SSSR, Ser.Mat., 1952, v.16, p.325-352.

70. B.M.L e v i t a n , On the asymptotical behaviour of the spectral function of selfadjoint second-order elliptic equation, II (in Russian), Izvestija Acad.Nauk SSSR, Ser.Mat., 1955, v.19. p.33-58.

71. B.M.L e v i t a n , Asymptotical behaviour of the spectral function of an elliptic equation (in Russian), Uspechi Matem.Nauk, 1971, v.26, No 6, p.151-212; English translation: Russian Math. Surveys, 1971, v.26, No 6, p.165-232.

72. B.M.L e v i t a n , Certain problems of the spectral theory of the differential operators, Proc.of Intern.Congress of Math., Nice, 1970.

73. J.L.L i o n s , E.M a g e n e s , Problèmes aux limites non homogènes et applications, Dunod, Paris. 1968.

74. B.M a l g r a n g e , Ideals of differentiable functions, Oxford, Univ.Press, 1966.

75. V.P.M a s l o v , The asymptotic behaviour of the eigenvalues of the Schrödinger operator (in Russian), Uspechi Matem.Nauk, 1965, v.20, No 6, p.134-138; English translation: Russian Math.Surveys, 1965, v.20, No 6.

76. R.M e l r o s e , Weyl's conjecture for manifolds with concave boundary, Proc.Symp. on Pure Math., 1980, v.36, Amer.Math.Soc., p.257-273.

77. R. M e l r o s e , Hypoelliptic operators with characteristic variety of codimensiona two and the wave equation, Sem.Goulaouic - Schwartz, 1979/80, Exp.XI.
78. R. M e l r o s e , J. S j ö s t r a n d , Singularities of boundary value problems, I, Communs. on Pure and Appl.Math., 1978, v.31. p.593-617.
79. R. M e l r o s e , J.S. S j ö s t r a n d , Singularities of boundary value problems, II, Communs. on Pure and Appl.Math., 1982, v.35, No 2, p.129-168.
80. V. P e t k o v , Comportement asymptotique de la phase de diffusion pour des obstacles non-convexes. Sem.Goulaouic-Schwartz, 1980/81 Exp.XIII.
81. V. P e t k o v , G. P o p o v , Asymptotique de la phase de diffusion pour des domaines non convexes, C.r.Acad.Sci.Paris, 1981, t.292, Ser.I, p.275-277.
82. P h a m T h e L a i , Meilleures estimations asymptotiques de restes de la fonction spectrale et des valeurs propres relatives du Laplacien, Math.Scand., 1981, v.48, No 1, p.5-38.
83. S. R e m p e l , B.W. S c h u l z e , Parametrices and boundary symbolic calculus for elliptic boundary problems without the transmission property, Math.Nachr., 1982, b.105, s.45-149.
84. G.V. R o s e n b l u m , Spectral asymptotics for the elliptic systems (in Russian), Boundary problems of Math.physics and related problems of the function theory, 12 (LOMI proceedings, v.96), Nauka, Leningrad, 1980.
85. P. S c h a p i r a , Théorie des hyperfonctions, Lecture Notes in Math., 1970, v.126.
86. B.W. S c h u l z e , Elliptic operators on manifolds with boundary, Contr.School "Global analysis", Ludwigsfelde, Berlin, 1976.
87. R. S e e l y , A sharp asymptotic remainder estimate for the eigenvalues of the Laplacian in a domain of R^3 , Adv. in Math., 1978, v.29, p.144-269.
88. R. S e e l y , An estimate near the boundary for the spectral function of the Laplace operator, Amer.J.Math., 1980, v.102, p.869-902.
89. A.I. Š n i r e l ' m a n , On the asymptotical multiplicity of the spectrum of the Laplace operator (in Russian). Uspechi Matem. Nauk, 1975, v,30, No 4, p.265-266; English translation: Russian Math.Surveys, v.30, No 4,
90. M.A. Š u b i n , Pseudo-differential operators and spectral theory (in Russian), Nauka, Moscow, 1978.
91. M. T a y l o r , Pseudo-differential operators, Lecture Notes in

Math., 1974, v.416.

92. F.T r e v e s , Introduction to pseudo-differential and Fourier integral operators, I,II, Plenum Press, N.Y., 1980.

93. V.N.T u l o v s k i i , M.A.Š u b i n , On the asymptotical distribution of the eigenvalues in R^n (in Russian), Matem.Sbornik, 1973, v.92, p.571-588; English translation: Math.USSR Sbornik, 1973, v.21, p.565-583.

94. A.W e i n s t e i n , Fourier integral operators, quantization and the spectra of Riemannian manifolds, Proc. de C.N.R.S., Colloque de géométrie symplectique et physics mathématique, Aix de Provence, 1974.

95. A.W e i n s t e i n , Asymptotics of eigenvalue clusters for the Laplacian plus a potential, Duke Math.,J., 1977, v.44, p.883-892.

96. H.W e y l , Über die asymptotische Verteilung der Eigenwerte, Göttinger Nachr., 1911, s.110-117.

97. H.W e y l , Das asymptotische Verteilung der Eigenwerte linearer partieller Differentialgleichungen, Math.Ann., 1912, b.71, s.441-479.

98. H.W e y l , Über die Abhängigkeit der Eigenshwingungen einer Membran von der Begrenzung, J.Reine Agew.Math., 1912, b.141, s.1-11.

99. G.M e t i v e r , Estimation du reste en théorie spectrale, Journees "Equats.aux derivées partielles", Saint-Jean-De-Monts, 1982. Exp.I.

100. D.G.V a s i l j e v , The binomial asymptotics for the spectrum of boundary value problems (in Russian), Funkt.Analys; Ego Pril., 1983, v.17, N 4, p.79-81; English Translation will appear in Funct.Anal.Applic.

101. V.I v r i i , On precise quasiclassical asymptotics for h-pseudo-differential operators acting in fiberings (in Russian), Proc.Acad.S.L.Sobolev' seminar, 1983, N 1, p.30-54.

102. V.I v r i i , On the asymptotics of discrete spectrum for certain operators in R^d (in Russian), to appear in Funkt.Analys i Ego Pril., 1984, v.18.

103. V.I v r i i , On precise eigenvalue asymptotics for two classes of differential operators in R^d (in Russian), to appear in Dokl.AN SSSR, 1984.

104. V.I v r i i , On the asymptotics of discrete spectrum for h-particle Schrödinger operator (in Russian), to appear in Dokl.AN SSSR, 1984.

105. V.I v r i i , Precise classical and quasiclassical asymptotics for spectral problems on manifold with the boundary (in Russi-

an), to appear in Dokl.AN SSSR, 1984.

106. V.I v r i i , Global and partially global operators. Propagation of singularities and spectral asymptotics, to appear in Contemp.Math., 1984, X (Micro).

107. M.K a l k a , A.M e n i k o f f , The wave equation on a cone, Communs. in Part.Different.Equats., 1982, v.7, N 3, p.223-278.

108. A.N.K o z e v n i k o v , On the remainder term in the spectrum asymptotics of the Stokes problem (in Russian), Dokl.AN SSSR, 1983, v.272, N 2, p.294-296; English translation will appear in Soviet Math.Dokl.

109. R.M e l r o s e , The trace of the wave group. Preprint, 1983.

110. Pham The Lai, V.P e t k o v , Comportement asymptotique de la fonction spectrale de l'opérateur de Läplace - Beltrami: sur une variété ayant des singularités coniques. Sem."Equats aux derivées partielles". Univ.de Nantes, 1982-1983, p.117-126.

111. Yu.G.S a f a r o v , The asymptotic of spectrum of the Maxwell's operator (in Russian), Boundary problems of math.physics and related problems of the function theory, 15 (LOMI proceedings, v. 127); Nauka, Leningrad, 1983, p.169-180.

Vol. 1008: Algebraic Geometry. Proceedings, 1981. Edited by J. Dolgachev. V, 138 pages. 1983.

Vol. 1009: T. A. Chapman, Controlled Simple Homotopy Theory and Applications. III, 94 pages. 1983.

Vol. 1010: J.-E. Dies, Chaînes de Markov sur les permutations. IX, 226 pages. 1983.

Vol. 1011: J.M. Sigal. Scattering Theory for Many-Body Quantum Mechanical Systems. IV, 132 pages. 1983.

Vol. 1012: S. Kantorovitz, Spectral Theory of Banach Space Operators. V, 179 pages. 1983.

Vol. 1013: Complex Analysis – Fifth Romanian-Finnish Seminar. Part 1. Proceedings, 1981. Edited by C. Andreian Cazacu, N. Boboc, M. Jurchescu and I. Suciu. XX, 393 pages. 1983.

Vol. 1014: Complex Analysis – Fifth Romanian-Finnish Seminar. Part 2. Proceedings, 1981. Edited by C. Andreian Cazacu, N. Boboc, M. Jurchescu and I. Suciu. XX, 334 pages. 1983.

Vol. 1015: Equations différentielles et systèmes de Pfaff dans le champ complexe – II. Seminar. Edited by R. Gérard et J.P. Ramis. V, 411 pages. 1983.

Vol. 1016: Algebraic Geometry. Proceedings, 1982. Edited by M. Raynaud and T. Shioda. VIII, 528 pages. 1983.

Vol. 1017: Equadiff 82. Proceedings, 1982. Edited by H. W. Knobloch and K. Schmitt. XXIII, 666 pages. 1983.

Vol. 1018: Graph Theory, Łagów 1981. Proceedings, 1981. Edited by M. Borowiecki, J. W. Kennedy and M. M. Sysło. X, 289 pages. 1983.

Vol. 1019: Cabal Seminar 79–81. Proceedings, 1979–81. Edited by A. S. Kechris, D. A. Martin and Y. N. Moschovakis. V, 284 pages. 1983.

Vol. 1020: Non Commutative Harmonic Analysis and Lie Groups. Proceedings, 1982. Edited by J. Carmona and M. Vergne. V, 187 pages. 1983.

Vol. 1021: Probability Theory and Mathematical Statistics. Proceedings, 1982. Edited by K. Itô and J.V. Prokhorov. VIII, 747 pages. 1983.

Vol. 1022: G. Gentili, S. Salamon and J.-P. Vigué. Geometry Seminar "Luigi Bianchi", 1982. Edited by E. Vesentini. VI, 177 pages. 1983.

Vol. 1023: S. McAdam, Asymptotic Prime Divisors. IX, 118 pages. 1983.

Vol. 1024: Lie Group Representations I. Proceedings, 1982–1983. Edited by R. Herb, R. Lipsman and J. Rosenberg. IX, 369 pages. 1983.

Vol. 1025: D. Tanré, Homotopie Rationnelle: Modèles de Chen, Quillen, Sullivan. X, 211 pages. 1983.

Vol. 1026: W. Plesken, Group Rings of Finite Groups Over p-adic Integers. V, 151 pages. 1983.

Vol. 1027: M. Hasumi, Hardy Classes on Infinitely Connected Riemann Surfaces. XII, 280 pages. 1983.

Vol. 1028: Séminaire d'Analyse P. Lelong – P. Dolbeault – H. Skoda. Années 1981/1983. Edité par P. Lelong, P. Dolbeault et H. Skoda. VIII, 328 pages. 1983.

Vol. 1029: Séminaire d'Algèbre Paul Dubreil et Marie-Paule Malliavin. Proceedings, 1982. Edité par M.-P. Malliavin. V, 339 pages. 1983.

Vol. 1030: U. Christian, Selberg's Zeta-, L-, and Eisensteinseries. XII, 196 pages. 1983.

Vol. 1031: Dynamics and Processes. Proceedings, 1981. Edited by Ph. Blanchard and L. Streit. IX, 213 pages. 1983.

Vol. 1032: Ordinary Differential Equations and Operators. Proceedings, 1982. Edited by W. N. Everitt and R. T. Lewis. XV, 521 pages. 1983.

Vol. 1033: Measure Theory and its Applications. Proceedings, 1982. Edited by J. M. Belley, J. Dubois and P. Morales. XV, 317 pages. 1983.

Vol. 1034: J. Musielak, Orlicz Spaces and Modular
222 pages. 1983.

Vol. 1035: The Mathematics and Physics of Disord
Proceedings, 1983. Edited by B.D. Hughes and B.
VII, 432 pages. 1983.

Vol. 1036: Combinatorial Mathematics X. Proceedings, 1
by L. R. A. Casse. XI, 419 pages. 1983.

Vol. 1037: Non-linear Partial Differential Operators and
Procedures. Proceedings, 1981. Edited by S. I. And
H.-D. Doebner. VII, 334 pages. 1983.

Vol. 1038: F. Borceux, G. Van den Bossche, Algebra i
Topos with Applications to Ring Theory. IX, 240 pages.

Vol. 1039: Analytic Functions, Błażejewko 1982. Proceed
by J. Ławrynowicz. X, 494 pages. 1983

Vol. 1040: A. Good, Local Analysis of Selberg's Trace
128 pages. 1983.

Vol. 1041: Lie Group Representations II. Proceedings 1
Edited by R. Herb, S. Kudla, R. Lipsman and J. Ros
340 pages. 1984.

Vol. 1042: A. Gut, K. D. Schmidt, Amarts and Set Function
III, 258 pages. 1983.

Vol. 1043: Linear and Complex Analysis Problem Book
V. P. Havin, S. V. Hruščëv and N. K. Nikol'skii. XVIII, 721 pa

Vol. 1044: E. Gekeler, Discretization Methods for Stable
Problems. VIII, 201 pages. 1984.

Vol. 1045: Differential Geometry. Proceedings, 1982.
A. M. Naveira. VIII, 194 pages. 1984.

Vol. 1046: Algebraic K-Theory, Number Theory, Geometry
sis. Proceedings, 1982. Edited by A. Bak. IX, 464 pages

Vol. 1047: Fluid Dynamics. Seminar, 1982. Edited by H.
Veiga. VII, 193 pages. 1984.

Vol. 1048: Kinetic Theories and the Boltzmann Equatio
1981. Edited by C. Cercignani. VII, 248 pages. 1984.

Vol. 1049: B. Iochum, Cônes autopolaires et algèbres
VI, 247 pages. 1984.

Vol. 1050: A. Prestel, P. Roquette, Formally p-adic Fields. V,
1984.

Vol. 1051: Algebraic Topology, Aarhus 1982. Proceedings.
I. Madsen and B. Oliver. X, 665 pages. 1984.

Vol. 1052: Number Theory. Seminar, 1982. Edited by D
novsky, G.V. Chudnovsky, H. Cohn and M.B. Nathans
pages. 1984.

Vol. 1053: P. Hilton, Nilpotente Gruppen und nilpoten
V, 221 pages. 1984.

Vol. 1054: V. Thomée, Galerkin Finite Element Methods fo
Problems. VII, 237 pages. 1984.

Vol. 1055: Quantum Probability and Applications to the
Theory of Irreversible Processes. Proceedings, 1982.
L. Accardi, A. Frigerio and V. Gorini. VI, 411 pages. 1984.

Vol. 1056: Algebraic Geometry. Bucharest 1982. Proceedi
Edited by L. Bădescu and D. Popescu. VII, 380 pages. 1

Vol. 1057: Bifurcation Theory and Applications. Seminar, 19
by L. Salvadori. VII, 233 pages. 1984.

Vol. 1058: B. Aulbach, Continuous and Discrete Dyna
Manifolds of Equilibria. IX, 142 pages. 1984.

Vol. 1059: Séminaire de Probabilités XVIII, 1982/83. Pro
Edité par J. Azéma et M. Yor. IV, 518 pages. 1984.

Vol. 1060: Topology. Proceedings, 1982. Edited by L. D. Fa
A. A. Mal'cev. VI, 389 pages. 1984.

Vol. 1061: Séminaire de Théorie du Potentiel. Paris, N
ceedings. Directeurs: M. Brelot, G. Choquet et J. Deny. Ré
F. Hirsch et G. Mokobodzki. IV, 281 pages. 1984.